Access Your Online Resources

Don't miss out on the Online Resources included with your purchase!

Your purchase of this product unlocks access to our Online Resources page. Elevate your study experience with our **interactive practice test interface**, along with all of the additional resources that we couldn't include in this book.

Flip to the Online Resources section at the end of this book to find the link and a QR code to get started!

Mometrix
TEST PREPARATION

Praxis® II

Mathematics: Content Knowledge (5161) Exam Secrets Study Guide

Praxis® II Test Review for the Praxis® II: Subject Assessments

Mometrix
TEST PREPARATION

Written and edited by the Mometrix Teacher Certification Test Team

Paperback
ISBN 13: 978-1-63094-510-7
ISBN 10: 1-63094-510-2

Ebook
ISBN 13: 978-1-5167-0283-1
ISBN 10: 1-5167-0283-2

Hardback
ISBN 13: 978-1-5167-0829-1
ISBN 10: 1-5167-0829-6

DEAR FUTURE EXAM SUCCESS STORY

First of all, **THANK YOU** for purchasing Mometrix study materials!

Second, congratulations! You are one of the few determined test-takers who are committed to doing whatever it takes to excel on your exam. **You have come to the right place.** We developed these study materials with one goal in mind: to deliver you the information you need in a format that's concise and easy to use.

In addition to optimizing your guide for the content of the test, we've outlined our recommended steps for breaking down the preparation process into small, attainable goals so you can make sure you stay on track.

We've also analyzed the entire test-taking process, identifying the most common pitfalls and showing how you can overcome them and be ready for any curveball the test throws you.

Standardized testing is one of the biggest obstacles on your road to success, which only increases the importance of doing well in the high-pressure, high-stakes environment of test day. Your results on this test could have a significant impact on your future, and this guide provides the information and practical advice to help you achieve your full potential on test day.

Your success is our success

We would love to hear from you! If you would like to share the story of your exam success or if you have any questions or comments in regard to our products, please contact us at **800-673-8175** or **support@mometrix.com**.

Thanks again for your business and we wish you continued success!

Sincerely,
The Mometrix Test Preparation Team

TABLE OF CONTENTS

Introduction

Thank you for purchasing this resource! You have made the choice to prepare yourself for a test that could have a huge impact on your future, and this guide is designed to help you be fully ready for test day. Obviously, it's important to have a solid understanding of the test material, but you also need to be prepared for the unique environment and stressors of the test, so that you can perform to the best of your abilities.

For this purpose, the first section that appears in this guide is the **Secret Keys**. We've devoted countless hours to meticulously researching what works and what doesn't, and we've boiled down our findings to the five most impactful steps you can take to improve your performance on the test. We start at the beginning with study planning and move through the preparation process, all the way to the testing strategies that will help you get the most out of what you know when you're finally sitting in front of the test.

We recommend that you start preparing for your test as far in advance as possible. However, if you've bought this guide as a last-minute study resource and only have a few days before your test, we recommend that you skip over the first two Secret Keys since they address a long-term study plan.

If you struggle with **test anxiety**, we strongly encourage you to check out our recommendations for how you can overcome it. Test anxiety is a formidable foe, but it can be beaten, and we want to make sure you have the tools you need to defeat it.

Secret Key #1 – Plan Big, Study Small

There's a lot riding on your performance. If you want to ace this test, you're going to need to keep your skills sharp and the material fresh in your mind. You need a plan that lets you review everything you need to know while still fitting in your schedule. We'll break this strategy down into three categories.

Information Organization

Start with the information you already have: the official test outline. From this, you can make a complete list of all the concepts you need to cover before the test. Organize these concepts into groups that can be studied together, and create a list of any related vocabulary you need to learn so you can brush up on any difficult terms. You'll want to keep this vocabulary list handy once you actually start studying since you may need to add to it along the way.

Time Management

Once you have your set of study concepts, decide how to spread them out over the time you have left before the test. Break your study plan into small, clear goals so you have a manageable task for each day and know exactly what you're doing. Then just focus on one small step at a time. When you manage your time this way, you don't need to spend hours at a time studying. Studying a small block of content for a short period each day helps you retain information better and avoid stressing over how much you have left to do. You can relax knowing that you have a plan to cover everything in time. In order for this strategy to be effective though, you have to start studying early and stick to your schedule. Avoid the exhaustion and futility that comes from last-minute cramming!

Study Environment

The environment you study in has a big impact on your learning. Studying in a coffee shop, while probably more enjoyable, is not likely to be as fruitful as studying in a quiet room. It's important to keep distractions to a minimum. You're only planning to study for a short block of time, so make the most of it. Don't pause to check your phone or get up to find a snack. It's also important to **avoid multitasking**. Research has consistently shown that multitasking will make your studying dramatically less effective. Your study area should also be comfortable and well-lit so you don't have the distraction of straining your eyes or sitting on an uncomfortable chair.

The time of day you study is also important. You want to be rested and alert. Don't wait until just before bedtime. Study when you'll be most likely to comprehend and remember. Even better, if you know what time of day your test will be, set that time aside for study. That way your brain will be used to working on that subject at that specific time and you'll have a better chance of recalling information.

Finally, it can be helpful to team up with others who are studying for the same test. Your actual studying should be done in as isolated an environment as possible, but the work of organizing the information and setting up the study plan can be divided up. In between study sessions, you can discuss with your teammates the concepts that you're all studying and quiz each other on the details. Just be sure that your teammates are as serious about the test as you are. If you find that your study time is being replaced with social time, you might need to find a new team.

Secret Key #2 – Make Your Studying Count

You're devoting a lot of time and effort to preparing for this test, so you want to be absolutely certain it will pay off. This means doing more than just reading the content and hoping you can remember it on test day. It's important to make every minute of study count. There are two main areas you can focus on to make your studying count.

Retention

It doesn't matter how much time you study if you can't remember the material. You need to make sure you are retaining the concepts. To check your retention of the information you're learning, try recalling it at later times with minimal prompting. Try carrying around flashcards and glance at one or two from time to time or ask a friend who's also studying for the test to quiz you.

To enhance your retention, look for ways to put the information into practice so that you can apply it rather than simply recalling it. If you're using the information in practical ways, it will be much easier to remember. Similarly, it helps to solidify a concept in your mind if you're not only reading it to yourself but also explaining it to someone else. Ask a friend to let you teach them about a concept you're a little shaky on (or speak aloud to an imaginary audience if necessary). As you try to summarize, define, give examples, and answer your friend's questions, you'll understand the concepts better and they will stay with you longer. Finally, step back for a big picture view and ask yourself how each piece of information fits with the whole subject. When you link the different concepts together and see them working together as a whole, it's easier to remember the individual components.

Finally, practice showing your work on any multi-step problems, even if you're just studying. Writing out each step you take to solve a problem will help solidify the process in your mind, and you'll be more likely to remember it during the test.

Modality

Modality simply refers to the means or method by which you study. Choosing a study modality that fits your own individual learning style is crucial. No two people learn best in exactly the same way, so it's important to know your strengths and use them to your advantage.

For example, if you learn best by visualization, focus on visualizing a concept in your mind and draw an image or a diagram. Try color-coding your notes, illustrating them, or creating symbols that will trigger your mind to recall a learned concept. If you learn best by hearing or discussing information, find a study partner who learns the same way or read aloud to yourself. Think about how to put the information in your own words. Imagine that you are giving a lecture on the topic and record yourself so you can listen to it later.

For any learning style, flashcards can be helpful. Organize the information so you can take advantage of spare moments to review. Underline key words or phrases. Use different colors for different categories. Mnemonic devices (such as creating a short list in which every item starts with the same letter) can also help with retention. Find what works best for you and use it to store the information in your mind most effectively and easily.

Secret Key #3 – Practice the Right Way

Your success on test day depends not only on how many hours you put into preparing, but also on whether you prepared the right way. It's good to check along the way to see if your studying is paying off. One of the most effective ways to do this is by taking practice tests to evaluate your progress. Practice tests are useful because they show exactly where you need to improve. Every time you take a practice test, pay special attention to these three groups of questions:

- The questions you got wrong
- The questions you had to guess on, even if you guessed right
- The questions you found difficult or slow to work through

This will show you exactly what your weak areas are, and where you need to devote more study time. Ask yourself why each of these questions gave you trouble. Was it because you didn't understand the material? Was it because you didn't remember the vocabulary? Do you need more repetitions on this type of question to build speed and confidence? Dig into those questions and figure out how you can strengthen your weak areas as you go back to review the material.

Additionally, many practice tests have a section explaining the answer choices. It can be tempting to read the explanation and think that you now have a good understanding of the concept. However, an explanation likely only covers part of the question's broader context. Even if the explanation makes perfect sense, **go back and investigate** every concept related to the question until you're positive you have a thorough understanding.

As you go along, keep in mind that the practice test is just that: practice. Memorizing these questions and answers will not be very helpful on the actual test because it is unlikely to have any of the same exact questions. If you only know the right answers to the sample questions, you won't be prepared for the real thing. **Study the concepts** until you understand them fully, and then you'll be able to answer any question that shows up on the test.

It's important to wait on the practice tests until you're ready. If you take a test on your first day of study, you may be overwhelmed by the amount of material covered and how much you need to learn. Work up to it gradually.

On test day, you'll need to be prepared for answering questions, managing your time, and using the test-taking strategies you've learned. It's a lot to balance, like a mental marathon that will have a big impact on your future. Like training for a marathon, you'll need to start slowly and work your way up. When test day arrives, you'll be ready.

Start with the strategies you've read in the first two Secret Keys—plan your course and study in the way that works best for you. If you have time, consider using multiple study resources to get different approaches to the same concepts. It can be helpful to see difficult concepts from more than one angle. Then find a good source for practice tests. Many times, the test website will suggest potential study resources or provide sample tests.

Practice Test Strategy

If you're able to find at least three practice tests, we recommend this strategy:

Untimed and Open-Book Practice

Take the first test with no time constraints and with your notes and study guide handy. Take your time and focus on applying the strategies you've learned.

Timed and Open-Book Practice

Take the second practice test open-book as well, but set a timer and practice pacing yourself to finish in time.

Timed and Closed-Book Practice

Take any other practice tests as if it were test day. Set a timer and put away your study materials. Sit at a table or desk in a quiet room, imagine yourself at the testing center, and answer questions as quickly and accurately as possible.

Keep repeating timed and closed-book tests on a regular basis until you run out of practice tests or it's time for the actual test. Your mind will be ready for the schedule and stress of test day, and you'll be able to focus on recalling the material you've learned.

Secret Key #4 – Pace Yourself

Once you're fully prepared for the material on the test, your biggest challenge on test day will be managing your time. Just knowing that the clock is ticking can make you panic even if you have plenty of time left. Work on pacing yourself so you can build confidence against the time constraints of the exam. Pacing is a difficult skill to master, especially in a high-pressure environment, so **practice is vital**.

Set time expectations for your pace based on how much time is available. For example, if a section has 60 questions and the time limit is 30 minutes, you know you have to average 30 seconds or less per question in order to answer them all. Although 30 seconds is the hard limit, set 25 seconds per question as your goal, so you reserve extra time to spend on harder questions. When you budget extra time for the harder questions, you no longer have any reason to stress when those questions take longer to answer.

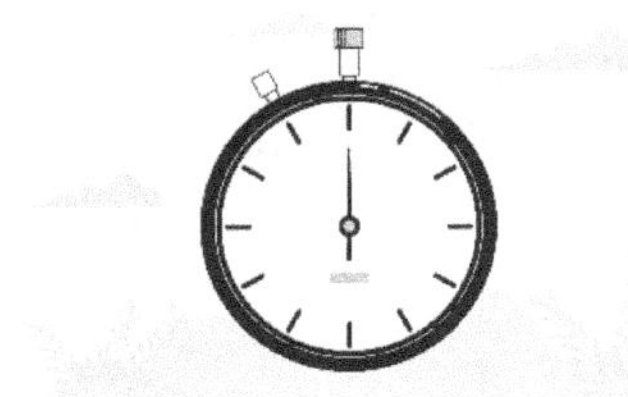

Don't let this time expectation distract you from working through the test at a calm, steady pace, but keep it in mind so you don't spend too much time on any one question. Recognize that taking extra time on one question you don't understand may keep you from answering two that you do understand later in the test. If your time limit for a question is up and you're still not sure of the answer, mark it and move on, and come back to it later if the time and the test format allow. If the testing format doesn't allow you to return to earlier questions, just make an educated guess; then put it out of your mind and move on.

On the easier questions, be careful not to rush. It may seem wise to hurry through them so you have more time for the challenging ones, but it's not worth missing one if you know the concept and just didn't take the time to read the question fully. Work efficiently but make sure you understand the question and have looked at all of the answer choices, since more than one may seem right at first.

Even if you're paying attention to the time, you may find yourself a little behind at some point. You should speed up to get back on track, but do so wisely. Don't panic; just take a few seconds less on each question until you're caught up. Don't guess without thinking, but do look through the answer choices and eliminate any you know are wrong. If you can get down to two choices, it is often worthwhile to guess from those. Once you've chosen an answer, move on and don't dwell on any that you skipped or had to hurry through. If a question was taking too long, chances are it was one of the harder ones, so you weren't as likely to get it right anyway.

On the other hand, if you find yourself getting ahead of schedule, it may be beneficial to slow down a little. The more quickly you work, the more likely you are to make a careless mistake that will affect your score. You've budgeted time for each question, so don't be afraid to spend that time. Practice an efficient but careful pace to get the most out of the time you have.

Secret Key #5 – Have a Plan for Guessing

When you're taking the test, you may find yourself stuck on a question. Some of the answer choices seem better than others, but you don't see the one answer choice that is obviously correct. What do you do?

The scenario described above is very common, yet most test takers have not effectively prepared for it. Developing and practicing a plan for guessing may be one of the single most effective uses of your time as you get ready for the exam.

In developing your plan for guessing, there are three questions to address:

- When should you start the guessing process?
- How should you narrow down the choices?
- Which answer should you choose?

When to Start the Guessing Process

Unless your plan for guessing is to select C every time (which, despite its merits, is not what we recommend), you need to leave yourself enough time to apply your answer elimination strategies. Since you have a limited amount of time for each question, that means that if you're going to give yourself the best shot at guessing correctly, you have to decide quickly whether or not you will guess.

Of course, the best-case scenario is that you don't have to guess at all, so first, see if you can answer the question based on your knowledge of the subject and basic reasoning skills. Focus on the key words in the question and try to jog your memory of related topics. Give yourself a chance to bring the knowledge to mind, but once you realize that you don't have (or you can't access) the knowledge you need to answer the question, it's time to start the guessing process.

It's almost always better to start the guessing process too early than too late. It only takes a few seconds to remember something and answer the question from knowledge. Carefully eliminating wrong answer choices takes longer. Plus, going through the process of eliminating answer choices can actually help jog your memory.

Summary: Start the guessing process as soon as you decide that you can't answer the question based on your knowledge.

How to Narrow Down the Choices

The next chapter in this book (**Test-Taking Strategies**) includes a wide range of strategies for how to approach questions and how to look for answer choices to eliminate. You will definitely want to read those carefully, practice them, and figure out which ones work best for you. Here though, we're going to address a mindset rather than a particular strategy.

Your odds of guessing an answer correctly depend on how many options you are choosing from.

Number of options left	5	4	3	2	1
Odds of guessing correctly	20%	25%	33%	50%	100%

You can see from this chart just how valuable it is to be able to eliminate incorrect answers and make an educated guess, but there are two things that many test takers do that cause them to miss out on the benefits of guessing:

- Accidentally eliminating the correct answer
- Selecting an answer based on an impression

We'll look at the first one here, and the second one in the next section.

To avoid accidentally eliminating the correct answer, we recommend a thought exercise called **the $5 challenge**. In this challenge, you only eliminate an answer choice from contention if you are willing to bet $5 on it being wrong. Why $5? Five dollars is a small but not insignificant amount of money. It's an amount you could afford to lose but wouldn't want to throw away. And while losing $5 once might not hurt too much, doing it twenty times will set you back $100. In the same way, each small decision you make—eliminating a choice here, guessing on a question there—won't by itself impact your score very much, but when you put them all together, they can make a big difference. By holding each answer choice elimination decision to a higher standard, you can reduce the risk of accidentally eliminating the correct answer.

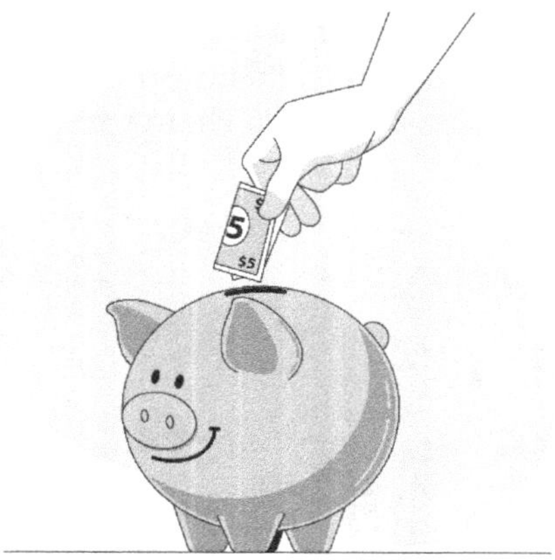

The $5 challenge can also be applied in a positive sense: If you are willing to bet $5 that an answer choice *is* correct, go ahead and mark it as correct.

Summary: Only eliminate an answer choice if you are willing to bet $5 that it is wrong.

Which Answer to Choose

You're taking the test. You've run into a hard question and decided you'll have to guess. You've eliminated all the answer choices you're willing to bet $5 on. Now you have to pick an answer. Why do we even need to talk about this? Why can't you just pick whichever one you feel like when the time comes?

The answer to these questions is that if you don't come into the test with a plan, you'll rely on your impression to select an answer choice, and if you do that, you risk falling into a trap. The test writers know that everyone who takes their test will be guessing on some of the questions, so they intentionally write wrong answer choices to seem plausible. You still have to pick an answer though, and if the wrong answer choices are designed to look right, how can you ever be sure that you're not falling for their trap? The best solution we've found to this dilemma is to take the decision out of your hands entirely. Here is the process we recommend:

Once you've eliminated any choices that you are confident (willing to bet $5) are wrong, select the first remaining choice as your answer.

Whether you choose to select the first remaining choice, the second, or the last, the important thing is that you use some preselected standard. Using this approach guarantees that you will not be enticed into selecting an answer choice that looks right, because you are not basing your decision on how the answer choices look.

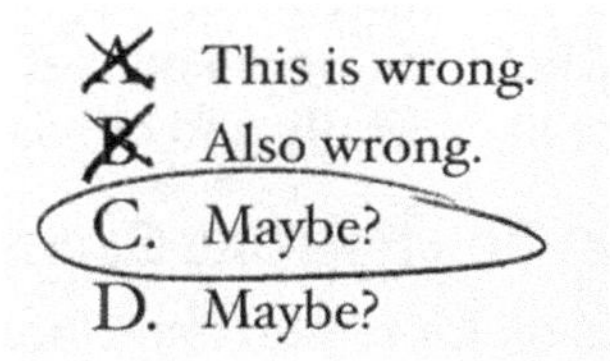

This is not meant to make you question your knowledge. Instead, it is to help you recognize the difference between your knowledge and your impressions. There's a huge difference between thinking an answer is right because of what you know, and thinking an answer is right because it looks or sounds like it should be right.

Summary: To ensure that your selection is appropriately random, make a predetermined selection from among all answer choices you have not eliminated.

Test-Taking Strategies

This section contains a list of test-taking strategies that you may find helpful as you work through the test. By taking what you know and applying logical thought, you can maximize your chances of answering any question correctly!

It is very important to realize that every question is different and every person is different: no single strategy will work on every question, and no single strategy will work for every person. That's why we've included all of them here, so you can try them out and determine which ones work best for different types of questions and which ones work best for you.

Question Strategies

✓ Read Carefully

Read the question and the answer choices carefully. Don't miss the question because you misread the terms. You have plenty of time to read each question thoroughly and make sure you understand what is being asked. Yet a happy medium must be attained, so don't waste too much time. You must read carefully and efficiently.

✓ Contextual Clues

Look for contextual clues. If the question includes a word you are not familiar with, look at the immediate context for some indication of what the word might mean. Contextual clues can often give you all the information you need to decipher the meaning of an unfamiliar word. Even if you can't determine the meaning, you may be able to narrow down the possibilities enough to make a solid guess at the answer to the question.

✓ Prefixes

If you're having trouble with a word in the question or answer choices, try dissecting it. Take advantage of every clue that the word might include. Prefixes can be a huge help. Usually, they allow you to determine a basic meaning. *Pre-* means before, *post-* means after, *pro-* is positive, *de-* is negative. From prefixes, you can get an idea of the general meaning of the word and try to put it into context.

✓ Hedge Words

Watch out for critical hedge words, such as *likely, may, can, often, almost, mostly, usually, generally, rarely*, and *sometimes*. Question writers insert these hedge phrases to cover every possibility. Often an answer choice will be wrong simply because it leaves no room for exception. Be on guard for answer choices that have definitive words such as *exactly* and *always*.

✓ Switchback Words

Stay alert for *switchbacks*. These are the words and phrases frequently used to alert you to shifts in thought. The most common switchback words are *but, although*, and *however*. Others include *nevertheless, on the other hand, even though, while, in spite of, despite*, and *regardless of*. Switchback words are important to catch because they can change the direction of the question or an answer choice.

✓ FACE VALUE

When in doubt, use common sense. Accept the situation in the problem at face value. Don't read too much into it. These problems will not require you to make wild assumptions. If you have to go beyond creativity and warp time or space in order to have an answer choice fit the question, then you should move on and consider the other answer choices. These are normal problems rooted in reality. The applicable relationship or explanation may not be readily apparent, but it is there for you to figure out. Use your common sense to interpret anything that isn't clear.

Answer Choice Strategies

✓ ANSWER SELECTION

The most thorough way to pick an answer choice is to identify and eliminate wrong answers until only one is left, then confirm it is the correct answer. Sometimes an answer choice may immediately seem right, but be careful. The test writers will usually put more than one reasonable answer choice on each question, so take a second to read all of them and make sure that the other choices are not equally obvious. As long as you have time left, it is better to read every answer choice than to pick the first one that looks right without checking the others.

✓ ANSWER CHOICE FAMILIES

An answer choice family consists of two (in rare cases, three) answer choices that are very similar in construction and cannot all be true at the same time. If you see two answer choices that are direct opposites or parallels, one of them is usually the correct answer. For instance, if one answer choice says that quantity x increases and another either says that quantity x decreases (opposite) or says that quantity y increases (parallel), then those answer choices would fall into the same family. An answer choice that doesn't match the construction of the answer choice family is more likely to be incorrect. Most questions will not have answer choice families, but when they do appear, you should be prepared to recognize them.

✓ ELIMINATE ANSWERS

Eliminate answer choices as soon as you realize they are wrong, but make sure you consider all possibilities. If you are eliminating answer choices and realize that the last one you are left with is also wrong, don't panic. Start over and consider each choice again. There may be something you missed the first time that you will realize on the second pass.

✓ AVOID FACT TRAPS

Don't be distracted by an answer choice that is factually true but doesn't answer the question. You are looking for the choice that answers the question. Stay focused on what the question is asking for so you don't accidentally pick an answer that is true but incorrect. Always go back to the question and make sure the answer choice you've selected actually answers the question and is not merely a true statement.

✓ EXTREME STATEMENTS

In general, you should avoid answers that put forth extreme actions as standard practice or proclaim controversial ideas as established fact. An answer choice that states the "process should be used in certain situations, if…" is much more likely to be correct than one that states the "process should be discontinued completely." The first is a calm rational statement and doesn't even make a definitive, uncompromising stance, using a hedge word *if* to provide wiggle room, whereas the second choice is far more extreme.

⊘ Benchmark

As you read through the answer choices and you come across one that seems to answer the question well, mentally select that answer choice. This is not your final answer, but it's the one that will help you evaluate the other answer choices. The one that you selected is your benchmark or standard for judging each of the other answer choices. Every other answer choice must be compared to your benchmark. That choice is correct until proven otherwise by another answer choice beating it. If you find a better answer, then that one becomes your new benchmark. Once you've decided that no other choice answers the question as well as your benchmark, you have your final answer.

⊘ Predict the Answer

Before you even start looking at the answer choices, it is often best to try to predict the answer. When you come up with the answer on your own, it is easier to avoid distractions and traps because you will know exactly what to look for. The right answer choice is unlikely to be word-for-word what you came up with, but it should be a close match. Even if you are confident that you have the right answer, you should still take the time to read each option before moving on.

General Strategies

⊘ Tough Questions

If you are stumped on a problem or it appears too hard or too difficult, don't waste time. Move on! Remember though, if you can quickly check for obviously incorrect answer choices, your chances of guessing correctly are greatly improved. Before you completely give up, at least try to knock out a couple of possible answers. Eliminate what you can and then guess at the remaining answer choices before moving on.

⊘ Check Your Work

Since you will probably not know every term listed and the answer to every question, it is important that you get credit for the ones that you do know. Don't miss any questions through careless mistakes. If at all possible, try to take a second to look back over your answer selection and make sure you've selected the correct answer choice and haven't made a costly careless mistake (such as marking an answer choice that you didn't mean to mark). This quick double check should more than pay for itself in caught mistakes for the time it costs.

⊘ Pace Yourself

It's easy to be overwhelmed when you're looking at a page full of questions; your mind is confused and full of random thoughts, and the clock is ticking down faster than you would like. Calm down and maintain the pace that you have set for yourself. Especially as you get down to the last few minutes of the test, don't let the small numbers on the clock make you panic. As long as you are on track by monitoring your pace, you are guaranteed to have time for each question.

⊘ Don't Rush

It is very easy to make errors when you are in a hurry. Maintaining a fast pace in answering questions is pointless if it makes you miss questions that you would have gotten right otherwise. Test writers like to include distracting information and wrong answers that seem right. Taking a little extra time to avoid careless mistakes can make all the difference in your test score. Find a pace that allows you to be confident in the answers that you select.

✓ Keep Moving

Panicking will not help you pass the test, so do your best to stay calm and keep moving. Taking deep breaths and going through the answer elimination steps you practiced can help to break through a stress barrier and keep your pace.

Final Notes

The combination of a solid foundation of content knowledge and the confidence that comes from practicing your plan for applying that knowledge is the key to maximizing your performance on test day. As your foundation of content knowledge is built up and strengthened, you'll find that the strategies included in this chapter become more and more effective in helping you quickly sift through the distractions and traps of the test to isolate the correct answer.

Now that you're preparing to move forward into the test content chapters of this book, be sure to keep your goal in mind. As you read, think about how you will be able to apply this information on the test. If you've already seen sample questions for the test and you have an idea of the question format and style, try to come up with questions of your own that you can answer based on what you're reading. This will give you valuable practice applying your knowledge in the same ways you can expect to on test day.

Good luck and good studying!

Number and Quantity

Transform passive reading into active learning! After immersing yourself in this chapter, put your comprehension to the test by taking a quiz. The insights you gained will stay with you longer this way. Scan the QR code to go directly to the chapter quiz interface for this study guide. If you're using a computer, simply visit the online resources page at **mometrix.com/resources719/priimath5161** and click the Chapter Quizzes link.

Number Basics

Classifications of Numbers

Numbers are the basic building blocks of mathematics. Specific features of numbers are identified by the following terms:

Integer – any positive or negative whole number, including zero. Integers do not include fractions $\left(\frac{1}{3}\right)$, decimals (0.56), or mixed numbers $\left(7\frac{3}{4}\right)$.

Prime number – any whole number greater than 1 that has only two factors, itself and 1; that is, a number that can be divided evenly only by 1 and itself.

Composite number – any whole number greater than 1 that has more than two different factors; in other words, any whole number that is not a prime number. For example: The composite number 8 has the factors of 1, 2, 4, and 8.

Even number – any integer that can be divided by 2 without leaving a remainder. For example: 2, 4, 6, 8, and so on.

Odd number – any integer that cannot be divided evenly by 2. For example: 3, 5, 7, 9, and so on.

Decimal number – any number that uses a decimal point to show the part of the number that is less than one. Example: 1.234.

Decimal point – a symbol used to separate the ones place from the tenths place in decimals or dollars from cents in currency.

Decimal place – the position of a number to the right of the decimal point. In the decimal 0.123, the 1 is in the first place to the right of the decimal point, indicating tenths; the 2 is in the second place, indicating hundredths; and the 3 is in the third place, indicating thousandths.

The **decimal**, or base 10, system is a number system that uses ten different digits (0, 1, 2, 3, 4, 5, 6, 7, 8, 9). An example of a number system that uses something other than ten digits is the **binary**, or base 2, number system, used by computers, which uses only the numbers 0 and 1. It is thought that the decimal system originated because people had only their 10 fingers for counting.

Rational numbers include all integers, decimals, and fractions. Any terminating or repeating decimal number is a rational number.

Irrational numbers cannot be written as fractions or decimals because the number of decimal places is infinite and there is no recurring pattern of digits within the number. For example, pi (π) begins with 3.141592 and continues without terminating or repeating, so pi is an irrational number.

Real numbers are the set of all rational and irrational numbers.

Review Video: Classification of Numbers
Visit mometrix.com/academy and enter code: 461071

Review Video: Prime and Composite Numbers
Visit mometrix.com/academy and enter code: 565581

NUMBERS IN WORD FORM AND PLACE VALUE

When writing numbers out in word form or translating word form to numbers, it is essential to understand how a place value system works. In the decimal or base-10 system, each digit of a number represents how many of the corresponding place value—a specific factor of 10—are contained in the number being represented. To make reading numbers easier, every three digits to the left of the decimal place is preceded by a comma. The following table demonstrates some of the place values:

Power of 10	10^3	10^2	10^1	10^0	10^{-1}	10^{-2}	10^{-3}
Value	1,000	100	10	1	0.1	0.01	0.001
Place	thousands	hundreds	tens	ones	tenths	hundredths	thousandths

For example, consider the number 4,546.09, which can be separated into each place value like this:

4: thousands
5: hundreds
4: tens
6: ones
0: tenths
9: hundredths

This number in word form would be *four thousand five hundred forty-six and nine hundredths.*

Review Video: Place Value
Visit mometrix.com/academy and enter code: 205433

Number Lines

A number line is a graph to see the distance between numbers. Basically, this graph shows the relationship between numbers. So a number line may have a point for zero and may show negative numbers on the left side of the line. Any positive numbers are placed on the right side of the line. For example, consider the points labeled on the following number line:

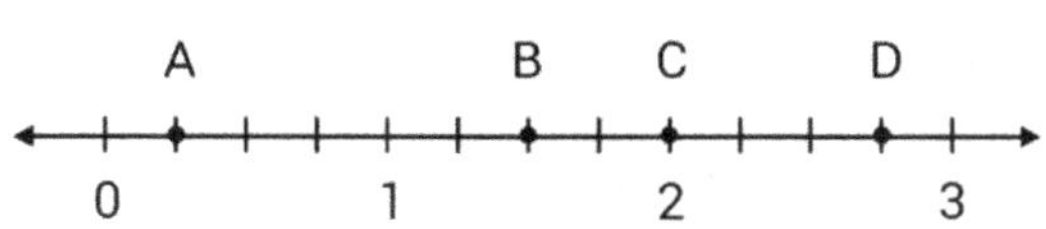

We can use the dashed lines on the number line to identify each point. Each dashed line between two whole numbers is $\frac{1}{4}$. The line halfway between two numbers is $\frac{1}{2}$.

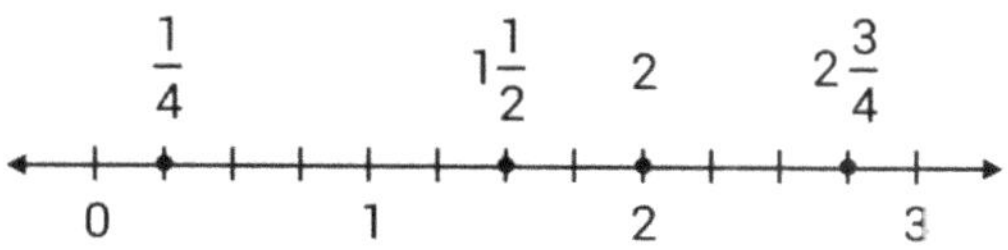

Review Video: The Number Line
Visit mometrix.com/academy and enter code: 816439

Comparing Numbers

Inequality Notation

The symbols $<$ and $>$ mean "is less than" and "is greater than," respectively. For instance, $3 < 5$ means "3 is less than 5," and $7 > 4$ means "7 is greater than 4." Statements like $3 < 5$ and $7 > 4$ are **inequalities**, and the symbols $<$ and $>$ are **inequality symbols**.

Whole Numbers and Decimal Numbers

To compare whole or decimal numbers, we look at the most significant place (the leftmost digit) at which they differ. The number with the larger digit in that place is larger. For instance, $0.3\underline{8}74$ and $0.3\underline{9}$ differ in the hundredths place (underlined). Since 8 is smaller than 9, we see $0.3874 < 0.39$. This is clearer if we make the decimals equal in length by writing extra zeroes: $0.3874 < 0.3900$. Similarly, $2\underline{3}.984 < 2\underline{5}.112$ because 3 is smaller than 5, or 23 is smaller than 25.

Fractions

If fractions have the same denominator, the fraction with the larger numerator is larger. For instance, $\frac{2}{7} < \frac{5}{7}$ since $2 < 5$. We compare fractions with different denominators by finding a common denominator. When comparing the fractions with a common denominator we only compare the numerator, so as a shortcut, we can multiply each numerator by the denominator of the other fraction. The numerator that produces the larger product belongs to the larger fraction. For example, to compare $\frac{7}{8}$ and $\frac{5}{6}$, we note that $7 \cdot 6 = 42$ is larger than $5 \cdot 8 = 40$. Since the numerator 7 produces the larger product, we see $\frac{7}{8} > \frac{5}{6}$. We can also compare fractions by converting them to decimals. For instance, since $\frac{3}{4} = 0.75$ and $\frac{4}{5} = 0.8$ and $0.75 < 0.8$, we conclude $\frac{3}{4} < \frac{4}{5}$.

Mixed Numbers

To compare mixed numbers we compare their whole number parts. If those are equal, then we compare their fractional parts. For instance, $5\frac{3}{8} > 4\frac{7}{8}$ because $5 > 4$, but $3\frac{5}{9} < 3\frac{8}{9}$ because $\frac{5}{9} < \frac{8}{9}$.

Square Roots

To compare square roots, we convert it to a decimal, usually with a calculator. To compare two square roots, we compare their radicands. For instance, $\sqrt{11} < \sqrt{14}$ because $11 < 14$.

NEGATIVE NUMBERS

A negative number is always less than a positive number. Two negative numbers compare in the reverse order of their opposites. For instance, $-6 < -2$ (that is, -6 is smaller, more negative, than -2) because $6 > 2$.

Absolute Value

A precursor to working with negative numbers is understanding what **absolute values** are. A number's absolute value is simply the distance away from zero a number is on the number line. The absolute value of a number is always positive and is written $|x|$. For example, the absolute value of 3, written as $|3|$, is 3 because the distance between 0 and 3 on a number line is three units. Likewise, the absolute value of -3, written as $|-3|$, is 3 because the distance between 0 and -3 on a number line is three units. So $|3| = |-3|$.

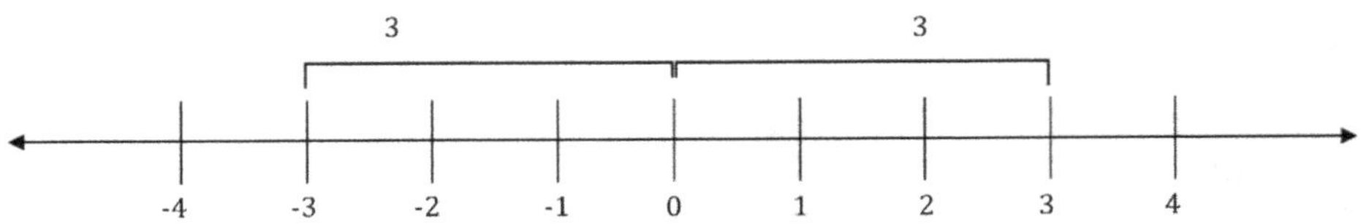

> **Review Video: Absolute Value**
> Visit mometrix.com/academy and enter code: 314669

Operations

An **operation** is simply a mathematical process that takes some value(s) as input(s) and produces an output. Elementary operations are often written in the following form: *value operation value*. For instance, in the expression $1 + 2$ the values are 1 and 2 and the operation is addition. Performing the operation gives the output of 3. In this way we can say that $1 + 2$ and 3 are equal, or $1 + 2 = 3$.

ADDITION

Addition increases the value of one quantity by the value of another quantity (both called **addends**). Example: $2 + 4 = 6$ or $8 + 9 = 17$. The result is called the **sum**. With addition, the order does not matter, $4 + 2 = 2 + 4$.

When adding signed numbers, if the signs are the same simply add the absolute values of the addends and apply the original sign to the sum. For example, $(+4) + (+8) = +12$ and $(-4) + (-8) = -12$. When the original signs are different, take the absolute values of the addends and subtract the smaller value from the larger value, then apply the original sign of the larger value to the difference. Example: $(+4) + (-8) = -4$ and $(-4) + (+8) = +4$.

SUBTRACTION

Subtraction is the opposite operation to addition; it decreases the value of one quantity (the **minuend**) by the value of another quantity (the **subtrahend**). For example, $6 - 4 = 2$ or $17 - 8 = 9$. The result is called the **difference**. Note that with subtraction, the order does matter, $6 - 4 \neq 4 - 6$.

For subtracting signed numbers, change the sign of the subtrahend and then follow the same rules used for addition. Example: $(+4) - (+8) = (+4) + (-8) = -4$

Multiplication

Multiplication can be thought of as repeated addition. One number (the **multiplier**) indicates how many times to add the other number (the **multiplicand**) to itself. Example: $3 \times 2 = 2 + 2 + 2 = 6$. With multiplication, the order does not matter, $2 \times 3 = 3 \times 2$ or $3 + 3 = 2 + 2 + 2$, either way the result (the **product**) is the same.

If the signs are the same, the product is positive when multiplying signed numbers. Example: $(+4) \times (+8) = +32$ and $(-4) \times (-8) = +32$. If the signs are opposite, the product is negative. Example: $(+4) \times (-8) = -32$ and $(-4) \times (+8) = -32$. When more than two factors are multiplied together, the sign of the product is determined by how many negative factors are present. If there are an odd number of negative factors then the product is negative, whereas an even number of negative factors indicates a positive product. Example: $(+4) \times (-8) \times (-2) = +64$ and $(-4) \times (-8) \times (-2) = -64$.

Division

Division is the opposite operation to multiplication; one number (the **divisor**) tells us how many parts to divide the other number (the **dividend**) into. The result of division is called the **quotient**. Example: $20 \div 4 = 5$. If 20 is split into 4 equal parts, each part is 5. With division, the order of the numbers does matter, $20 \div 4 \neq 4 \div 20$.

The rules for dividing signed numbers are similar to multiplying signed numbers. If the dividend and divisor have the same sign, the quotient is positive. If the dividend and divisor have opposite signs, the quotient is negative. Example: $(-4) \div (+8) = -0.5$.

> **Review Video: Mathematical Operations**
> Visit mometrix.com/academy and enter code: 208095

Parentheses

Parentheses are used to designate which operations should be done first when there are multiple operations. Example: $4 - (2 + 1) = 1$; the parentheses tell us that we must add 2 and 1, and then subtract the sum from 4, rather than subtracting 2 from 4 and then adding 1 (this would give us an answer of 3).

> **Review Video: Mathematical Parentheses**
> Visit mometrix.com/academy and enter code: 978600

Exponents

An **exponent** is a superscript number placed next to another number at the top right. It indicates how many times the base number is to be multiplied by itself. Exponents provide a shorthand way to write what would be a longer mathematical expression, Example $2^4 = 2 \times 2 \times 2 \times 2$. A number with an exponent of 2 is said to be "squared," while a number with an exponent of 3 is said to be "cubed." The value of a number raised to an exponent is called its power. So 8^4 is read as "8 to the 4th power," or "8 raised to the power of 4."

Review Video: Exponents
Visit mometrix.com/academy and enter code: 600998

Roots

A **root**, such as a square root, is another way of writing a fractional exponent. Instead of using a superscript, roots use the radical symbol ($\sqrt{\ }$) to indicate the operation. A radical will have a number underneath the bar, and may sometimes have a number in the upper left: $\sqrt[n]{a}$, read as "the nth root of a." The relationship between radical notation and exponent notation can be described by this equation:

$$\sqrt[n]{a} = a^{\frac{1}{n}}$$

The two special cases of $n = 2$ and $n = 3$ are called square roots and cube roots. If there is no number to the upper left, the radical is understood to be a square root ($n = 2$). Nearly all of the roots you encounter will be square roots. A square root is the same as a number raised to the one-half power. When we say that a is the square root of b ($a = \sqrt{b}$), we mean that a multiplied by itself equals b: ($a \times a = b$).

A **perfect square** is a number that has an integer for its square root. There are 10 perfect squares from 1 to 100: 1, 4, 9, 16, 25, 36, 49, 64, 81, 100 (the squares of integers 1 through 10).

Review Video: Roots
Visit mometrix.com/academy and enter code: 795655

Review Video: Perfect Squares and Square Roots
Visit mometrix.com/academy and enter code: 648063

Word Problems and Mathematical Symbols

When working on word problems, you must be able to translate verbal expressions or "math words" into math symbols. This chart contains several "math words" and their appropriate symbols:

Phrase	Symbol
equal, is, was, will be, has, costs, gets to, is the same as, becomes	$=$
times, of, multiplied by, product of, twice, doubles, halves, triples	$\times$
divided by, per, ratio of/to, out of	$\div$
plus, added to, sum, combined, and, more than, totals of	$+$
subtracted from, less than, decreased by, minus, difference between	$-$
what, how much, original value, how many, a number, a variable	x, n, etc.

Review Video: Understanding Word Problems
Visit mometrix.com/academy and enter code: 499199

Examples of Translated Mathematical Phrases

- The phrase four more than twice a number can be written algebraically as $2x + 4$.
- The phrase half a number decreased by six can be written algebraically as $\frac{1}{2}x - 6$.
- The phrase the sum of a number and the product of five and that number can be written algebraically as $x + 5x$.

- You may see a test question that says, "Olivia is constructing a bookcase from seven boards. Two of them are for vertical supports and five are for shelves. The height of the bookcase is twice the width of the bookcase. If the seven boards total 36 feet in length, what will be the height of Olivia's bookcase?" You would need to make a sketch and then create the equation to determine the width of the shelves. The height can be represented as double the width. (If x represents the width of the shelves in feet, then the height of the bookcase is $2x$. Since the seven boards total 36 feet, $2x + 2x + x + x + x + x + x = 36$ or $9x = 36$; $x = 4$. The height is twice the width, or 8 feet.)

Subtraction with Regrouping

A great way to make use of some of the features built into the decimal system would be regrouping when attempting longform subtraction operations. When subtracting within a place value, sometimes the minuend is smaller than the subtrahend, **regrouping** enables you to 'borrow' a unit from a place value to the left in order to get a positive difference. For example, consider subtracting 189 from 525 with regrouping.

First, set up the subtraction problem in vertical form:

$$\begin{array}{r} 525 \\ -\ 189 \\ \hline \end{array}$$

Notice that the numbers in the ones and tens columns of 525 are smaller than the numbers in the ones and tens columns of 189. This means you will need to use regrouping to perform subtraction:

$$\begin{array}{rrrr} & 5 & 2 & 5 \\ - & 1 & 8 & 9 \\ \hline \end{array}$$

To subtract 9 from 5 in the ones column you will need to borrow from the 2 in the tens columns:

$$\begin{array}{rrrr} & 5 & 1 & 15 \\ - & 1 & 8 & 9 \\ \hline & & & 6 \end{array}$$

Next, to subtract 8 from 1 in the tens column you will need to borrow from the 5 in the hundreds column:

$$\begin{array}{rrrr} & 4 & 11 & 15 \\ - & 1 & 8 & 9 \\ \hline & & 3 & 6 \end{array}$$

Last, subtract the 1 from the 4 in the hundreds column:

$$\begin{array}{rrrr} & 4 & 11 & 15 \\ - & 1 & 8 & 9 \\ \hline & 3 & 3 & 6 \end{array}$$

Review Video: Subtracting Large Numbers
Visit mometrix.com/academy and enter code: 603350

Order of Operations

The **order of operations** is a set of rules that dictates the order in which we must perform each operation in an expression so that we will evaluate it accurately. If we have an expression that includes multiple different operations, the order of operations tells us which operations to do first. The most common mnemonic for the order of operations is **PEMDAS**, or "Please Excuse My Dear Aunt Sally." PEMDAS stands for parentheses, exponents, multiplication, division, addition, and subtraction. It is important to understand that multiplication and division have equal precedence, as do addition and subtraction, so those pairs of operations are simply worked from left to right in order.

For example, evaluating the expression $5 + 20 \div 4 \times (2 + 3)^2 - 6$ using the correct order of operations would be done like this:

- **P:** Perform the operations inside the parentheses: $(2 + 3) = 5$
- **E:** Simplify the exponents: $(5)^2 = 5 \times 5 = 25$
    - The expression now looks like this: $5 + 20 \div 4 \times 25 - 6$
- **MD:** Perform multiplication and division from left to right: $20 \div 4 = 5$; then $5 \times 25 = 125$
    - The expression now looks like this: $5 + 125 - 6$
- **AS:** Perform addition and subtraction from left to right: $5 + 125 = 130$; then $130 - 6 = 124$

Review Video: Order of Operations
Visit mometrix.com/academy and enter code: 259675

Properties of Operations

The Commutative Property

The commutative property applies to addition and multiplication and states that these operations can be completed in any order. The **commutative property of addition** states that numbers and terms can be added together in any order to still get the same value. For example, $3 + 4 = 7$ and $4 + 3 = 7$. Also, we can use the commutative property of addition to show that $3x + 4 + 2^2$ is equivalent to $4 + 3x + 2^2$ and $2^2 + 4 + 3x$. When adding terms, you can add in any order and get the same value.

The **commutative property of multiplication** states that numbers and terms can be multiplied in any order to get the same value. For example, 12×3 is equivalent to 3×12. Additionally, we can use the commutative property of multiplication to assume that $(5 + 3) \times (36 - 6)$ is equivalent to $(36 - 6) \times (5 + 3)$. You can multiply terms in any order and still get the same value.

The Associative Property

The **associative property of addition** states that if three or more terms are being added together, the value is the same regardless of the groupings.

For example, given the expression $3 + 4 + 6$, these terms can be grouped and added in any form. $3 + 4 + 6$ is equivalent to $(3 + 4) + 6$ and is also equivalent to $3 + (4 + 6)$. This can be applied to write equivalent expressions in a variety of ways.

For example, suppose we are given the expression $5 + (y + 2) + 4$. We can generate equivalent expressions knowing the associative property. Knowing that when three or more terms are added, the grouping is irrelevant, we can say that this expression is equivalent to $5 + y + (2 + 4)$, and it is equivalent to $(5 + y) + (2 + 4)$. It is even equivalent to $5 + y + 2 + 4$.

The **associative property of multiplication** states that if three or more terms are being multiplied together, the value is the same regardless of the grouping. We can use this property to identify and generate equivalent expressions.

For example, given the expression $2 \times 7 \times 3$, these terms can be grouped in any way and still get the same value. $2 \times 7 \times 3$ is equivalent to $(2 \times 7) \times 3$ or $2 \times (7 \times 3)$.

The Identity Property

The **identity property of multiplication** states that when a number is multiplied by 1, you get the same number. That is, anything multiplied by 1 is itself. For example, $2 \times 1 = 2$, or $1 \times -36 = -36$. Using the identity property of multiplication, we can identify and generate equivalent expressions. Let's say that we are given the expression $15 - (3 \times 4)$. We can generate equivalent expressions using the identity property. One equivalent expression example would be $(15 \times 1) - (3 \times 4)$. Another example would be $15 - (1 \times 3 \times 4)$. We can say these expressions are equivalent because the identity property of multiplication states that we can multiply any portion of an expression by 1 to get the same value.

The **identity property of addition** states that when 0 is added to a number, you get the same number. For example, $2 + 0 = 2$, or $0 + -3 = -3$. We can also use this property to identify and generate equivalent expressions. For example, if we are given the expression $2 \times (1 + 2)$, we could write the equivalent expressions $2 \times (0 + 1 + 2)$ or $(2 + 0) \times (1 + 2)$.

The Inverse Property

The **inverse property of addition** states that the sum of a number and its opposite is always equal to 0. Remember, the opposite of a number is a number that is opposite on the number line from zero, or the same number with the opposite sign. For example, −4 is opposite to 4, and 1,726.9 is opposite to −1,726.9. So, the inverse property of addition states that if you add opposite numbers, their sum is zero. For example, $5 + (-5) = 0$ and $-5 + 5 = 0$.

The **inverse property of multiplication** states that a number multiplied by its reciprocal is always equal to 1. The **reciprocal** of a number is its "flipped" fraction. For example, the reciprocal of 5 is $\frac{1}{5}$, or the reciprocal of $\frac{2}{3}$ is $\frac{3}{2}$. The inverse property of multiplication can be applied for these values, $5 \times \frac{1}{5} = 1$ and $\frac{2}{3} \times \frac{3}{2} = 1$. This is because when you multiply across, you get a fraction that is equal to 1.

$$\frac{2}{3} \times \frac{3}{2} = \frac{6}{6} = 1$$

The Distributive Property

The **distributive property** explains how multiplication and addition interact. It says that when multiplying one number by the sum of two other numbers, the same result can also be obtained by multiplying the one number by each of the numbers individually and then adding the products. For example, to multiply 2 by the sum of 7 and 3, the direct approach says, "the sum of 7 and 3 is 10, and 2 times 10 is 20." This would be expressed as $2 \times (7 + 3) = 2 \times 10 = 20$. On the other hand, the distributive property states that the same answer can be achieved by multiplying each number inside the parentheses and adding the products. That is, "the product of 2 and 7 is 14, the product of

2 and 3 is 6, and the sum of 14 and 6 is 20." This would be expressed as $2 \times (7 + 3) = 2 \times 7 + 2 \times 3 = 14 + 6 = 20$, and it is demonstrated below.

$$2 \times (7 + 3) = 2 \times 7 + 2 \times 3$$

This same concept can be used when multiplying a number by the difference of two numbers. For example, $5 \times (10 - 4) = 5 \times 10 - 5 \times 4$. Since $5 \times 10 = 50$ and $5 \times 4 = 20$, the result is $50 - 20 = 30$. This answer can be checked by subtracting inside the parentheses first and then multiplying: $5 \times (10 - 4) = 5 \times 6 = 30$.

Review Video: Commutative, Associative, and Distributive Properties
Visit mometrix.com/academy and enter code: 483176

Properties of Exponents

The properties of exponents are as follows:

Property	Description
$a^1 = a$	Any number to the power of 1 is equal to itself
$1^n = 1$	The number 1 raised to any power is equal to 1
$a^0 = 1$	Any number raised to the power of 0 is equal to 1
$a^n \times a^m = a^{n+m}$	Add exponents to multiply powers of the same base number
$a^n \div a^m = a^{n-m}$	Subtract exponents to divide powers of the same base number
$(a^n)^m = a^{n \times m}$	When a power is raised to a power, the exponents are multiplied
$(a \times b)^n = a^n \times b^n$ $(a \div b)^n = a^n \div b^n$	Multiplication and division operations inside parentheses can be raised to a power. This is the same as each term being raised to that power.
$a^{-n} = \frac{1}{a^n}$	A negative exponent is the same as the reciprocal of a positive exponent

Note that exponents do not have to be integers. Fractional or decimal exponents follow all the rules above as well. Example: $5^{\frac{1}{4}} \times 5^{\frac{3}{4}} = 5^{\frac{1}{4}+\frac{3}{4}} = 5^1 = 5$.

Review Video: Properties of Exponents
Visit mometrix.com/academy and enter code: 532558

Factors and Multiples

Factors and Greatest Common Factor

A whole number a is a **factor** (or **divisor**) of a whole number b if a divides b evenly. In other words, a is a factor of b if the quotient $b \div a$ is a whole number with a remainder of 0. For instance, 3 is a factor of 12 because $12 \div 3 = 4$ with no remainder. Another way to say this is that a is a factor of b if we can multiply a by another whole number to get b. So, we can also show that 3 is a factor of 12 by noting that $3 \times 4 = 12$.

Every positive whole number has 1 and itself as factors. If a whole number greater than one has *only* 1 and itself as factors, we call it a **prime number**. For instance, 5 is a prime number because its only factors are 1 and 5. The first several prime numbers are 2, 3, 5, 7, 11, and 13.

If a whole number greater than 1 is not prime—that is, if it has factors besides 1 and itself—then it is a **composite number.** For instance, 10 is a composite number because it has factors 2 and 5 in addition to 1 and 10. The first several composite numbers are 4, 6, 8, 9, 10, 12, 14, and 15.

A **prime factor** of a whole number is a factor that is also a prime number. For example, the prime factors of 12 are 2 and 3. The prime factors of 15 are 3 and 5.

A **common factor** of two (or more) whole numbers is a number that is a factor of both (or all) of them. For example, the factors of 12 are <u>1</u>, 2, <u>3</u>, 4, 6, and 12, while the factors of 15 are <u>1</u>, <u>3</u>, 5, and 15. The common factors (underlined) of 12 and 15 are 1 and 3.

The **greatest common factor** (**GCF**) of two (or more) whole numbers is the largest number that is a factor of both (or all) of them. For example, the factors of 15 are <u>1</u>, 3, <u>5</u>, and 15; the factors of 35 are <u>1</u>, <u>5</u>, 7, and 35. Therefore, the greatest common factor of 15 and 35 is 5.

Review Video: Factors
Visit mometrix.com/academy and enter code: 920086

Review Video: Prime Numbers and Factorization
Visit mometrix.com/academy and enter code: 760669

Multiples and Least Common Multiple

A whole number b is a **multiple** of a whole number a when a is a factor of b. This means that b is the product of a and another whole number. For example, the multiples of 7 are $0 \times 7 = 0, 1 \times 7 = 7, 2 \times 7 = 14, 3 \times 7 = 21, 4 \times 7 = 28, 5 \times 7 = 35$, ... Dividing 0, 7, 14, 21, 28, and 35 by 7 results in the whole numbers 0, 1, 2, 3, 4, and 5, respectively, showing that 7 is a factor of these numbers.

The least common multiple (**LCM**) of two (or more) whole numbers is the smallest number that is a multiple of both (or all) of them. For example, the multiples of 3 are 3, 6, 9, 12, <u>15</u>, ...; the multiples of 5 are 5, 10, <u>15</u>, 20, ... The smallest number that appears in both lists is 15, so the least common multiple of 3 and 5 is 15.

Review Video: Multiples
Visit mometrix.com/academy and enter code: 626738

Review Video: Greatest Common Factor and Least Common Multiple
Visit mometrix.com/academy and enter code: 838699

Fractions

A **fraction** is a number that is expressed as one integer written above another integer, with a dividing line between them $\left(\frac{x}{y}\right)$. It represents the **quotient** of the two numbers "x divided by y." It can also be thought of as x out of y equal parts.

The top number of a fraction is called the **numerator**, and it represents the number of parts under consideration. The 1 in $\frac{1}{4}$ means that 1 part out of the whole is being considered in the calculation.

The bottom number of a fraction is called the **denominator**, and it represents the total number of equal parts. The 4 in $\frac{1}{4}$ means that the whole consists of 4 equal parts. A fraction cannot have a denominator of zero; this is referred to as "*undefined.*"

Fractions can be manipulated, without changing the value of the fraction, by multiplying or dividing (but not adding or subtracting) both the numerator and denominator by the same number. If you divide both numbers by a common factor, you are **reducing** or simplifying the fraction. Two fractions that have the same value but are expressed differently are known as **equivalent fractions**. For example, $\frac{2}{10}, \frac{3}{15}, \frac{4}{20}$, and $\frac{5}{25}$ are all equivalent fractions. They can also all be reduced or simplified to $\frac{1}{5}$.

When two fractions are manipulated so that they have the same denominator, this is known as finding a **common denominator**. The number chosen to be that common denominator should be the least common multiple of the two original denominators. Example: $\frac{3}{4}$ and $\frac{5}{6}$; the least common multiple of 4 and 6 is 12. Manipulating to achieve the common denominator: $\frac{3}{4} = \frac{9}{12}; \frac{5}{6} = \frac{10}{12}$.

Review Video: Overview of Fractions
Visit mometrix.com/academy and enter code: 262335

Proper Fractions and Mixed Numbers

A fraction whose denominator is greater than its numerator is known as a **proper fraction**, while a fraction whose numerator is greater than its denominator is known as an **improper fraction**. Proper fractions have values *less than one* and improper fractions have values *greater than one.*

A **mixed number** is a number that contains both an integer and a fraction. Any improper fraction can be rewritten as a mixed number. Example: $\frac{8}{3} = \frac{6}{3} + \frac{2}{3} = 2 + \frac{2}{3} = 2\frac{2}{3}$. Similarly, any mixed number can be rewritten as an improper fraction. Example: $1\frac{3}{5} = 1 + \frac{3}{5} = \frac{5}{5} + \frac{3}{5} = \frac{8}{5}$.

Review Video: Proper and Improper Fractions and Mixed Numbers
Visit mometrix.com/academy and enter code: 211077

Adding and Subtracting Fractions

If two fractions have a common denominator, they can be added or subtracted simply by adding or subtracting the two numerators and retaining the same denominator. If the two fractions do not already have the same denominator, one or both of them must be manipulated to achieve a common denominator before they can be added or subtracted. Example: $\frac{1}{2} + \frac{1}{4} = \frac{2}{4} + \frac{1}{4} = \frac{3}{4}$.

Review Video: Adding and Subtracting Fractions
Visit mometrix.com/academy and enter code: 378080

Multiplying Fractions

Two fractions can be multiplied by multiplying the two numerators to find the new numerator and the two denominators to find the new denominator. Example: $\frac{1}{3} \times \frac{2}{3} = \frac{1 \times 2}{3 \times 3} = \frac{2}{9}$.

Dividing Fractions

Two fractions can be divided by flipping the numerator and denominator of the second fraction and then proceeding as though it were a multiplication problem. Example: $\frac{2}{3} \div \frac{3}{4} = \frac{2}{3} \times \frac{4}{3} = \frac{8}{9}$.

> **Review Video: Multiplying and Dividing Fractions**
> Visit mometrix.com/academy and enter code: 473632

Multiplying a Mixed Number by a Whole Number or a Decimal

When multiplying a mixed number by something, it is usually best to convert it to an improper fraction first. Additionally, if the multiplicand is a decimal, it is most often simplest to convert it to a fraction. For instance, to multiply $4\frac{3}{8}$ by 3.5, begin by rewriting each quantity as a whole number plus a proper fraction. Remember, a mixed number is a fraction added to a whole number and a decimal is a representation of the sum of fractions, specifically tenths, hundredths, thousandths, and so on:

$$4\frac{3}{8} \times 3.5 = \left(4 + \frac{3}{8}\right) \times \left(3 + \frac{1}{2}\right)$$

Next, the quantities being added need to be expressed with the same denominator. This is achieved by multiplying and dividing the whole number by the denominator of the fraction. Recall that a whole number is equivalent to that number divided by 1:

$$= \left(\frac{4}{1} \times \frac{8}{8} + \frac{3}{8}\right) \times \left(\frac{3}{1} \times \frac{2}{2} + \frac{1}{2}\right)$$

When multiplying fractions, remember to multiply the numerators and denominators separately:

$$= \left(\frac{4 \times 8}{1 \times 8} + \frac{3}{8}\right) \times \left(\frac{3 \times 2}{1 \times 2} + \frac{1}{2}\right)$$
$$= \left(\frac{32}{8} + \frac{3}{8}\right) \times \left(\frac{6}{2} + \frac{1}{2}\right)$$

Now that the fractions have the same denominators, they can be added:

$$= \frac{35}{8} \times \frac{7}{2}$$

Finally, perform the last multiplication and then simplify:

$$= \frac{35 \times 7}{8 \times 2} = \frac{245}{16} = \frac{240}{16} - \frac{5}{16} = 15\frac{5}{16}$$

Comparing Fractions

It is important to master the ability to compare and order fractions. This skill is relevant to many real-world scenarios. For example, carpenters often compare fractional construction nail lengths when preparing for a project, and bakers often compare fractional measurements to have the correct ratio of ingredients. There are three commonly used strategies when comparing fractions. These strategies are referred to as the common denominator approach, the decimal approach, and the cross-multiplication approach.

Using a Common Denominator to Compare Fractions

The fractions $\frac{2}{3}$ and $\frac{4}{7}$ have different denominators. $\frac{2}{3}$ has a denominator of 3, and $\frac{4}{7}$ has a denominator of 7. In order to precisely compare these two fractions, it is necessary to use a common denominator. A common denominator is a common multiple that is shared by both denominators. In this case, the denominators 3 and 7 share a multiple of 21. In general, it is most efficient to select the least common multiple for the two denominators.

Rewrite each fraction with the common denominator of 21. Then, calculate the new numerators as illustrated below.

$$\frac{2}{3} \overset{\times 7}{=} \frac{\mathbf{14}}{21} \qquad \frac{4}{7} \overset{\times 3}{=} \frac{\mathbf{12}}{21}$$

For $\frac{2}{3}$, multiply the numerator and denominator by 7. The result is $\frac{14}{21}$.

For $\frac{4}{7}$, multiply the numerator and denominator by 3. The result is $\frac{12}{21}$.

Now that both fractions have a denominator of 21, the fractions can accurately be compared by comparing the numerators. Since 14 is greater than 12, the fraction $\frac{14}{21}$ is greater than $\frac{12}{21}$. This means that $\frac{2}{3}$ is greater than $\frac{4}{7}$.

Using Decimals to Compare Fractions

Sometimes decimal values are easier to compare than fraction values. For example, $\frac{5}{8}$ is equivalent to 0.625 and $\frac{3}{5}$ is equivalent to 0.6. This means that the comparison of $\frac{5}{8}$ and $\frac{3}{5}$ can be determined by comparing the decimals 0.625 and 0.6. When both decimal values are extended to the thousandths place, they become 0.625 and 0.600, respectively. It becomes clear that 0.625 is greater than 0.600 because 625 thousandths is greater than 600 thousandths. In other words, $\frac{5}{8}$ is greater than $\frac{3}{5}$ because 0.625 is greater than 0.6.

Using Cross-Multiplication to Compare Fractions

Cross-multiplication is an efficient strategy for comparing fractions. This is a shortcut for the common denominator strategy. Start by writing each fraction next to one another. Multiply the numerator of the fraction on the left by the denominator of the fraction on the right. Write down the result next to the fraction on the left. Now multiply the numerator of the fraction on the right by the denominator of the fraction on the left. Write down the result next to the fraction on the right. Compare both products. The fraction with the larger result is the larger fraction.

Consider the fractions $\frac{4}{7}$ and $\frac{5}{9}$.

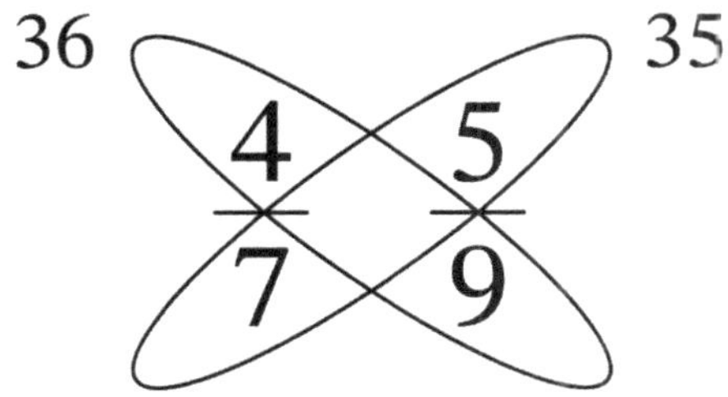

36 is greater than 35. Therefore, $\frac{4}{7}$ is greater than $\frac{5}{9}$.

Decimals

Decimals are one way to represent parts of a whole. Using the place value system, each digit to the right of a decimal point denotes the number of units of a corresponding *negative* power of ten. For example, consider the decimal 0.24. We can use a model to represent the decimal. Since a dime is worth one-tenth of a dollar and a penny is worth one-hundredth of a dollar, one possible model to represent this fraction is to have 2 dimes representing the 2 in the tenths place and 4 pennies representing the 4 in the hundredths place:

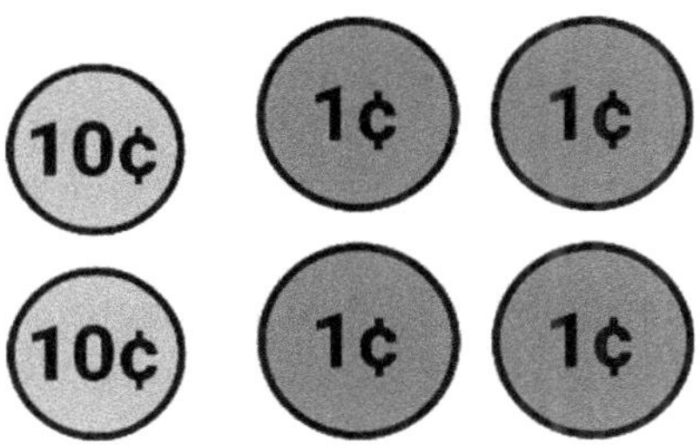

To write the decimal as a fraction, put the decimal in the numerator with 1 in the denominator. Multiply the numerator and denominator by tens until there are no more decimal places. Then simplify the fraction to lowest terms. For example, converting 0.24 to a fraction:

$$0.24 = \frac{0.24}{1} = \frac{0.24 \times 100}{1 \times 100} = \frac{24}{100} = \frac{6}{25}$$

Review Video: Decimals
Visit mometrix.com/academy and enter code: 837268

Operations with Decimals

Adding and Subtracting Decimals

When adding and subtracting decimals, the decimal points must always be aligned. Adding decimals is just like adding regular whole numbers. Example: $4.5 + 2.0 = 6.5$.

If the problem-solver does not properly align the decimal points, an incorrect answer of 4.7 may result. An easy way to add decimals is to align all of the decimal points in a vertical column visually. This will allow you to see exactly where the decimal should be placed in the final answer. Begin adding from right to left. Add each column in turn, making sure to carry the number to the left if a column adds up to more than 9. The same rules apply to the subtraction of decimals.

Review Video: Adding and Subtracting Decimals
Visit mometrix.com/academy and enter code: 381101

MULTIPLYING DECIMALS

A simple multiplication problem has two components: a **multiplicand** and a **multiplier**. When multiplying decimals, work as though the numbers were whole rather than decimals. Once the final product is calculated, count the number of places to the right of the decimal in both the multiplicand and the multiplier. Then, count that number of places from the right of the product and place the decimal in that position.

For example, 12.3×2.56 has a total of three places to the right of the respective decimals. Multiply 123×256 to get 31,488. Now, beginning on the right, count three places to the left and insert the decimal. The final product will be 31.488.

Review Video: How to Multiply Decimals
Visit mometrix.com/academy and enter code: 731574

DIVIDING DECIMALS

Every division problem has a **divisor** and a **dividend**. The dividend is the number that is being divided. In the problem $14 \div 7$, 14 is the dividend and 7 is the divisor. In a division problem with decimals, the divisor must be converted into a whole number. Begin by moving the decimal in the divisor to the right until a whole number is created. Next, move the decimal in the dividend the same number of spaces to the right. For example, 4.9 into 24.5 would become 49 into 245. The decimal was moved one space to the right to create a whole number in the divisor, and then the same was done for the dividend. Once the whole numbers are created, the problem is carried out normally: $245 \div 49 = 5$.

Review Video: Dividing Decimals
Visit mometrix.com/academy and enter code: 560690

Review Video: Dividing Decimals by Whole Numbers
Visit mometrix.com/academy and enter code: 535669

Percentages

Percentages can be thought of as fractions that are based on a whole of 100; that is, one whole is equal to 100%. The word **percent** means "per hundred." Percentage problems are often presented in three main ways:

- Find what percentage of some number another number is.
  - Example: What percentage of 40 is 8?
- Find what number is some percentage of a given number.
  - Example: What number is 20% of 40?
- Find what number another number is a given percentage of.
  - Example: What number is 8 20% of?

There are three components in each of these cases: a **whole** (W), a **part** (P), and a **percentage** (%). These are related by the equation: $P = W \times \%$. This can easily be rearranged into other forms that may suit different questions better: $\% = \frac{P}{W}$ and $W = \frac{P}{\%}$. Percentage problems are often also word problems. As such, a large part of solving them is figuring out which quantities are what. For example, consider the following word problem:

In a school cafeteria, 7 students choose pizza, 9 choose hamburgers, and 4 choose tacos. What percentage of student choose tacos?

To find the whole, you must first add all of the parts: $7 + 9 + 4 = 20$. The percentage can then be found by dividing the part by the whole $\left(\% = \frac{P}{W}\right)$: $\frac{4}{20} = \frac{20}{100} = 20\%$.

Review Video: Computation with Percentages
Visit mometrix.com/academy and enter code: 693099

Calculating Percent Change

Suppose a quantity has a particular value (the *old value*) and then we add something (the *change*) to it to get another value (the *new value*). We can describe this process by the simple equation (old value) + change = (new value). If we know the old and new values, we can rearrange this equation to find the change, getting change = (new value) − (old value). For instance, if a store's price for a box of computer paper goes from \$20 last week to \$25 this week, this is a change of (new value) − (old value) = \$25 − \$20 = \$5. Or, if the size of the freshman class at a college goes from 500 students one year to 440 students the next year, this is a change of (new value) − (old value) = 440 − 500 = −60 students. So, we see that change can be positive or negative.

Instead of the word *change*, we sometimes use the words *increase* or *decrease* to specify whether the value goes up or down, respectively. In the examples above, the price of computer paper increases by \$5 and the freshman class decreases by 60 students. Note that the decrease is 60 students and not -60 because the word *decrease* already means that the value goes down. So, *increase* is the same as positive change and *decrease* is the opposite or negative change.

If the changing quantity represents an amount (how much of something there is), we can also calculate the **percent change**. This is the change expressed as a percentage of the old amount. To calculate this, we divide the change by the old amount and express the quotient as a percent. That is, we use the formula percent change $= \frac{\text{change}}{\text{old value}}$, converting the resulting decimal answer to a percent. In the examples above, the price of a box of computer paper has a percent change of $\frac{\text{change in price}}{\text{old price}} = \frac{\$5}{\$20} = 0.25 = 25\%$, and the size of the freshman class at the college has a percent change of $\frac{\text{change in enrollment}}{\text{old enrollment}} = \frac{-60}{500} = -0.12 = -12\%$. We can also use the terms *percent increase* and *percent decrease*, saying that the price of computer paper increases by 25% and the size of the freshman class decreases by 12%. Note that the denominator is always the old amount, never the new amount.

Example: Your landlord raises your rent from \$1,500 to \$1,700 per month. To find the percent change in your rent (rounded to the nearest tenth of a percent), you calculate as follows.

$$\text{percent change in rent} = \frac{\text{change in rent}}{\text{old rent}} = \frac{(\text{new rent}) - (\text{old rent})}{\text{old rent}}$$
$$= \frac{\$1{,}700 - \$1{,}500}{\$1{,}500} = \frac{\$200}{\$1{,}500} = 0.1333\ldots \approx 13.3\%$$

Therefore, the percent change in your rent is approximately 13.3%.

Review Video: Percent Change
Visit mometrix.com/academy and enter code: 907890

Converting Between Percentages, Fractions, and Decimals

Converting decimals to percentages and percentages to decimals is as simple as moving the decimal point. To *convert from a decimal to a percentage*, move the decimal point **two places to the right**. To *convert from a percentage to a decimal*, move it **two places to the left**. It may be helpful to remember that the percentage number will always be larger than the equivalent decimal number. Example:

$$0.23 = 23\% \qquad 5.34 = 534\% \qquad 0.007 = 0.7\%$$
$$700\% = 7.00 \qquad 86\% = 0.86 \qquad 0.15\% = 0.0015$$

To convert a fraction to a decimal, simply divide the numerator by the denominator in the fraction. To convert a decimal to a fraction, put the decimal in the numerator with 1 in the denominator. Multiply the numerator and denominator by tens until there are no more decimal places. Then simplify the fraction to lowest terms. For example, converting 0.24 to a fraction:

$$0.24 = \frac{0.24}{1} = \frac{0.24 \times 100}{1 \times 100} = \frac{24}{100} = \frac{6}{25}$$

Fractions can be converted to a percentage by finding equivalent fractions with a denominator of 100. Example:

$$\frac{7}{10} = \frac{70}{100} = 70\% \quad \frac{1}{4} = \frac{25}{100} = 25\%$$

To convert a percentage to a fraction, divide the percentage number by 100 and reduce the fraction to its simplest possible terms. Example:

$$60\% = \frac{60}{100} = \frac{3}{5} \quad 96\% = \frac{96}{100} = \frac{24}{25}$$

Review Video: Converting Fractions to Percentages and Decimals
Visit mometrix.com/academy and enter code: 306233

Review Video: Converting Percentages to Decimals and Fractions
Visit mometrix.com/academy and enter code: 287297

Review Video: Converting Decimals to Fractions and Percentages
Visit mometrix.com/academy and enter code: 986765

Review Video: Converting Decimals, Improper Fractions, and Mixed Numbers
Visit mometrix.com/academy and enter code: 696924

Rational and Irrational Numbers

The term **rational** means that the number can be expressed as a ratio or fraction. That is, a number, r, is rational if and only if it can be represented by a fraction $\frac{a}{b}$ where a and b are integers and b does not equal 0. The set of rational numbers includes integers and decimals. If there is no finite way to represent a value with a fraction of integers, then the number is **irrational**. Common irrational numbers are π and the square roots of whole numbers that are not perfect squares (e.g., $\sqrt{5}$ or $\sqrt{21}$). The sum or product of an integer and an irrational number is always irrational (e.g., 3π or $7 + \sqrt{6}$).

Review Video: Rational and Irrational Numbers
Visit mometrix.com/academy and enter code: 280645

Review Video: Ordering Rational Numbers
Visit mometrix.com/academy and enter code: 419578

Review Video: Irrational Numbers on a Number Line
Visit mometrix.com/academy and enter code: 433866

Chapter Quiz

Ready to see how well you retained what you just read? Scan the QR code to go directly to the chapter quiz interface for this study guide. If you're using a computer, simply visit the online resources page at **mometrix.com/resources719/priimath5161** and click the Chapter Quizzes link.

Algebra and Functions

Transform passive reading into active learning! After immersing yourself in this chapter, put your comprehension to the test by taking a quiz. The insights you gained will stay with you longer this way. Scan the QR code to go directly to the chapter quiz interface for this study guide. If you're using a computer, simply visit the online resources page at **mometrix.com/resources719/priimath5161** and click the Chapter Quizzes link.

Proportionality

Proportions and Ratios

Proportions

There is a **proportion** between two variable quantities if there is a constant relationship between their products or quotients, a relationship that does not change as the quantities themselves change.

Given variable quantities x and y, we say that they are **directly proportional** (or that y **varies directly with** x) if their quotient or *ratio* is constant—that is, if there is a constant k such that $\frac{y}{x} = k$ is always true. Another way of saying this is that y is a constant multiple of x, so that $y = kx$ is always true. We call the number k the **constant of proportionality**. For example, if you drive at a constant 50 miles per hour, then the distance, y, that you travel in miles is 50 times the number of hours, x, that you drive. In symbols, $y = 50x$ miles (or $\frac{y}{x} = 50$ mph). So, the distance you travel, y, is directly proportional to (or varies directly with) the time you travel, x, with constant of proportionality $k = 50$ mph.

The quantities x and y are **inversely proportional** (or y varies inversely with x) if their product is constant—that is, if there is a constant k such that $xy = k$ is always true. Another way of saying this is to say that y is a constant multiple of the reciprocal of x so that $y = \frac{k}{x}$ is always true. For instance, suppose you drive at speed (rate) y mph for x hours, going a total of 120 miles. Since rate × time = distance, we get $xy = 120$ miles (or $y = \frac{120}{x}$ miles per hour). Thus, your driving speed, y, is inversely proportional to (or varies inversely with) your drive time, x, with constant of proportionality $k = 120$ miles.

Review Video: Proportions
Visit mometrix.com/academy and enter code: 505355

Ratios

A **ratio** expresses the sizes of two quantities relative to each other. For instance, suppose we have 3 copies of sheet music to share among 6 singers. We can divide the singers into groups of 2 and give each group 1 copy of the music. Thus, there is 1 copy of the music for every 2 singers, and we say that the **ratio** of sheet music to singers is 1 to 2, which we write either as a fraction $\frac{1}{2}$ or using a colon $1 : 2$. Of course, it is also true there are 3 copies for every 6 singers so that the ratio of sheet music to singers is also 3 to 6, which we write as $\frac{3}{6}$ or $3 : 6$. So, the ratios $\frac{1}{2}$ and $\frac{3}{6}$ express the same relative quantities of music and singers. We say that these ratios are equal or **equivalent**, and we note that ratios are equal precisely when their fractions are equal (so, in this case, $\frac{1}{2} = \frac{3}{6}$ as fractions). We can also express the quantities in the other order and say that the ratio of singers to music is $\frac{2}{1}$ or 2 : 1 (or $\frac{6}{3}$ or 6 : 3).

Review Video: Ratios
Visit mometrix.com/academy and enter code: 996914

Constant of Proportionality

If variable quantities x and y are proportional and we know a pair of corresponding values for them, then we can find their constant of proportionality. If they are directly proportional, we use the formula $\frac{y}{x} = k$. If they are inversely proportional, we use the formula $xy = k$

Example: The cost in dollars, y, of buying fence posts is directly proportional to the number, x, that you buy. If it costs $51 to buy 17 fence posts, what is the constant of proportionality? Because of direct proportionality, we know that $\frac{y}{x} = k$. Since this works for every pair of corresponding x- and y-values, it also works for $x = 17$ and $y = 51$. This gives us $\frac{51}{17} = k$, which simplifies to $k = 3$. Note also that this is the unit price, namely $3 per fence post.

Work/Unit Rate

Unit rate expresses a quantity of one thing in terms of one unit of another. For example, if you travel 30 miles every two hours, a unit rate expresses this comparison in terms of one hour: in one hour you travel 15 miles, so your unit rate is 15 miles per hour. Other examples are how much one ounce of food costs (price per ounce) or figuring out how much one egg costs out of the dozen (price per 1 egg, instead of price per 12 eggs). The denominator of a unit rate is always 1. Unit rates are used to compare different situations to solve problems. For example, to make sure you get the best deal when deciding which kind of soda to buy, you can find the unit rate of each. If soda #1 costs $1.50 for a 1-liter bottle, and soda #2 costs $2.75 for a 2-liter bottle, it would be a better deal to buy soda #2, because its unit rate is only $1.375 per 1-liter, which is cheaper than soda #1. Unit rates can also help determine the length of time a given event will take. For example, if you can paint 2 rooms in 4.5 hours, you can determine how long it will take you to paint 5 rooms by solving for the unit rate per room and then multiplying that by 5.

Review Video: Rates and Unit Rates
Visit mometrix.com/academy and enter code: 185363

Cross Multiplication

Finding an Unknown in Equivalent Expressions

It is often necessary to apply information given about a rate or proportion to a new scenario. For example, if you know that Jedha can run a marathon (26.2 miles) in 3 hours, how long would it take her to run 10 miles at the same pace? Start by setting up equivalent expressions:

$$\frac{26.2\text{ mi}}{3\text{ hr}} = \frac{10\text{ mi}}{x\text{ hr}}$$

Now, cross multiply and solve for x:

$$26.2x = 30$$
$$x = \frac{30}{26.2} = \frac{15}{13.1}$$
$$x \approx 1.15 \text{ hrs } or \text{ } 1 \text{ hr } 9 \text{ min}$$

So, at this pace, Jedha could run 10 miles in about 1.15 hours or about 1 hour and 9 minutes.

Review Video: Cross Multiplying Fractions
Visit mometrix.com/academy and enter code: 893904

Linear Expressions and Equations

Linear Expressions

Terms and Coefficients

Mathematical expressions consist of a combination of one or more values arranged in terms that are added together. As such, an expression could be just a single number, including zero. A **variable term** is the product of a real number, also called a **coefficient**, and one or more variables, each of which may be raised to an exponent. Expressions may also include numbers without a variable, called **constants** or **constant terms**. The expression $6s^2$, for example, is a single term where the coefficient is the real number 6 and the variable term is s^2. Note that if a term is written as simply a variable to some exponent, like t^2, then the coefficient is 1, because $t^2 = 1t^2$.

Linear Expressions

A **single variable linear expression** is the sum of a single variable term, where the variable has no exponent, and a constant, which may be zero. For instance, the expression $2w + 7$ has $2w$ as the variable term and 7 as the constant term. It is important to realize that terms are separated by addition or subtraction. Since an expression is a sum of terms, expressions such as $5x - 3$ can be written as $5x + (-3)$ to emphasize that the constant term is negative. A real-world example of a single variable linear expression is the perimeter of a square, four times the side length, often expressed: $4s$.

In general, a **linear expression** is the sum of any number of variable terms so long as none of the variables have an exponent and none of the terms have two variables multiplied together. For example, $3m + 8n - \frac{1}{4}p + 5.5q - 1$ is a linear expression, but $3y^3$ and $5xy$ are not. In the same way, the expression for the perimeter of a general triangle $(a + b + c)$ is linear, but the expression for the area of a square (s^2) is not.

Slope

Finding Slope Given Graph or Table

On a graph with two points, (x_1, y_1) and (x_2, y_2), the **slope** is found with the formula $m = \frac{y_2 - y_1}{x_2 - x_1}$; where $x_1 \neq x_2$ and m stands for slope. If the value of the slope is **positive**, the line has an *upward direction* from left to right. If the value of the slope is **negative**, the line has a *downward direction* from left to right. Consider the following example:

A new book goes on sale in bookstores and online stores. In the first month, 5,000 copies of the book are sold. Over time, the book continues to grow in popularity. The data for the number of copies sold is in the table below.

# of Months on Sale	1	2	3	4	5
# of Copies Sold (In Thousands)	5	10	15	20	25

So, the number of copies that are sold and the time that the book is on sale is a proportional relationship. In this example, an equation can be used to show the data: $y = 5x$, where x is the number of months that the book is on sale. Also, y is the number of copies sold. So, the slope of the corresponding line is $\frac{\text{rise}}{\text{run}} = \frac{5}{1} = 5$.

FINDING SLOPE GIVEN AN EQUATION

When given an equation of a line, it is necessary to solve for y to determine the slope of the line. Given the equation $6x + 2y = 8$, find the slope. First, subtract $6x$ from both sides of the equation, resulting in $2y = -6x + 8$. Then divide both sides of the equation by 2, resulting in $y = -3x + 4$. This then allows us to conclude that the slope of the line is $m = -3$, the coefficient of x. Once an equation is in the form $y = mx + b$, the slope and y-intercept can easily be determined. For this reason, we refer to the equation $y = mx + b$ as "slope-intercept form" of the equation of a line.

Review Video: Finding the Slope of a Line
Visit mometrix.com/academy and enter code: 766664

LINEAR EQUATIONS

Equations like $5x = 100$ and $8x - 120 = 200$ and $6x + 4y = 240$ are **linear equations**. Linear equations are named based off the number of distinct variables they include. For example, the equation $3x + 30 = 8x$ is a **one-variable linear equation** because it involves only the single variable x. It does not matter that x appears more than once. Any equations that can be written as $ax + b = 0$, where $a \neq 0$, falls into this category. Furthermore, the equation $3x - 5y = 14 + 9y$ is a **two-variable linear equation** because it involves the two variables x and y. The equation $7x + 8y - 12z + 14w = 56$ is a linear equation in four variables.

SATISFYING THE EQUATION

When given a one-variable linear equation, the goal is typically to solve it. This means that we want to find the number that makes the equation true if we substitute it for the variable. That number is the **solution,** or root, of the equation. For instance, the equation $5x = 10$ has the solution $x = 2$. This is true because when 2 is substituted for x, the result is $5 \cdot 2 = 10$, which is true. On the other hand, $x = 6$ can not be a solution because $5 \cdot 6 \neq 10$, so it is false. Two equations with the same solution are **equivalent equations**. For example, the equations $5x = 10$ and $5x + 3 = 13$ are equivalent because both have the same solution of $x = 2$.

DETERMINING A SOLUTION SET

The **solution set** is the set of all solutions of an equation. In the previous example, the solution set would be 2. Solutions to a linear equation in two variables consist of pairs of numbers. For instance, the equation $6x + 4y = 240$ has the solution $x = 20$ and $y = 30$ since $6 \cdot 20 + 4 \cdot 30 = 240$ is true. We can write this solution as the ordered pair (20,30) and plot it as a point on the coordinate plane. Such equations usually have infinitely many solutions; and if we plot the points for all these solutions we get a line, which is a picture of all the solutions. We call this **graphing the equation**. When an equation has no true solutions, it is referred to as an **empty set**.

Linear Equation Forms

Linear equations can be written many ways. Below is a list of some forms linear equations can take:

- **Standard Form**: $Ax + By = C$; the slope is $\frac{-A}{B}$ and the y-intercept is $\frac{C}{B}$
- **Slope Intercept Form**: $y = mx + b$, where m is the slope and b is the y-intercept
- **Point-Slope Form**: $y - y_1 = m(x - x_1)$, where m is the slope and (x_1, y_1) is a point on the line
- **Two-Point Form**: $\frac{y-y_1}{x-x_1} = \frac{y_2-y_1}{x_2-x_1}$, where (x_1, y_1) and (x_2, y_2) are two points on the given line
- **Intercept Form**: $\frac{x}{x_1} + \frac{y}{y_1} = 1$, where $(x_1, 0)$ is the point at which a line intersects the x-axis, and $(0, y_1)$ is the point at which the same line intersects the y-axis

> **Review Video: Slope-Intercept and Point-Slope Forms**
> Visit mometrix.com/academy and enter code: 113216
>
> **Review Video: Converting Between Standard and Slope-Intercept Forms**
> Visit mometrix.com/academy and enter code: 982828
>
> **Review Video: Linear Equations Basics**
> Visit mometrix.com/academy and enter code: 793005

Solving Equations

Manipulating Equations

Like Terms

Like terms are terms in an equation that have the same variable, regardless of whether they also have the same coefficient. This includes terms that *lack* a variable; all constants (i.e., numbers without variables) are considered like terms. If the equation involves terms with a variable raised to different powers, the like terms are those that have the variable raised to the same power.

For example, consider the equation $x^2 + 3x + 2 = 2x^2 + x - 7 + 2x$. In this equation, 2 and -7 are like terms; they are both constants. The terms $3x$, x, and $2x$ are like terms, they all include the variable x raised to the first power. The terms x^2 and $2x^2$ are like terms, they both include the variable x, raised to the second power. The terms $2x$ and $2x^2$ are not like terms; although they both involve the variable x, the variable is not raised to the same power in both terms. The fact that they have the same coefficient, 2, is not relevant.

> **Review Video: Rules for Manipulating Equations**
> Visit mometrix.com/academy and enter code: 838871

Carrying Out the Same Operation on Both Sides of an Equation

When solving an equation, the general procedure is to carry out a series of operations on both sides of an equation, choosing operations that simplify the equation when doing so. The reason why the same operation must be carried out on both sides of the equation is because that leaves the meaning of the equation unchanged, and yields a result that is equivalent to the original equation. This would not be the case if we carried out an operation on one side of an equation and not the other. Consider what an equation means: it is a statement that two values or expressions are equal. If we carry out the same operation on both sides of the equation—add 3 to both sides, for example—then the two sides of the equation are changed in the same way, and so remain equal. If we do that to only one side of the equation—add 3 to one side but not the other—then that wouldn't be true; if we change one side of the equation but not the other then the two sides are no longer equal.

Combining Like Terms

Combining like terms refers to adding or subtracting like terms—terms with the same variable—and therefore reducing sets of like terms to a single term. The main advantage of doing this is that it simplifies the equation. Often, combining like terms can be done as the first step in solving an equation, though it can also be done later, such as after distributing terms in a product.

For example, consider the equation $2(x + 3) + 3(2 + x + 3) = -4$. The 2 and the 3 in the second set of parentheses are like terms, and we can combine them, yielding $2(x + 3) + 3(x + 5) = -4$. Now we can carry out the multiplications implied by the parentheses, distributing the outer 2 and 3 accordingly: $2x + 6 + 3x + 15 = -4$. The $2x$ and the $3x$ are like terms, and we can add them together: $5x + 6 + 15 = -4$. Now, the constants 6, 15, and -4 are also like terms, and we can combine them as well: subtracting 6 and 15 from both sides of the equation, we get $5x = -4 - 6 - 15$, or $5x = -25$, which simplifies further to $x = -5$.

Review Video: Solving Equations by Combining Like Terms
Visit mometrix.com/academy and enter code: 668506

Canceling Terms on Opposite Sides of an Equation

Two terms on opposite sides of an equation can be canceled if and only if they *exactly* match each other. They must have the same variable raised to the same power and the same coefficient. For example, in the equation $3x + 2x^2 + 6 = 2x^2 - 6$, $2x^2$ appears on both sides of the equation and can be canceled, leaving $3x + 6 = -6$. The 6 on each side of the equation *cannot* be canceled, because it is added on one side of the equation and subtracted on the other. While they cannot be canceled, however, the 6 and -6 are like terms and can be combined, yielding $3x = -12$, which simplifies further to $x = -4$.

It's also important to note that the terms to be canceled must be independent terms and cannot be part of a larger term. For example, consider the equation $2(x + 6) = 3(x + 4) + 1$. We cannot cancel the x's, because even though they match each other they are part of the larger terms $2(x + 6)$ and $3(x + 4)$. We must first distribute the 2 and 3, yielding $2x + 12 = 3x + 12 + 1$. Now we see that the terms with the x's do not match, but the 12s do, and can be canceled, leaving $2x = 3x + 1$, which simplifies to $x = -1$.

Isolating Variables

To isolate a variable means to manipulate the equation so that the variable appears by itself on one side of the equation, and does not appear at all on the other side. Generally, an equation or inequality is considered to be solved once the variable is isolated and the other side of the equation or inequality is simplified as much as possible. In the case of a two-variable equation or inequality, only one variable needs to be isolated; it will not usually be possible to simultaneously isolate both variables.

For a linear equation—an equation in which the variable only appears raised to the first power—isolating a variable can be done by first moving all the terms with the variable to one side of the equation and all other terms to the other side. (*Moving* a term really means adding the inverse of the term to both sides; when a term is *moved* to the other side of the equation its sign is flipped.)

Then combine like terms on each side. Finally, divide both sides by the coefficient of the variable, if applicable. The steps need not necessarily be done in this order, but this order will always work.

Review Video: Solving Equations for Specific Variables
Visit mometrix.com/academy and enter code: 130695

Review Video: Solving Equations Involving Algebraic Fractions
Visit mometrix.com/academy and enter code: 237770

Review Video: Solving One-Step Equations
Visit mometrix.com/academy and enter code: 777004

SOLVING ONE-VARIABLE LINEAR EQUATIONS

EQUATIONS WITH ONE SOLUTION (THE USUAL CASE)

To solve a one-variable linear equation, we use the techniques above to isolate the variable.

1. If any coefficients or constants are fractions, it is often helpful first to multiply both sides of the equation by the least common denominator (of all fractions) to clear the fractions.
2. Simplify both sides of the equation by combining any like terms.
3. Put all terms with the variable on one side of the equation and all constant terms on the other side, by adding or subtracting the same terms on both sides of the equation.
4. Divide both sides by the coefficient of the variable (or multiply both sides by its reciprocal).
5. When we have a value for the variable, we can check it by substituting the value into the original equation to make sure it produces a true result.

Consider the following example for solving the equation $\frac{2}{3}x + 8 = 14$:

$3 \cdot \left(\frac{2}{3}x + 8\right) = 3 \cdot 14$	Clear fractions by multiplying both sides by 3.
$2x + 24 = 42$	Simplify, remembering to apply the distributive property.
$2x + 24 - 24 = 42 - 24$	Subtract 24 from both sides to isolate $2x$.
$2x = 18$	Simplify by combining like terms.
$\frac{2x}{2} = \frac{18}{2}$	Divide both sides by 2 to isolate x.
$x = 9$	Simplify

Finally, we check this answer by substituting $x = 9$ into the original equation to make sure we get a true result.

$$\frac{2}{3}x + 8 = \frac{2}{3}(9) + 8 = 6 + 8 = 14$$

This is correct, so the value of x is 9.

Review Video: Solving Equations Using the Distributive Property
Visit mometrix.com/academy and enter code: 765499

Equations with More Than One Solution

Some types of non-linear equations, such as equations involving squares of variables, may have more than one solution. For example, the equation $x^2 = 4$ has two solutions: 2 and -2. Equations with absolute values can also have multiple solutions: $|x| = 1$ has the solutions $x = 1$ and $x = -1$.

It is possible for a linear equation to have more than one solution but only if the equation is true regardless of the value of the variable. We call such an equation an **identity**. In this case, the equation has infinitely many solutions, because every possible value of the variable is a solution. We discover that a linear equation is an identity when our attempts to isolate the variable cause the variable to disappear, leaving a *true* equation involving only constants. For example, consider the equation $2(3x + 5) = x + 5(x + 2)$. Distributing, we get $6x + 10 = x + 5x + 10$; combining like terms gives $6x + 10 = 6x + 10$, and the $6x$-terms cancel to leave $10 = 10$. This is clearly true, so the original equation is an identity. We could also cancel the 10's leaving $0 = 0$, which is also is clearly true—in general if both sides of the equation can be reduced to match one another exactly, the original equation is an identity.

Equations with No Solution

Some types of non-linear equations, such as equations involving squares of variables, may have no solution. For example, the equation $x^2 = -2$ has no solutions in the real numbers because the square of a real number must be positive. Similarly, $|x| = -1$ has no solution because the absolute value of a number is always positive.

It is also possible for a linear equation to have no solution. We call such an equation a **contradiction**. We discover that a linear equation is a contradiction when our attempts to isolate the variable cause the variable to disappear, leaving a *false* equation involving only constants. For example, the equation $2(x + 3) + x = 3x$ has no solution. We can see this by trying to solve it: first we distribute, leaving $2x + 6 + x = 3x$. Combining like terms gives us $3x + 6 = 3x$, and cancelling the term $3x$ on both sides leaves us with $6 = 0$. This is clearly false, so the original equation is a contradiction, having no solutions.

Features of Equations That Require Special Treatment

A linear equation is an equation in which variables only appear by themselves: not multiplied together, not with exponents other than one, and not inside absolute value signs or any other functions. For example, the equation $x + 1 - 3x = 5 - x$ is a linear equation; while x appears multiple times, it never appears with an exponent other than one, or inside any function. The two-variable equation $2x - 3y = 5 + 2x$ is also a linear equation. In contrast, the equation $x^2 - 5 = 3x$ is *not* a linear equation, because it involves the term x^2. The equation $\sqrt{x} = 5$ is not linear, because it involves a square root. The equation $(x - 1)^2 = 4$ is not linear because even though there's no exponent on the x directly, it appears as part of an expression that is squared. The two-variable equation $x + xy - y = 5$ is not linear because it includes the term xy, where two variables are multiplied together.

As we see above, linear equations can always be solved (or shown to have no solution) by combining like terms and performing simple operations on both sides of the equation. Some non-linear equations can be solved by similar methods, but others may require more advanced methods of solution, if they can be solved analytically at all.

Solving Equations Involving Roots

In an equation involving roots, the first step is to isolate the term with the root, if possible, and then raise both sides of the equation to the appropriate power to eliminate it. Consider an example equation, $2\sqrt{x+1}-1=3$. In this case, begin by adding 1 to both sides, yielding $2\sqrt{x+1}=4$, and then dividing both sides by 2, yielding $\sqrt{x+1}=2$. Now square both sides, yielding $x+1=4$. Finally, subtracting 1 from both sides yields $x=3$.

Squaring both sides of an equation (or raising both sides to any *even* power) may, however, yield a spurious solution—a solution to the squared equation that is *not* a solution of the original equation. It's therefore necessary to plug the solution back into the original equation to make sure it works. In this case, it does: $2\sqrt{3+1}-1=2\sqrt{4}-1=2(2)-1=4-1=3$.

The same procedure applies for other roots as well. For example, given the equation $3+\sqrt[3]{2x}=5$, we can first subtract 3 from both sides, yielding $\sqrt[3]{2x}=2$ and isolating the root. Raising both sides to the third power yields $2x=2^3$; i.e., $2x=8$. We can now divide both sides by 2 to get $x=4$.

Review Video: Solving Equations Involving Roots
Visit mometrix.com/academy and enter code: 297670

Solving Equations with Exponents

In solving an equation with powers of a variable, sometimes it is possible to eliminate all but one term involving the variable. In that case, we can isolate the power of the variable and then take the appropriate root of both sides to eliminate the exponent. For instance, for the equation $2x^3+17=5x^3-7$, we can subtract $5x^3$ from both sides to get $-3x^3+17=-7$, and then subtract 17 from both sides to get $-3x^3=-24$. Finally, we can divide both sides by -3 to get $x^3=8$. Since this isolates the cube of the variable, we can take the cube root of both sides to get $x=\sqrt[3]{8}=2$.

One important but often overlooked point is that equations with an exponent greater than 1 may have more than one answer. The solution to $x^2=9$ isn't simply $x=3$; it's $x=\pm3$ (that is, $x=3$ or $x=-3$). For a slightly more complicated example, consider the equation $(x-1)^2-1=3$. Adding 1 to both sides yields $(x-1)^2=4$; taking the square root of both sides yields $x-1=2$. We can then add 1 to both sides to get $x=3$. However, there's a second solution. We also have the possibility that $x-1=-2$, in which case $x=-1$. Both $x=3$ and $x=-1$ are valid solutions, as can be verified by substituting them both into the original equation.

Review Video: Solving Equations with Exponents
Visit mometrix.com/academy and enter code: 514557

Review Video: Adding and Subtracting with Exponents
Visit mometrix.com/academy and enter code: 875756

Solving Equations with Absolute Values

When solving an equation with an absolute value, the first step is to isolate the absolute value term. We then consider two possibilities: when the expression inside the absolute value is positive or when it is negative. In the former case, the expression in the absolute value equals the expression on the other side of the equation; in the latter, it equals the additive inverse of that expression—the expression times negative one. We consider each case separately and finally check for spurious solutions.

For instance, consider solving $|2x-1|+x=5$ for x. We can first isolate the absolute value by moving the x to the other side: $|2x-1|=-x+5$. Now, we have two possibilities. First, that $2x-1$ is positive, and hence $2x-1=-x+5$. Rearranging and combining like terms yields $3x=6$, and hence $x=2$. The other possibility is that $2x-1$ is negative, and hence $2x-1=-(-x+5)=x-$

5. In this case, rearranging and combining like terms yields $x = -4$. Substituting $x = 2$ and $x = -4$ back into the original equation, we see that they are both valid solutions.

Note that the absolute value of a sum or difference applies to the sum or difference as a whole, not to the individual terms; in general, $|2x - 1|$ is not equal to $|2x + 1|$ or to $|2x| - 1$.

Review Video: Solving Absolute Value Equations
Visit mometrix.com/academy and enter code: 501208

EXTRANEOUS SOLUTIONS

An **extraneous solution** may arise when we square both sides of an equation (or raise both sides to an even power) as a step in solving it or under certain other operations on the equation. It is a solution to the squared or otherwise modified equation that is *not* a solution of the original equation. To identify an extraneous solution, it's useful when you solve an equation involving roots or absolute values to plug the solution back into the original equation to make sure it's valid.

TWO-VARIABLE EQUATIONS

Similar to methods for a one-variable equation, solving a two-variable equation involves isolating a variable: manipulating the equation so that a variable appears by itself on one side of the equation, and not at all on the other side. However, in a two-variable equation, you will usually only be able to isolate one of the variables; the other variable may appear on the other side along with constant terms, or with exponents or other functions. If an equation has multiple variables, the problem should tell you which variable to isolate.

Review Video: Solving Equations with Variables on Both Sides
Visit mometrix.com/academy and enter code: 402497

GRAPHING EQUATIONS

GRAPHICAL SOLUTIONS TO EQUATIONS

When equations are shown graphically, they are usually shown on a **Cartesian coordinate plane**. The Cartesian coordinate plane consists of two number lines placed perpendicular to each other and intersecting at the zero point, also known as the origin. The horizontal number line is known as the x-axis, with positive values to the right of the origin, and negative values to the left of the origin. The vertical number line is known as the y-axis, with positive values above the origin, and negative values below the origin. Any point on the plane can be identified by an ordered pair in the form (x, y), called coordinates. The x-value of the coordinate is called the abscissa, and the y-value of the coordinate is called the ordinate. The two number lines divide the plane into **four quadrants**: I, II, III, and IV.

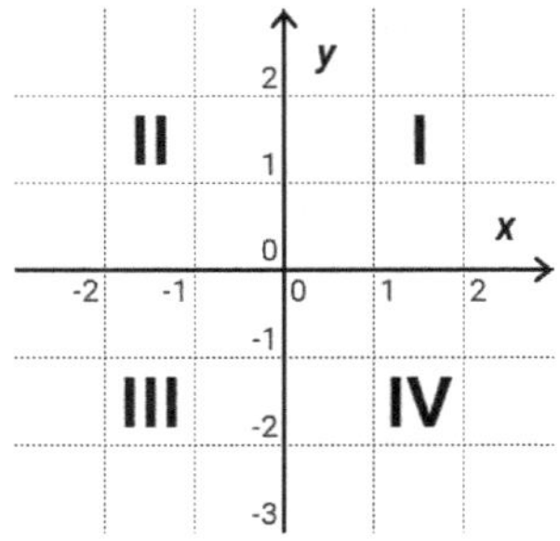

Note that in quadrant I $x > 0$ and $y > 0$, in quadrant II $x < 0$ and $y > 0$, in quadrant III $x < 0$ and $y < 0$, and in quadrant IV $x > 0$ and $y < 0$.

Recall that if the value of the slope of a line is positive, the line slopes upward from left to right. If the value of the slope is negative, the line slopes downward from left to right. If the y-coordinates are the same for two points on a line, the slope is 0 and the line is a **horizontal line**. If the x-coordinates are the same for two points on a line, there is no slope and the line is a **vertical line**. Two or more lines that have equivalent slopes are **parallel lines**. **Perpendicular lines** have slopes that are negative reciprocals of each other, such as $\frac{a}{b}$ and $\frac{-b}{a}$.

Review Video: Cartesian Coordinate Plane and Graphing
Visit mometrix.com/academy and enter code: 115173

Graphing Equations in Two Variables

One way of graphing an equation in two variables is to plot enough points to get an idea for its shape and then draw the appropriate curve through those points. A point can be plotted by substituting in a value for one variable and solving for the other. If the equation is linear, we only need two points and can then draw a straight line between them.

For example, consider the equation $y = 2x - 1$. This is a linear equation—both variables only appear raised to the first power—so we only need two points. When $x = 0, y = 2(0) - 1 = -1$. When $x = 2, y = 2(2) - 1 = 3$. We can therefore choose the points $(0, -1)$ and $(2, 3)$, and draw a line between them:

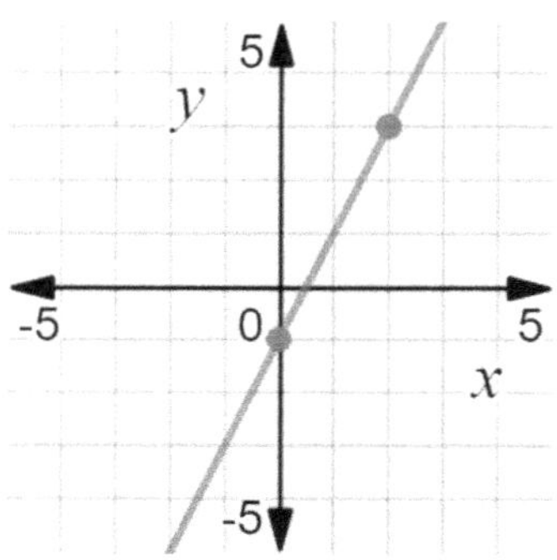

Inequalities

INEQUALITIES

Commonly in algebra and other upper-level fields of math you find yourself working with mathematical expressions that do not equal each other. The statement comparing such expressions with symbols such as $<$ (less than) or $>$ (greater than) is called an *inequality*. An example of an inequality is $7x > 5$. To solve for x, simply divide both sides by 7 and the solution is shown to be $x > \frac{5}{7}$. Graphs of the solution set of inequalities are represented on a number line. Open circles are used to show that an expression approaches a number but is never quite equal to that number.

Review Video: Solving One-Step Inequalities
Visit mometrix.com/academy and enter code: 229684

Review Video: Solving Multi-Step Inequalities
Visit mometrix.com/academy and enter code: 347842

Review Video: Solving Inequalities Using All 4 Basic Operations
Visit mometrix.com/academy and enter code: 401111

TYPES OF INEQUALITIES

Conditional inequalities are those with certain values for the variable that will make the condition true and other values for the variable where the condition will be false. **Absolute inequalities** can have any real number as the value for the variable to make the condition true, while there is no real number value for the variable that will make the condition false. Solving inequalities is done by following the same rules for solving equations with the exception that when multiplying or dividing by a negative number the direction of the inequality sign must be flipped or reversed. **Double inequalities** are situations where two inequality statements apply to the same variable expression. Example: $-c < ax + b < c$.

Review Video: Conditional and Absolute Inequalities
Visit mometrix.com/academy and enter code: 980164

SOLVING INEQUALITIES

DETERMINING SOLUTIONS TO INEQUALITIES

To determine whether a coordinate is a solution of an inequality, you can substitute the values of the coordinate into the inequality, simplify, and check whether the resulting statement holds true. For instance, to determine whether $(-2,4)$ is a solution of the inequality $y \geq -2x + 3$, substitute the values into the inequality, $4 \geq -2(-2) + 3$. Simplify the right side of the inequality and the result is $4 \geq 7$, which is a false statement. Therefore, the coordinate is not a solution of the inequality. You can also use this method to determine which part of the graph of an inequality is shaded. The graph of $y \geq -2x + 3$ includes the solid line $y = -2x - 3$ and, since it excludes the point $(-2,4)$ to the left of the line, it is shaded to the right of the line.

Review Video: Graphing Linear Inequalities
Visit mometrix.com/academy and enter code: 439421

Review Video: Graphing Solutions to Inequalities
Visit mometrix.com/academy and enter code: 391281

FLIPPING INEQUALITY SIGNS

When given an inequality, we can always turn the entire inequality around, swapping the two sides of the inequality and changing the inequality sign. For instance, $x + 2 > 2x - 3$ is equivalent to $2x - 3 < x + 2$. Aside from that, normally the inequality does not change if we carry out the same operation on both sides of the inequality. There is, however, one principal exception: if we *multiply* or *divide* both sides of the inequality by a *negative number*, the inequality is flipped. For example, if we take the inequality $-2x < 6$ and divide both sides by -2, the inequality flips and we are left with $x > -3$. This *only* applies to multiplication and division, and only with negative numbers. Multiplying or dividing both sides by a positive number, or adding or subtracting any number regardless of sign, does not flip the inequality. Another special case that flips the inequality sign is when reciprocals are used. For instance, $3 > 2$ but the relation of the reciprocals is $\frac{1}{3} < \frac{1}{2}$.

COMPOUND INEQUALITIES

A **compound inequality** is an equality that consists of two inequalities combined with *and* or *or*. The two components of a proper compound inequality must be of opposite type: that is, one must be greater than (or greater than or equal to), the other less than (or less than or equal to). For instance, "$x + 1 < 2$ or $x + 1 > 3$" is a compound inequality, as is "$2x \geq 4$ and $2x \leq 6$." An *and* inequality can be written more compactly by having one inequality on each side of the common part: "$2x \geq 1$ and $2x \leq 6$," can also be written as $1 \leq 2x \leq 6$.

In order for the compound inequality to be meaningful, the two parts of an *and* inequality must overlap; otherwise, no numbers satisfy the inequality. On the other hand, if the two parts of an *or* inequality overlap, then *all* numbers satisfy the inequality and as such the inequality is usually not meaningful.

Solving a compound inequality requires solving each part separately. For example, given the compound inequality "$x + 1 < 2$ or $x + 1 > 3$," the first inequality, $x + 1 < 2$, reduces to $x < 1$, and the second part, $x + 1 > 3$, reduces to $x > 2$, so the whole compound inequality can be written as "$x < 1$ or $x > 2$." Similarly, $1 \leq 2x \leq 6$ can be solved by dividing each term by 2, yielding $\frac{1}{2} \leq x \leq 3$.

Review Video: Compound Inequalities Visit mometrix.com/academy and enter code: 786318

SOLVING INEQUALITIES INVOLVING ABSOLUTE VALUES

To solve an inequality involving an absolute value, first isolate the term with the absolute value. Then proceed to treat the two cases separately as with an absolute value equation, but flipping the inequality in the case where the expression in the absolute value is negative (since that essentially involves multiplying both sides by −1.) The two cases are then combined into a compound inequality; if the absolute value is on the greater side of the inequality, then it is an *or* compound inequality, if on the lesser side, then it's an *and*.

Consider the inequality $2 + |x - 1| \geq 3$. We can isolate the absolute value term by subtracting 2 from both sides: $|x - 1| \geq 1$. Now, we're left with the two cases $x - 1 \geq 1$ or $x - 1 \leq -1$: note that in the latter, negative case, the inequality is flipped. $x - 1 \geq 1$ reduces to $x \geq 2$, and $x - 1 \leq -1$ reduces to $x \leq 0$. Since in the inequality $|x - 1| \geq 1$ the absolute value is on the greater side, the two cases combine into an *or* compound inequality, so the final, solved inequality is "$x \leq 0$ or $x \geq 2$."

Review Video: Solving Absolute Value Inequalities Visit mometrix.com/academy and enter code: 997008

Solving Inequalities Involving Square Roots

Solving an inequality with a square root involves two parts. First, we solve the inequality as if it were an equation, isolating the square root and then squaring both sides of the equation. Second, we restrict the solution to the set of values of x for which the value inside the square root sign is non-negative.

For example, in the inequality, $\sqrt{x-2}+1<5$, we can isolate the square root by subtracting 1 from both sides, yielding $\sqrt{x-2}<4$. Squaring both sides of the inequality yields $x-2<16$, so $x<18$. Since we can't take the square root of a negative number, we also require the part inside the square root to be non-negative. In this case, that means $x-2\geq 0$. Adding 2 to both sides of the inequality yields $x\geq 2$. Our final answer is a compound inequality combining the two simple inequalities: $x\geq 2$ and $x<18$, or $2\leq x<18$.

Note that we only get a compound inequality if the two simple inequalities are in opposite directions; otherwise, we take the one that is more restrictive.

The same technique can be used for other even roots, such as fourth roots. It is *not*, however, used for cube roots or other odd roots—negative numbers *do* have cube roots, so the condition that the quantity inside the root sign cannot be negative does not apply.

Review Video: Solving Inequalities Involving Square Roots
Visit mometrix.com/academy and enter code: 800288

Special Circumstances

Sometimes an inequality involving an absolute value or an even exponent is true for all values of x, and we don't need to do any further work to solve it. This is true if the inequality, once the absolute value or exponent term is isolated, says that term is greater than a negative number (or greater than or equal to zero). Since an absolute value or a number raised to an even exponent is *always* non-negative, this inequality is always true.

Graphing Inequalities

Graphing Simple Inequalities

To graph a simple inequality, we first mark on the number line the value that signifies the end point of the inequality. If the inequality is strict (involves a less than or greater than), we use a hollow circle; if it is not strict (less than or equal to or greater than or equal to), we use a solid circle. We then fill in the part of the number line that satisfies the inequality: to the left of the marked point for less than (or less than or equal to), to the right for greater than (or greater than or equal to).

For example, we would graph the inequality $x<5$ by putting a hollow circle at 5 and filling in the part of the line to the left:

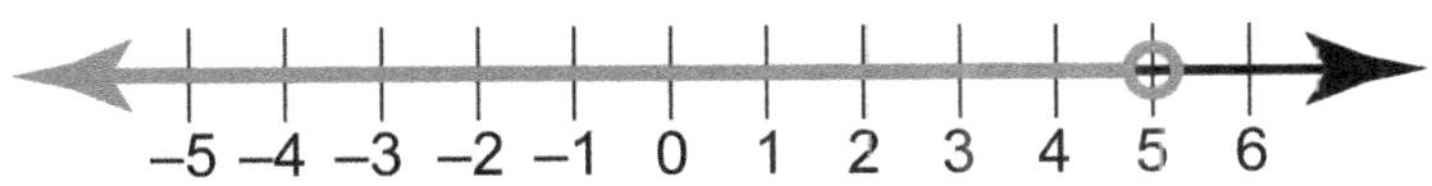

Graphing Compound Inequalities

To graph a compound inequality, we fill in both parts of the inequality for an *or* inequality, or the overlap between them for an *and* inequality. More specifically, we start by plotting the endpoints of each inequality on the number line. For an *or* inequality, we then fill in the appropriate side of the line for each inequality. Typically, the two component inequalities do not overlap, which means the shaded part is *outside* the two points. For an *and* inequality, we instead fill in the part of the line that meets both inequalities.

For the inequality "$x \leq -3$ or $x > 4$," we first put a solid circle at –3 and a hollow circle at 4. We then fill the parts of the line *outside* these circles:

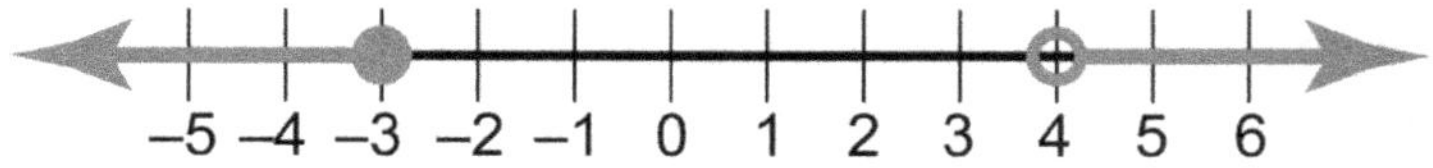

Graphing Inequalities Including Absolute Values

An inequality with an absolute value can be converted to a compound inequality. To graph the inequality, first convert it to a compound inequality, and then graph that normally. If the absolute value is on the greater side of the inequality, we end up with an *or* inequality; we plot the endpoints of the inequality on the number line and fill in the part of the line *outside* those points. If the absolute value is on the smaller side of the inequality, we end up with an *and* inequality; we plot the endpoints of the inequality on the number line and fill in the part of the line *between* those points.

For example, the inequality $|x + 1| \geq 4$ can be rewritten as $x \geq 3$ or $x \leq -5$. We place solid circles at the points 3 and –5 and fill in the part of the line *outside* them:

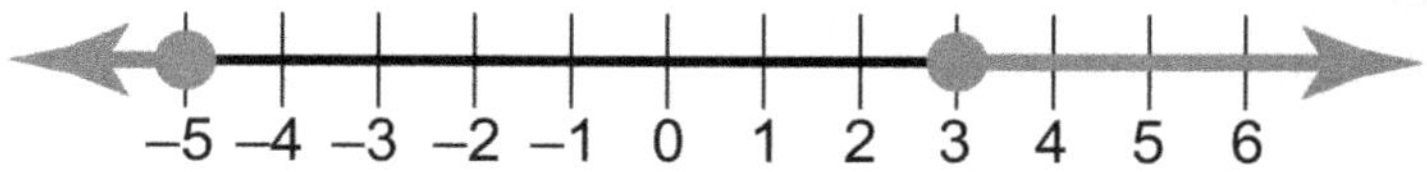

Graphing Inequalities in Two Variables

To graph an inequality in two variables, we first graph the border of the inequality. This means graphing the equation that we get if we replace the inequality sign with an equals sign. If the inequality is strict (> or <), we graph the border with a dashed or dotted line; if it is not strict ($\geq$ or $\leq$), we use a solid line. We can then test any point not on the border to see if it satisfies the inequality. If it does, we shade in that side of the border; if not, we shade in the other side. As an example, consider $y > 2x + 2$. To graph this inequality, we first graph the border, $y = 2x + 2$. Since it is a strict inequality, we use a dashed line. Then, we choose a test point. This can be any point not on the border; in this case, we will choose the origin, (0,0). (This makes the calculation easy and is generally a good choice unless the border passes through the origin.) Putting this into the original inequality, we get $0 > 2(0) + 2$, i.e., $0 > 2$. This is *not* true, so we shade in the side of the border that does *not* include the point (0,0):

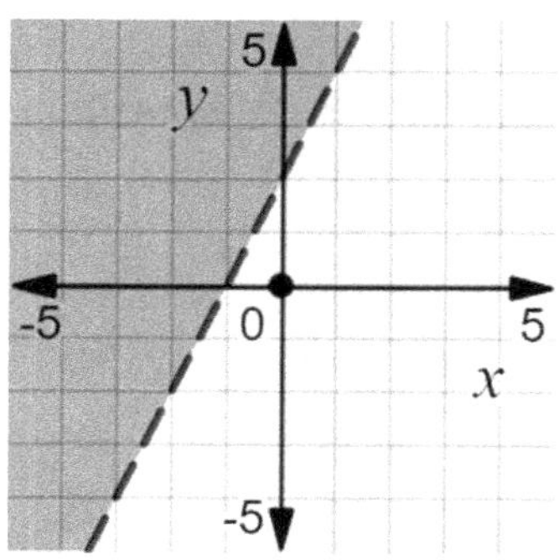

GRAPHING COMPOUND INEQUALITIES IN TWO VARIABLES

One way to graph a compound inequality in two variables is to first graph each of the component inequalities. For an *and* inequality, we then shade in only the parts where the two graphs overlap; for an *or* inequality, we shade in any region that pertains to either of the individual inequalities.

Consider the graph of "$y \geq x - 1$ and $y \leq -x$":

We first shade in the individual inequalities:

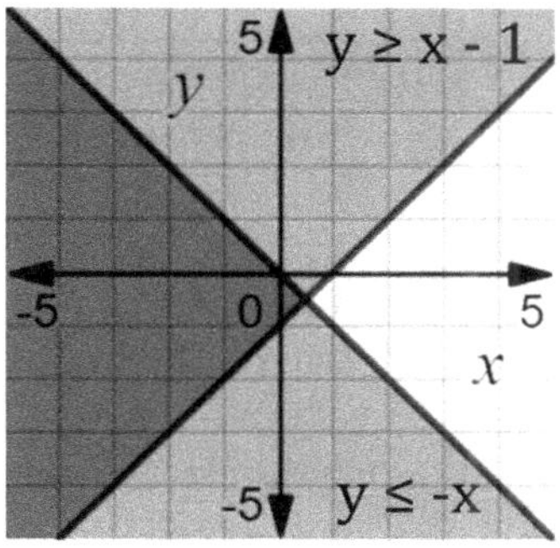

Now, since the compound inequality has an *and*, we only leave shaded the overlap—the part that pertains to *both* inequalities:

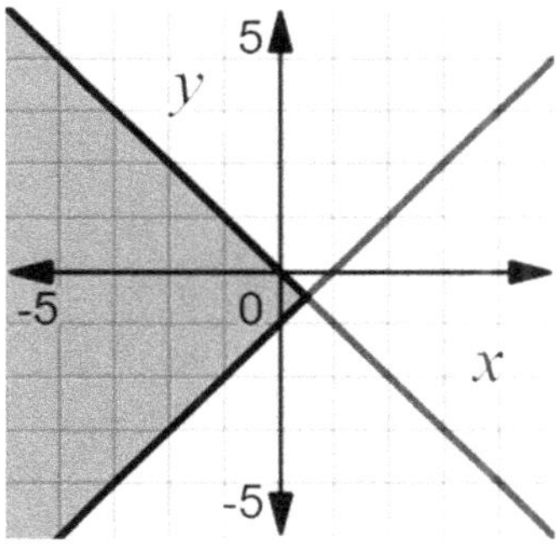

If instead the inequality had been "$y \geq x - 1$ or $y \leq -x$," our final graph would involve the *total* shaded area:

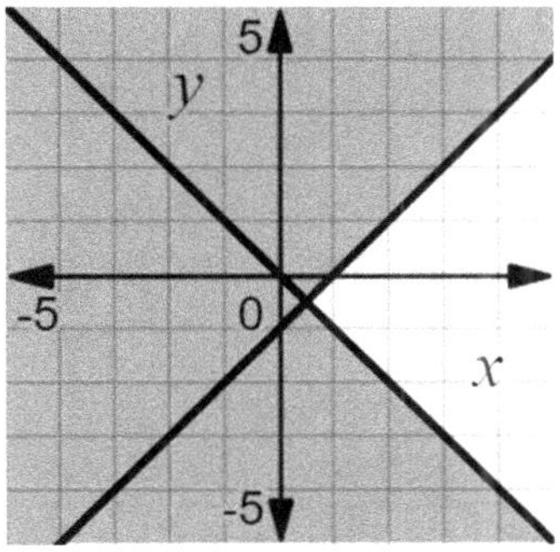

Review Video: Graphing Solutions to Inequalities
Visit mometrix.com/academy and enter code: 391281

Advanced Expressions and Equations

Systems of Equations

Solving Systems of Equations

A **system of equations** is a set of simultaneous equations that all use the same variables. A solution to a system of equations must be true for each equation in the system. **Consistent systems** are those with at least one solution. **Inconsistent systems** are systems of equations that have no solution.

> **Review Video: Solving Systems of Linear Equations**
> Visit mometrix.com/academy and enter code: 746745

Substitution

To solve a system of linear equations by **substitution**, start with the easier equation and solve for one of the variables. Express this variable in terms of the other variable. Substitute this expression in the other equation and solve for the other variable. The solution should be expressed in the form (x, y). Substitute the values into both of the original equations to check your answer. Consider the following system of equations:

$$x + 6y = 15$$
$$3x - 12y = 18$$

Solving the first equation for x: $x = 15 - 6y$

Substitute this value in place of x in the second equation, and solve for y:

$$3(15 - 6y) - 12y = 18$$
$$45 - 18y - 12y = 18$$
$$30y = 27$$
$$y = \frac{27}{30} = \frac{9}{10} = 0.9$$

Plug this value for y back into the first equation to solve for x:

$$x = 15 - 6(0.9) = 15 - 5.4 = 9.6$$

Check both equations if you have time:

$$9.6 + 6(0.9) = 15 \qquad 3(9.6) - 12(0.9) = 18$$
$$9.6 + 5.4 = 15 \qquad 28.8 - 10.8 = 18$$
$$15 = 15 \qquad 18 = 18$$

Therefore, the solution is (9.6,0.9).

> **Review Video: The Substitution Method**
> Visit mometrix.com/academy and enter code: 565151
>
> **Review Video: Substitution and Elimination**
> Visit mometrix.com/academy and enter code: 958611

Elimination

To solve a system of equations using **elimination**, begin by rewriting both equations in standard form $Ax + By = C$. Check to see if the coefficients of one pair of like variables add to zero. If not, multiply one or both of the equations by a non-zero number to make one set of like variables add to zero. Add the two equations to solve for one of the variables. Substitute this value into one of the original equations to solve for the other variable. Check your work by substituting into the other equation. Now, let's look at solving the following system using the elimination method:

$$5x + 6y = 4$$
$$x + 2y = 4$$

If we multiply the second equation by -3, we can eliminate the y-terms:

$$5x + 6y = 4$$
$$-3x - 6y = -12$$

Add the equations together and solve for x:

$$2x = -8$$
$$x = \frac{-8}{2} = -4$$

Plug the value for x back in to either of the original equations and solve for y:

$$-4 + 2y = 4$$
$$y = \frac{4+4}{2} = 4$$

Check both equations if you have time:

$$5(-4) + 6(4) = 4 \qquad -4 + 2(4) = 4$$
$$-20 + 24 = 4 \qquad -4 + 8 = 4$$
$$4 = 4 \qquad 4 = 4$$

Therefore, the solution is $(-4,4)$.

Review Video: The Elimination Method
Visit mometrix.com/academy and enter code: 449121

Graphically

To solve a system of linear equations **graphically**, plot both equations on the same graph. The solution of the equations is the point where both lines cross. If the lines do not cross (are parallel), then there is **no solution**.

For example, consider the following system of equations:

$$y = 2x + 7$$
$$y = -x + 1$$

Since these equations are given in slope-intercept form, they are easy to graph; the y-intercepts of the lines are (0,7) and (0,1). The respective slopes are 2 and −1, thus the graphs look like this:

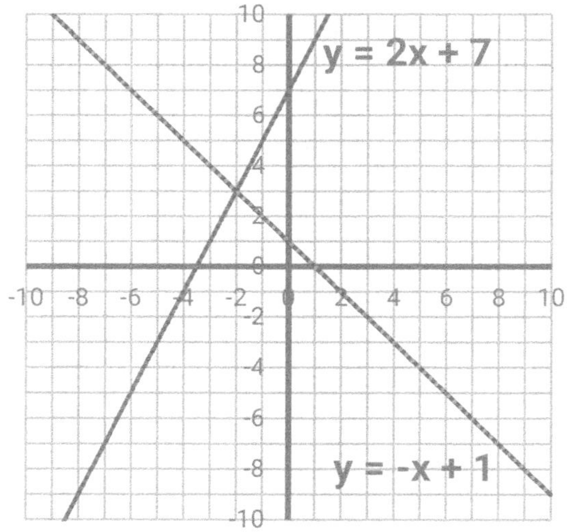

The two lines intersect at the point (−2,3), thus this is the solution to the system of equations.

Solving a system graphically is generally only practical if both coordinates of the solution are integers; otherwise the intersection will lie between gridlines on the graph and the coordinates will be difficult or impossible to determine exactly. It also helps if, as in this example, the equations are in slope-intercept form or some other form that makes them easy to graph. Otherwise, another method of solution (by substitution or elimination) is likely to be more useful.

Review Video: Solving Systems by Graphing
Visit mometrix.com/academy and enter code: 634812

Solving Systems of Equations Using the Trace Feature

Using the trace feature on a calculator requires that you rewrite each equation, isolating the y-variable on one side of the equal sign. Enter both equations in the graphing calculator and plot the graphs simultaneously. Use the trace cursor to find where the two lines cross. Use the zoom feature if necessary to obtain more accurate results. Always check your answer by substituting into the original equations. The trace method is likely to be less accurate than other methods due to the resolution of graphing calculators but is a useful tool to provide an approximate answer.

Advanced Systems of Equations

Solving a System of Equations with a Linear Equation and a Quadratic Equation

Algebraically

Generally, the simplest way to solve a system of equations consisting of a linear equation and a quadratic equation algebraically is through the method of substitution. One possible strategy is to solve the linear equation for y and then substitute that expression into the quadratic equation. After expansion and combining like terms, this will result in a new quadratic equation for x, which, like all quadratic equations, may have zero, one, or two solutions. Plugging each solution for x back into one of the original equations will then produce the corresponding value of y.

For example, consider the following system of equations:

$$x + y = 1$$
$$y = (x + 3)^2 - 2$$

We can solve the linear equation for y to yield $y = -x + 1$. Substituting this expression into the quadratic equation produces $-x + 1 = (x + 3)^2 - 2$. We can simplify this equation:

$$\begin{aligned} -x + 1 &= (x + 3)^2 - 2 \\ -x + 1 &= x^2 + 6x + 9 - 2 \\ -x + 1 &= x^2 + 6x + 7 \\ 0 &= x^2 + 7x + 6 \end{aligned}$$

This quadratic equation can be factored as $(x + 1)(x + 6) = 0$. It therefore has two solutions: $x_1 = -1$ and $x_2 = -6$. Plugging each of these back into the original linear equation yields $y_1 = -x_1 + 1 = -(-1) + 1 = 2$ and $y_2 = -x_2 + 1 = -(-6) + 1 = 7$. Thus, this system of equations has two solutions, $(-1,2)$ and $(-6,7)$.

It may help to check your work by putting each x- and y-value back into the original equations and verifying that they do provide a solution.

Graphically

To solve a system of equations consisting of a linear equation and a quadratic equation graphically, plot both equations on the same graph. The linear equation will, of course, produce a straight line, while the quadratic equation will produce a parabola. These two graphs will intersect at zero, one, or two points; each point of intersection is a solution of the system.

For example, consider the following system of equations:

$$y = -2x + 2$$
$$y = -2x^2 + 4x + 2$$

The linear equation describes a line with a y-intercept of $(0,2)$ and a slope of -2.

To graph the quadratic equation, we can first find the vertex of the parabola: the x-coordinate of the vertex is $h = -\frac{b}{2a} = -\frac{4}{2(-2)} = 1$, and the y-coordinate is $k = -2(1)^2 + 4(1) + 2 = 4$. Thus, the vertex lies at $(1,4)$. To get a feel for the rest of the parabola, we can plug in a few more values of x to find more points; by putting in $x = 2$ and $x = 3$ in the quadratic equation, we find that the points

(2,2) and $(3,-4)$ lie on the parabola; by symmetry, so must $(0,2)$ and $(-1,-4)$. We can now plot both equations:

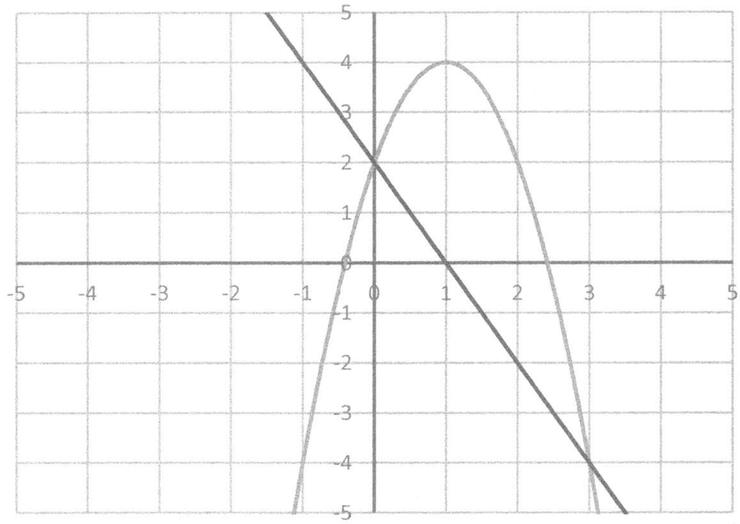

These two curves intersect at the points (0,2) and $(3,-4)$, thus these are the solutions of the equation.

Review Video: **Parabolas**
Visit mometrix.com/academy and enter code: 129187

Review Video: **Vertex of a Parabola**
Visit mometrix.com/academy and enter code: 272300

Review Video: Solving a System of Linear and Quadratic Equations
Visit mometrix.com/academy and enter code: 194870

Midpoint and Distance Formulas

If you know the coordinates of the endpoints of a line segment, you can calculate the midpoint and length of the line segment. Conveniently, the length of the line segment is also the distance between the two endpoints.

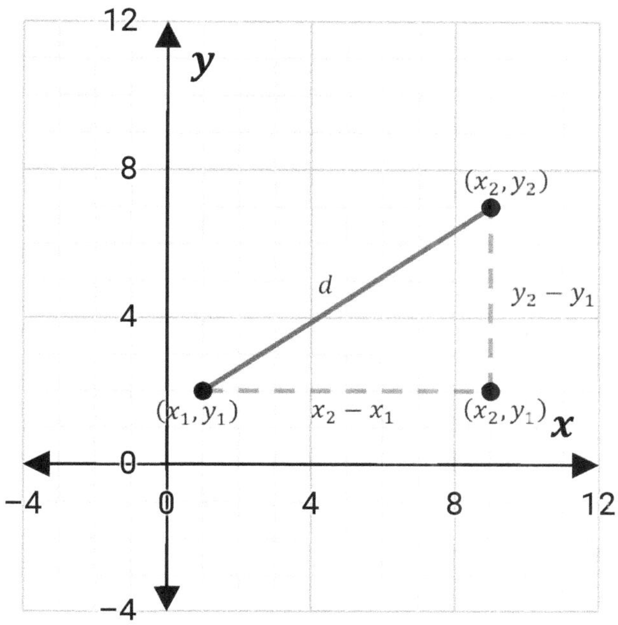

To find the **midpoint** of the line segment with endpoints (x_1, y_1) and (x_2, y_2), average the x-coordinates to get the x-coordinate of the midpoint, and average the y-coordinates to get the y-coordinate of the midpoint. Thus, the **midpoint formula** is:

$$\left(\frac{x_1 + x_2}{2}, \frac{y_1 + y_2}{2}\right)$$

The **distance** between points (x_1, y_1) and (x_2, y_2) is the same as the length of the hypotenuse of a right triangle with the two given points as endpoints, and the two sides of the right triangle parallel to the x-axis and y-axis, respectively. The length of the segment parallel to the x-axis is the difference between the x-coordinates of the two points. The length of the segment parallel to the y-axis is the difference between the y-coordinates of the two points. Use the Pythagorean theorem $a^2 + b^2 = c^2$ or $c = \sqrt{a^2 + b^2}$ to find the distance. Thus, the **distance formula** is:

$$d = \sqrt{(x_2 - x_1)^2 + (y_2 - y_1)^2}$$

> **Review Video: Calculations Using Points on a Graph**
> Visit mometrix.com/academy and enter code: 883228

POLYNOMIALS

MONOMIALS AND POLYNOMIALS

A **monomial** is a single constant, variable, or product of constants and variables, such as 7, x, $2x$, or x^3y. There will never be addition or subtraction symbols in a monomial. Like monomials have like variables, but they may have different coefficients. A **polynomial** is a monomial or the result of combining two or more monomials by sums or differences. In a polynomial we call each monomial a **term**. Two terms make a **binomial** (e.g., $2x + 3y$), three terms make a **trinomial** (e.g., $5x^2 - 4x + 9$). The **degree of a monomial** is the sum of the exponents of the variables. The **degree of a polynomial** is the highest degree of any individual term.

> **Review Video: Polynomials**
> Visit mometrix.com/academy and enter code: 305005

SIMPLIFYING POLYNOMIALS

Simplifying polynomials requires combining like terms. The like terms in a polynomial expression are those that have the same variables raised to the same powers. It is often helpful to connect the like terms with arrows or lines in order to separate them from the other monomials. Once you have determined the like terms, you can rearrange the polynomial by placing them together. Remember to include the sign that is in front of each term. Once the like terms are placed together, you can apply each operation and simplify. When adding and subtracting polynomials, only add and subtract the **coefficients**, or the number part; the variable and exponent stay the same.

ADDING POLYNOMIALS

To add polynomials, you need to add like terms. These terms have the same variable part. For example, the terms $4x^2$ and $3x^2$ both include x^2 terms. To find the sum of like terms, find the sum of the coefficients. Then, keep the same variable part. You can use the distributive property to distribute the plus sign to each term of the polynomial. For example:

$(4x^2 - 5x + 7) + (3x^2 + 2x + 1) =$
$(4x^2 - 5x + 7) + 3x^2 + 2x + 1 =$
$(4x^2 + 3x^2) + (-5x + 2x) + (7 + 1) =$
$7x^2 - 3x + 8$

SUBTRACTING POLYNOMIALS

To subtract polynomials, you need to subtract like terms. To find the difference of like terms, find the difference of the coefficients. Then, keep the same variable part. You can use the distributive property to distribute the minus sign to each term of the polynomial. For example:

$(-2x^2 - x + 5) - (3x^2 - 4x + 1) =$
$(-2x^2 - x + 5) - 3x^2 + 4x - 1 =$
$(-2x^2 - 3x^2) + (-x + 4x) + (5 - 1) =$
$-5x^2 + 3x + 4$

> **Review Video: Adding and Subtracting Polynomials**
> Visit mometrix.com/academy and enter code: 124088

MULTIPLYING POLYNOMIALS

In general, multiplying polynomials is done by multiplying each term in one polynomial by each term in the other and adding the results. In the specific case for multiplying binomials, there is a useful acronym, FOIL, that can help you make sure to cover each combination of terms. The **FOIL method** for $(Ax + By)(Cx + Dy)$ would be:

F	Multiply the *first* terms of each binomial	$(\overbrace{Ax}^{first} + By)(\overbrace{Cx}^{first} + Dy)$	ACx^2
O	Multiply the *outer* terms	$(\overbrace{Ax}^{outer} + By)(Cx + \overbrace{Dy}^{outer})$	$ADxy$
I	Multiply the *inner* terms	$(Ax + \overbrace{By}^{inner})(\overbrace{Cx}^{inner} + Dy)$	$BCxy$
L	Multiply the *last* terms of each binomial	$(Ax + \overbrace{By}^{last})(Cx + \overbrace{Dy}^{last})$	BDy^2

Then, add up the result of each and combine like terms: $ACx^2 + (AD + BC)xy + BDy^2$.

For example, using the FOIL method on binomials $(x + 2)$ and $(x - 3)$:

$$\begin{aligned}
&\text{First:} && (\boxed{x} + 2)(\boxed{x} + (-3)) && \rightarrow && (x)(x) &&= x^2 \\
&\text{Outer:} && (\boxed{x} + 2)(x + \boxed{(-3)}) && \rightarrow && (x)(-3) &&= -3x \\
&\text{Inner:} && (x + \boxed{2})(\boxed{x} + (-3)) && \rightarrow && (2)(x) &&= 2x \\
&\text{Last:} && (x + \boxed{2})(x + \boxed{(-3)}) && \rightarrow && (2)(-3) &&= -6
\end{aligned}$$

This results in: $(x^2) + (-3x) + (2x) + (-6)$

Combine like terms: $x^2 + (-3 + 2)x + (-6) = x^2 - x - 6$

> **Review Video: Multiplying Polynomials**
> Visit mometrix.com/academy and enter code: 598293
>
> **Review Video: Multiplying Terms Using the FOIL Method**
> Visit mometrix.com/academy and enter code: 854792

DIVIDING POLYNOMIALS

Use long division to divide a polynomial by either a monomial or another polynomial of equal or lesser degree.

When **dividing by a monomial**, divide each term of the polynomial by the monomial.

> **Review Video: Dividing Monomials**
> Visit mometrix.com/academy and enter code: 584409

When **dividing by a polynomial**, begin by arranging the terms of each polynomial in order of one variable. You may arrange in ascending or descending order, but be consistent with both polynomials. To get the first term of the quotient, divide the first term of the dividend by the first term of the divisor. Multiply the first term of the quotient by the entire divisor and subtract that product from the dividend. Repeat for the second and successive terms until you either get a remainder of zero or a remainder whose degree is less than the degree of the divisor. If the quotient has a remainder, write the answer as a mixed expression in the form:

$$\text{quotient} + \frac{\text{remainder}}{\text{divisor}}$$

For example, we can evaluate the following expression in the same way as long division:

$$\frac{x^3 - 3x^2 - 2x + 5}{x - 5}$$

$$\begin{array}{r} x^2 \;\; + 2x \;\; + 8 \\ x - 5 \overline{) \; x^3 - 3x^2 \;\; - 2x \;\; + 5} \\ \underline{-(x^3 - 5x^2)} \quad\quad\quad\quad\;\; \\ 2x^2 \;\; - 2x \quad\quad\; \\ \underline{-(2x^2 - 10x)} \quad\quad\; \\ 8x \;\; + 5 \\ \underline{-(8x - 40)} \\ 45 \end{array}$$

$$\frac{x^3 - 3x^2 - 2x + 5}{x - 5} = x^2 + 2x + 8 + \frac{45}{x - 5}$$

Review Video: Dividing Polynomials by Monomials
Visit mometrix.com/academy and enter code: 253551

Review Video: Dividing Trinomials by Binomials
Visit mometrix.com/academy and enter code: 651465

When **factoring** a polynomial, first see whether you can factor out a nontrivial greatest common factor (GCF). For example, the trinomial $3x^5 - 18x^4 + 15x^3$ has a GCF of $3x^3$ since the GCF of 3, 18, and 15 is 3 and the GCF of x^5, x^4, and x^3 is x^3. Factoring out the GCF simplifies the expression to $3x^3(x^2 - 6x + 5)$.

To factor a quadratic trinomial (this comes up frequently), first check whether it is a perfect square trinomial (see bulleted list below). If not, see if you can factor it by trial and error by making clever choices of values for a and b (or a, b, c, and d) in the formulas below (this amounts to trying to use the FOIL mnemonic backwards):

$$x^2 + (a + b)x + ab = (x + a)(x + b)$$
$$(ac)x^2 + (ad + bc)x + bd = (ax + b)(cx + d)$$

For instance, you would try to factor the trinomial $x^2 - 6x + 5$ using the equation $x^2 + (a + b)x + ab = (x + a)(x + b)$. This means that you need to find integers a and b such that $a + b = -6$ and $ab = 5$. Starting with $ab = 5$, you can see that the only ways to write 5 as a product of integers are $(1)(5) = 5$ and $(-1)(-5) = 5$. So, it is easy to see that you want $a = -1$ and $b = -5$ since $a + b = -1 + (-5) = -6$ and $ab = (-1)(-5) = 5$. This tells you that $x^2 - 6x + 5 = (x - 1)(x - 5)$.

For polynomials with four terms (usually a cubic polynomial), sometimes factoring by grouping works: You group the two higher-power terms and the two lower-power terms, factor the GCF out of each group, and then factor out the resulting common binomial factor, if there is one. For example, $x^3 + 5x^2 + 3x + 15 = (x^3 + 5x^2) + (3x + 15) = x^2(x + 5) + 3(x + 5) = (x^2 + 3)(x + 5)$.

Once you have found the factors, write the original polynomial as the product of all the factors. Make sure all of the factors are either monomials, or else linear or irreducible quadratic polynomials (*irreducible* means they have no real zeros, which is easy to check with the quadratic formula). Check your work by multiplying the factors to make sure you get the original polynomial.

Review Video: Factoring Out Common Monomial Factors
Visit mometrix.com/academy and enter code: 398578

Review Video: Factoring Trinomials of the Form x^2+bx+c
Visit mometrix.com/academy and enter code: 270556

Below are patterns of some special products to remember to help make factoring easier:

- Perfect square trinomials: $x^2 + 2xy + y^2 = (x + y)^2$ or $x^2 - 2xy + y^2 = (x - y)^2$. For example, $x^2 + 10x + 25 = (x + 5)^2$.
- Difference between two squares: $x^2 - y^2 = (x + y)(x - y)$. For example, $x^2 - 9 = (x + 3)(x - 3)$.
- Sum of two cubes: $x^3 + y^3 = (x + y)(x^2 - xy + y^2)$. For example, $x^3 + 27 = (x + 3)(x^2 - 3x + 9)$.
    - Note: the second factor is *not* the same as a perfect square trinomial, so do not try to factor it further.
- Difference between two cubes: $x^3 - y^3 = (x - y)(x^2 + xy + y^2)$. For example, $x^3 - 1000 = (x - 10)(x^2 + 10x + 100)$.
    - Again, the second factor is *not* the same as a perfect square trinomial.
- Perfect cubes: $x^3 + 3x^2y + 3xy^2 + y^3 = (x + y)^3$ and $x^3 - 3x^2y + 3xy^2 - y^3 = (x - y)^3$

Review Video: Factoring the Difference of Two Squares
Visit mometrix.com/academy and enter code: 128954

Algebra and Functions

Rational and Irrational Expressions

Rational Expressions

Rational expressions are fractions with polynomials in both the numerator and the denominator; the value of the polynomial in the denominator cannot be equal to zero. Be sure to keep track of values that make the denominator of the original expression zero as the final result inherits the same restrictions. For example, a denominator of $x - 3$ indicates that the expression is not defined when $x = 3$ and, as such, regardless of any operations done to the expression, it remains undefined there.

To **add or subtract** rational expressions, first find the common denominator, then rewrite each fraction as an equivalent fraction with the common denominator. Finally, add or subtract the numerators to get the numerator of the answer, and keep the common denominator as the denominator of the answer.

When **multiplying** rational expressions, factor each polynomial and cancel like factors (a factor which appears in both the numerator and the denominator). Then, multiply all remaining factors in the numerator to get the numerator of the product, and multiply the remaining factors in the denominator to get the denominator of the product. Remember: cancel entire factors, not individual terms.

To **divide** rational expressions, take the reciprocal of the divisor (the rational expression you are dividing by) and multiply by the dividend.

Review Video: Rational Expressions
Visit mometrix.com/academy and enter code: 415183

Simplifying Rational Expressions

To simplify a rational expression, factor the numerator and denominator completely. Factors that are the same and appear in the numerator and denominator have a ratio of 1. For example, look at the following expression:

$$\frac{x-1}{1-x^2}$$

The denominator, $(1-x^2)$, is a difference of squares. It can be factored as $(1-x)(1+x)$. The factor $1-x$ and the numerator $x-1$ are opposites and have a ratio of -1. Rewrite the numerator as $-1(1-x)$. So, the rational expression can be simplified as follows:

$$\frac{x-1}{1-x^2} = \frac{-1(1-x)}{(1-x)(1+x)} = \frac{-1}{1+x}$$

Note that since the original expression is only defined for $x \neq \{-1, 1\}$, the simplified expression has the same restrictions.

Review Video: Reducing Rational Expressions
Visit mometrix.com/academy and enter code: 788868

Review Video: Simplifying Algebraic Expressions with Parentheses
Visit mometrix.com/academy and enter code: 850843

Irrational Expressions

Irrational expressions are mathematical expressions that contain an irrational number and cannot be simplified into a rational form. Usually, this includes expressions that contain radicals or constants such as π. Most commonly, you will encounter these in the forms of expressions containing roots of non-perfect squares.

Basic Operations on Radical Expressions

To add or subtract radical numbers, the numbers within the radicals must match, similar to finding a common denominator in fractions:

$$a\sqrt{x} + b\sqrt{x} = (a+b)\sqrt{x}$$

$$a\sqrt{x} - b\sqrt{x} = (a-b)\sqrt{x}$$

To multiply radicals, the numbers outside the radical are multiplied together and the numbers inside the radical are multiplied together:

$$a\sqrt{x} \times b\sqrt{y} = ab\sqrt{xy}$$

To divide radicals, the radical must be eliminated from the denominator by multiplying both numerator and denominator by a value that will make the denominator a rational number:

$$\frac{a\sqrt{x}}{b\sqrt{y}} = \frac{a\sqrt{x}\sqrt{y}}{b\sqrt{y}\sqrt{y}} = \frac{a\sqrt{xy}}{by}$$

Review Video: Adding and Subtracting Radical Expressions
Visit mometrix.com/academy and enter code: 752176

Example

To solve $\frac{(3\sqrt{6})(2\sqrt{3})}{4\sqrt{5}}$, we first multiply the numerator, inside the radical and out: $3 \times 2\sqrt{6 \times 3} = 6\sqrt{18} = 18\sqrt{2}$. To divide, we multiply both numerator and denominator by a value that will eliminate the radical:

$$\frac{18\sqrt{2} \times \sqrt{5}}{4\sqrt{5} \times \sqrt{5}} = \frac{18\sqrt{10}}{4 \times 5} = \frac{18\sqrt{10}}{20} = \frac{9\sqrt{10}}{10}$$

Algebraic Theorems

According to the **fundamental theorem of algebra**, every non-constant, single-variable polynomial has exactly as many roots as the polynomial's highest exponent. For example, if x^4 is the largest exponent of a term, the polynomial will have exactly 4 roots. However, some of these roots may have multiplicity or be complex numbers. For instance, in the polynomial function $f(x) = x^4 - 4x + 3$, the only real root is 1, though it has multiplicity of 2 - that is, it occurs twice. The other two roots, $(-1 - i\sqrt{2})$ and $(-1 + i\sqrt{2})$, are complex, consisting of both real and non-real components.

The **remainder theorem** is useful for determining the remainder when a polynomial is divided by a binomial. The remainder theorem states that if a polynomial function $f(x)$ is divided by a binomial $x - a$, where a is a real number, the remainder of the division will be the value of $f(a)$. If $f(a) = 0$, then a is a root of the polynomial.

The **factor theorem** is related to the remainder theorem and states that if $f(a) = 0$ then $(x - a)$ is a factor of the function.

According to the **rational root theorem,** any rational root of a polynomial function $f(x) = a_nx^n + a_{n-1}x^{n-1} + \cdots + a_1x + a_0$ with integer coefficients will, when reduced to its lowest terms, be a positive or negative fraction such that the numerator is a factor of a_0 and the denominator is a factor of a_n. For instance, if the polynomial function $f(x) = x^3 + 3x^2 - 4$ has any rational roots, the numerators of those roots can only be factors of 4 (1, 2, 4), and the denominators can only be factors of 1 (1). The function in this example has roots of 1 (or $\frac{1}{1}$) and -2 (or $\frac{-2}{1}$).

Quadratics

Solving Quadratic Equations

A quadratic equation is an equation that can be written (possibly after simplification) in the form $ax^2 + bx + c = 0$. Thus, the **solutions** of this equation are precisely the **zeros** of the quadratic polynomial $P(x) = ax^2 + bx + c$. On the graph of this polynomial the zeros, if any, appear as

x-intercepts. There are several ways to find these solutions including the quadratic formula, factoring, completing the square, and graphing the function.

Review Video: Quadratic Equations Overview
Visit mometrix.com/academy and enter code: 476276

Review Video: Solutions of a Quadratic Equation on a Graph
Visit mometrix.com/academy and enter code: 328231

Quadratic Formula

The **quadratic formula** gives the zeros of a quadratic polynomial. It always works, but it is sometimes a little harder to use than other methods. To use it to solve a quadratic equation, rewrite the equation in the form $ax^2 + bx + c = 0$, where a, b, and c are coefficients. Now, as explained above, the solutions of the equation are the zeros of the quadratic polynomial $P(x) = ax^2 + bx + c$. To find them, substitute the values of a, b, and c into the Quadratic Formula:

$$x = \frac{-b \pm \sqrt{b^2 - 4ac}}{2a}$$

After simplification this formula produces two, one, or zero real solutions, depending on whether the **discriminant** (the number $b^2 - 4ac$ under the radical) is positive, zero, or negative. It is a good practice to check each solution by substituting it into the original equation. Incidentally, if the discriminant is negative, then the equation does have two complex solutions, but you often ignore these as meaningless in real-world settings.

Review Video: Using the Quadratic Formula
Visit mometrix.com/academy and enter code: 163102

Factoring

To solve a quadratic equation by factoring, begin by rewriting the equation in the standard form, $ax^2 + bx + c = 0$. In the important special case that $a = 1$, the goal of factoring is to find numbers f and g such that $x^2 + bx + c = (x + f)(x + g) = x^2 + (f + g)x + fg$. In other words, you want to choose f and g to make $fg = c$ and $f + g = b$. To do this, find pairs of numbers (factors) whose product is c and look for a pair whose sum is b.

For example, suppose you want to find the solutions of the equation $x^2 + 6x - 16 = 0$ by factoring. Here $b = 6$ and $c = -16$. First, you find the pairs of numbers whose product is -16. These are -4 and 4, -8 and 2, -2 and 8, -1 and 16, and 1 and -16. The pair -2 and 8 has a sum of 6. This means $f = -2$ and $g = 8$. So, the factorization is $x^2 + 6x - 16 = (x + f)(x + g) = (x - 2)(x + 8)$, allowing you to rewrite the original equation as $(x - 2)(x + 8) = 0$. The only way for a product to equal zero is for one of the factors to equal zero; so, either $x - 2 = 0$ (in which case $x = 2$) or $x + 8 = 0$ (in which case $x = -8$). Thus, the equation has the solution $x = 2$ or $x = -8$.

In the case that $a \neq 1$, you can attempt to factor the quadratic polynomial $ax^2 + bx + c$ in the form $(mx + f)(nx + g)$ by a similar trial-and-error procedure, but the work tends to be much harder.

Review Video: Factoring Quadratic Equations
Visit mometrix.com/academy and enter code: 336566

Completing the Square

The technique of completing the square comes from a simple observation: Suppose you have the expression $x^2 + bx$. If you take half the linear coefficient, b, square it, and add the result to the expression, the result is always a perfect square trinomial:

$$x^2 + bx + \left(\frac{b}{2}\right)^2 = \left(x + \frac{b}{2}\right)^2$$

For example, if you begin with $x^2 + 6x$ and add the square of half of 6 (half of 6 is 3, and $3^2 = 9$), then you get a perfect square trinomial:

$$x^2 + 6x + 9 = (x + 3)^2$$

This also works if b is negative. For instance, if you begin with the expression $x^2 - 10x$ and complete the square by adding 25 (half of -10 is -5, and $(-5)^2 = 25$), then you get another perfect square trinomial:

$$x^2 - 10x + 25 = (x - 5)^2$$

Now, suppose you want to solve the equation $x^2 + bx + c = 0$. Subtract c from both sides, to get the equation $x^2 + bx = -c$. Complete the square on the left side by adding $(b/2)^2$, but also add this same term to the right side so that the new equation is equivalent to the old one:

$$x^2 + bx + \left(\frac{b}{2}\right)^2 = -c + \left(\frac{b}{2}\right)^2$$
$$\left(x + \frac{b}{2}\right)^2 = -c + \left(\frac{b}{2}\right)^2$$

Take square roots of both sides:

$$x + \frac{b}{2} = \pm\sqrt{-c + \left(\frac{b}{2}\right)^2}$$

Remember to include $\pm$ since every positive number has both a positive and a negative square root. Subtract $b/2$ from both sides to isolate the variable x to finish the problem:

$$x = -\frac{b}{2} \pm \sqrt{-c + \left(\frac{b}{2}\right)^2}$$

This may sound complicated, but in practice it is not hard. For example, suppose you want to solve the equation $x^2 + 6x - 16 = 0$ by completing the square. First, add 16 to both sides:

$$x^2 + 6x = 16$$

Half of 6 is 3 and $3^2 = 9$, so complete the square by adding 9 to both sides of the equation:

$$x^2 + 6x + 9 = 16 + 9$$
$$(x + 3)^2 = 25$$

Now take square root of both sides, remembering to include the $\pm$:

$$\sqrt{(x+3)^2} = \pm\sqrt{25}$$
$$x + 3 = \pm 5$$
$$x = -3 \pm 5$$

So we see that the two solutions to the equation are $x = 2$ and $x = -8$.

Review Video: Completing the Square
Visit mometrix.com/academy and enter code: 982479

Using Given Solutions to Find a Quadratic Equation

To find a quadratic equation with given numbers as solutions, simply find a quadratic polynomial that has those numbers as zeros and set that polynomial equal to zero. This is easy because a polynomial has the number p as a zero precisely when it has the binomial $x - p$ as a factor. Thus, for instance, to find a quadratic polynomial with zeros at $x = 3$ and $x = -5$, construct the polynomial $P(x) = (x-3)\big(x-(-5)\big) = (x-3)(x+5) = x^2 + 2x - 15$. Setting this equal to zero produces an equation, $x^2 + 2x - 15 = 0$, whose solutions are $x = 3$ and $x = -5$.

Of course, any constant multiple $P(x) = a(x-3)(x+5)$ will also have the same zeros. For instance, if you choose $a = 4$, then you get the polynomial $P(x) = 4(x-3)(x+5) = 4x^2 + 8x - 60$, which also has zeros at $x = 3$ and $x = -5$. From this you can get another equation, $4x^2 + 8x - 60 = 0$, with solutions $x = 3$ and $x = -5$.

Functions

Basics of Functions

Definition of a Function

A function is a rule that assigns to every number in a given set (called the **domain**) exactly one corresponding value. For example, if our domain is the set $\{-2,1,2,3\}$, we can define a function by assigning to each number its square. This function assigns to -2 the value 4, to 1 the value 1, to 2 the value 4, and to 3 the value 9 (since $(-2)^2 = 4$, $1^2 = 1$, $2^2 = 4$, and $3^2 = 9$). The set of all the values assigned by a function is the **range** of the function. The range of the function in our example is the set $\{1, 4, 9\}$. We may think of a function as a kind of machine: we give it a number as an input, and it uses its rule to produce a number as an output. In the squaring function above, the input 3 produces the output 9.

Review Video: What is a Function?
Visit mometrix.com/academy and enter code: 784611

Function Notation

We usually name a function by a letter, often the letter f (for *function*—if we need to talk about more than one function, we name the second one g, the third one h, etc.). To specify the value (the output) corresponding to a particular number in the domain (the input), we write the function letter followed by the input number in parentheses. For instance, in the example above the notation $f(3)$ means the value that the function assigns to the number 3, namely 9—that is, $f(3) = 9$. We read the symbols $f(3)$ as, "f of 3," and we call 3 the **argument** of the function and 9 the **value** of the function (so *argument* means *input* and *value* means *output*).

Using function notation we can define the squaring function above by listing the values the function assigns to each argument in the domain: $f(-2) = 4$. $f(1) = 1$, $f(2) = 4$, and $f(3) = 9$. More efficiently, we can define the function by the single equation $f(x) = x^2$, which says that if x is a number from the domain, then we calculate the value assigned to it by substituting the number x in the formula x^2. For instance, we calculate $f(5) = 5^2 = 25$. Similarly, if we define a function g by the equation $g(x) = x^2 - 4x + 7$, then we calculate the value $g(3)$ by substituting 3 for each x in the formula: $g(3) = 3^2 - 4 \cdot 3 + 7 = 9 - 12 + 7 = 4$.

Other Ways to Define Functions

Instead of denoting the value of the function by $f(x)$, sometimes we simply use another letter, usually y. For instance, instead of defining the squaring function by the equation $f(x) = x^2$, we might use the equation $y = x^2$. In this case, we refer to x (the input) as the **independent variable** and y (the output) as the **dependent variable** because the value, y, depends on the number we choose for x.

A formula (with y or $f(x)$) is the most common way to define a function; but sometimes, if the domain is small enough, we prefer to list explicitly the possible inputs and their corresponding outputs. Some ways of doing this appear above, but a more common approach is to put the input-output pairs in a table. For instance, we can define the squaring function above by the table

x	-2	1	2	3
y	4	1	4	9

We see that the domain of this function is the set of all numbers in the x-row and the range is the set of all numbers in the y-row. We note that numbers cannot repeat in the x-row (because a function assigns exactly one value to each argument in the domain) but they can repeat in the y-row (because the function can assign the same value to multiple arguments—for instance, the number 4 appears twice in the y-row).

We can also define a function by writing the inputs and corresponding outputs as ordered pairs of x- and y-values. For instance, we can write the squaring function above as the set of ordered pairs $\{(-2,4), (1,1), (2,4), (3,9)\}$. Further, by treating these ordered pairs as coordinates and plotting the corresponding points on the coordinate plane, we get the **graph** of the function:

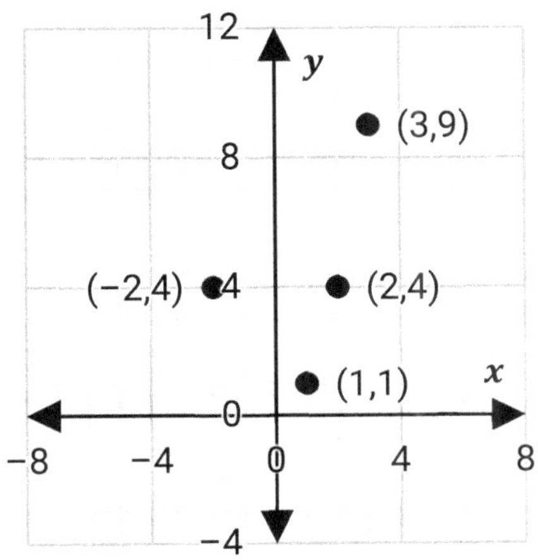

Turning this around, we can potentially use a graph to define a function, namely the function consisting of the coordinate pairs of all the points in the graph. This always works unless the graph

has two points with the same x-coordinate (because then the function would assign two different y-values to the same x). It is easy to detect such points: because they have the same x-coordinate, a vertical line passes through both of them. Thus, a graph always defines a function unless it is possible to draw a vertical line that intersects the graph in two or more points. We call this condition the **vertical line test**. For example, if our graph is a circle, then by the Vertical Line Test the graph does not define a function because there are vertical lines that will intersect the circle in two different points.

MORE ON DOMAINS AND RANGES

When we define a function by a formula and do not specify the domain, then by default the domain consists of all real numbers for which the formula produces an answer. For instance, suppose we define a function f by the formula $f(x) = 1/x$. If $x = 0$, then $1/x = 1/0$, which is undefined. But if x is any other real number, then we can calculate the value of $1/x$. So, the default domain of this function is all real numbers except zero. Because of this domain convention, the graph of a function defined by a formula usually consists of infinitely many points that "connect to" each other in a way that produces a line or curve (see examples below) rather than the isolated points we see in the squaring function above.

If we have the graph of a function, its domain consists of all numbers on the x-axis with corresponding points on the graph and its range consists of all numbers on the y-axis with corresponding points on the graph. For example, consider the function $f(x) = x^2 + 3$:

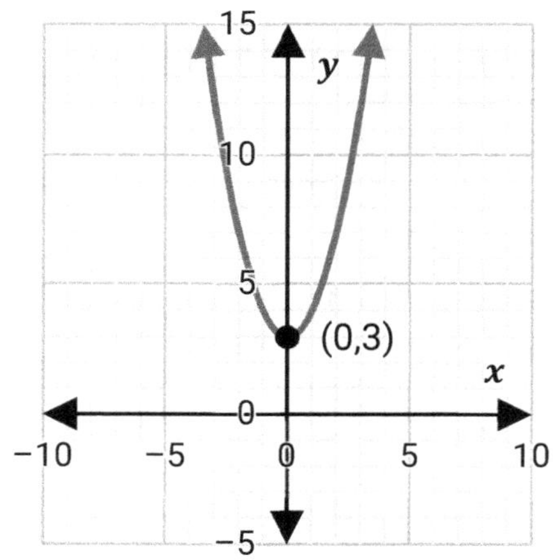

Since the graph continues infinitely to the left and right beyond what we can see, every point on the x-axis has a corresponding point on the graph; so, the domain of this function is all real numbers. On the other hand, the lowest point on this graph has a y-value of 3, and the graph passes through all higher y-values. So, the range of this function is all real numbers greater than or equal to 3, which we can denote algebraically by $y \geq 3$ or, using interval notation, by $[3, \infty)$.

Review Video: How to Find Domain and Range
Visit mometrix.com/academy and enter code: 778133

Review Video: Domain and Range of Quadratic Functions
Visit mometrix.com/academy and enter code: 331768

MONOTONIC AND EVEN/ODD FUNCTIONS

A function, f, is **increasing** if it always assigns larger values to larger arguments. It is **decreasing** if it always assigns smaller values to larger arguments. That is, f is increasing if $a < b$ always guarantees $f(a) < f(b)$, and it is decreasing if $a < b$ always guarantees $f(a) > f(b)$. The graph of an increasing function consistently rises from left to right, and the graph of a decreasing function consistently falls from left to right. For example, the function $f(x) = 2x$ is an increasing function because doubling a larger number always gives us a larger result than doubling a smaller number. The graph of $f(x) = 2x$ is a line with slope $m = 2$, which, as we expect, rises from left to right. We call a function **monotonic** if it is either increasing or decreasing.

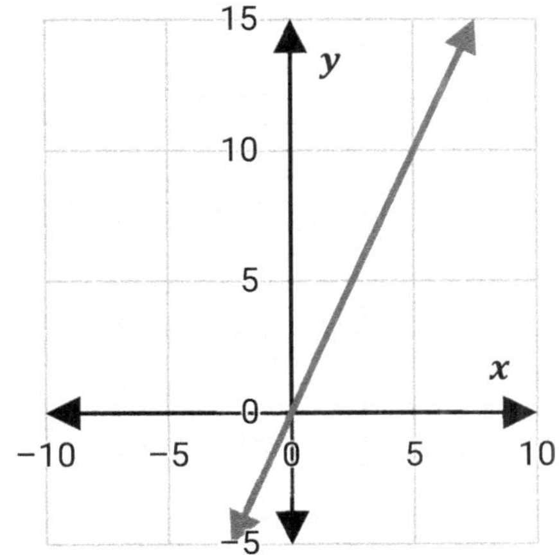

A function, f, is **even** if changing the sign of its argument produces the same value. It is **odd** if changing the sign of its argument produces the same value except with the opposite sign. That is, f is even if $f(-x) = f(x)$ and odd if $f(-x) = -f(x)$ for every argument x. The function $f(x) = x^2 + 3$ is even because substituting opposite arguments always produces the same value. For instance, $f(5) = 28$ and $f(-5) = 28$ because $5^2 + 3 = 25 + 3 = 28$ and $(-5)^2 + 3 = 25 + 3 = 28$. The function $f(x) = 2x$ is odd because substituting opposite arguments always produces opposite values. For instance, $f(10) = 20$ and $f(-10) = -20$ because $2(10) = 20$ and $2(-10) = -20$. The graph of an even function is always symmetric with respect to the y-axis, making the left and right halves of the graph mirror images of each other, as in the graph of the even function $f(x) = x^2 + 3$ above. The graph of an odd function is always symmetric with respect to the origin. This means that if we rotate the graph 180° around the origin (think of sticking a pin through the origin on a sheet of graph paper and rotating the paper halfway around) the graph looks the same, as in the graph of the odd function $f(x) = 2x$ above.

It is worth noting that most functions are neither increasing nor decreasing (that is, they are not monotonic) and most functions are neither even nor odd. For example, the function $f(x) = x^2 -$

x is neither increasing nor decreasing and neither even nor odd: its graph neither rises nor falls consistently, and it is symmetric with respect to neither the y-axis nor the origin.

Review Video: Even and Odd Functions
Visit mometrix.com/academy and enter code: 278985

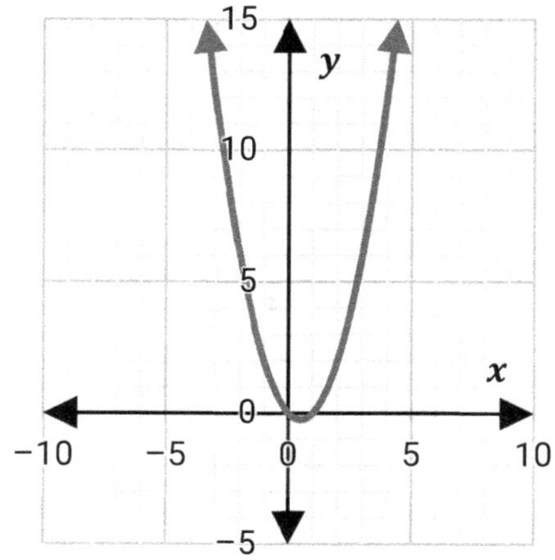

INVERTIBLE (ONE-TO-ONE) FUNCTIONS

A function, f, is one-to-one if it never assigns the same value to different arguments—that is, if $f(a)$ and $f(b)$ are different whenever a and b are different. The graph of a one-to-one function never has two points that lie on the same horizontal line because such points would have different x-values but the same y-value. Thus, a function is one-to-one if it is impossible to draw a horizontal line that intersects its graph in more than one point. We call this condition the **horizonal line test**. For example, the graph of the function $f(x) = 2x$ above is a line that rises from left to right. Every horizontal line intersects this line in exactly one point, so the function $f(x) = 2x$ is one-to-one. This is also clear without the graph because it is impossible to double two different numbers and get the same answer.

When a function, f, is one-to-one, it is possible to define its inverse function, f^{-1}, that "undoes" what f does, assigning to each output from f the input that produced it. That is, for each x in the domain of f, if $y = f(x)$, then $f^{-1}(y) = x$. For example, the inverse of the function $f(x) = 2x$ above is $f^{-1}(y) = y/2$. So, for instance, $f(5) = 2 \cdot 5 = 10$, and $f^{-1}(10) = 10/2 = 5$ (and similarly for every other value of x). Thus, the domain of f^{-1} is the range of f and vice versa. If a function, f, has an inverse, we say that f is **invertible**. Since a function has an inverse precisely when it is one-to-one, the terms *invertible* and *one-to-one* are synonyms.

If f is an invertible function defined by a formula, then to find its inverse we simply write the equation $y = f(x)$ and solve it for x (that is, we isolate the x). The result will be the equation $f^{-1}(y) = x$. For instance, starting with the function $f(x) = 2x$, we write $y = 2x$ and isolate the x by dividing both sides of the equation by 2. This gives us $y/2 = x$, so we know that $f^{-1}(y) = y/2$. Although this procedure is theoretically simple, in practice the algebra can be difficult.

COMMON FUNCTIONS

Certain functions and certain kinds of functions are particularly useful, coming up frequently in mathematics and its applications. Once we know some basic function terminology and concepts, it is useful to begin developing a mental library of the most common and useful functions.

Review Video: Common Functions
Visit mometrix.com/academy and enter code: 629798

CONSTANT FUNCTIONS

A function of the form $f(x) = a$, where a is a real number, is a **constant function**. This function assigns the same value, a, to every real argument x. For instance, given the constant function $f(x) = 5$, we have $f(2) = 5$, $f(100) = 5$, and $f(-7.1) = 5$.

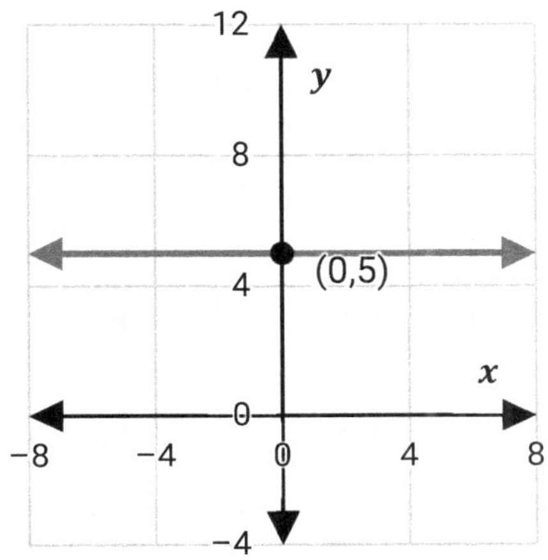

The domain of a constant function is the set of all real numbers, and the range is the set containing the single number a. Its graph is a horizontal line passing through the number $y = a$ on the y-axis (we call the number at which a function's graph intersects the y-axis the **y-intercept** of the function).

THE IDENTITY FUNCTION

The function $f(x) = x$ is the **identity function**. Its value always equals its argument. Thus, for instance, $f(2) = 2$, $f(100) = 100$, and $f(-7.1) = -7.1$.

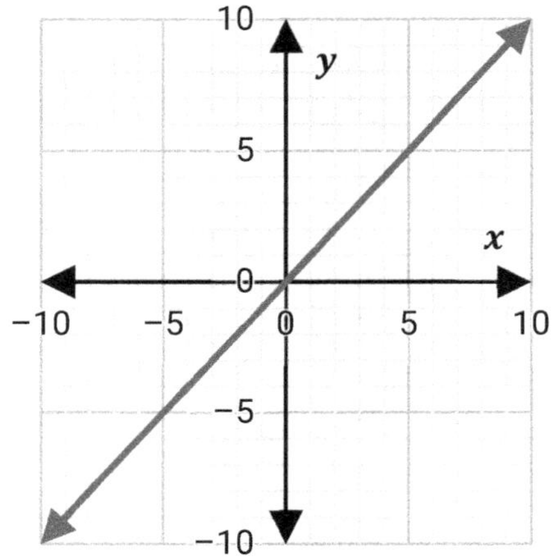

Its domain and range are the set of all real numbers. It is both an increasing function and an odd function. Its graph is a line that passes through the origin and rises from left to right at a 45° angle to the horizontal. Since it passes through the origin, its y intercept is $y = 0$ and it also has an ***x*-intercept** (a number at which the function's graph intersects the x-axis) of $x = 0$.

LINEAR FUNCTIONS

A function of the form $f(x) = ax + b$, where a and b are real numbers (with $a \neq 0$), is a **linear function** (the identity function is a linear function with $a = 1$ and $b = 0$). Its domain and range are the set of all real numbers. Its graph is a line (the word *linear* contains the root word *line*) with one x-intercept (at $x = -b/a$), with a y-intercept at $y = b$, and with a direction and steepness that depend on the coefficient a, which we call the **slope**. Specifically, the slope a is the amount the y-value increases for each increase of 1 in the x-value. Thus, for $a > 0$, the line rises from left to right (making f an increasing function), and larger values of a produce steeper ascents. Similarly, for $a < 0$, the line falls from left to right (making f a decreasing function), and smaller (more negative) values of a produce steeper descents. For instance, the graph of the linear function

$f(x) = (1/2)x + 3$ is a line that passes through the point $y = 3$ on the y-axis and that rises by $1/2$ unit for every unit that x increases.

> **Review Video: Linear Functions**
> Visit mometrix.com/academy and enter code: 200735
>
> **Review Video: Graphing Linear Functions**
> Visit mometrix.com/academy and enter code: 699478

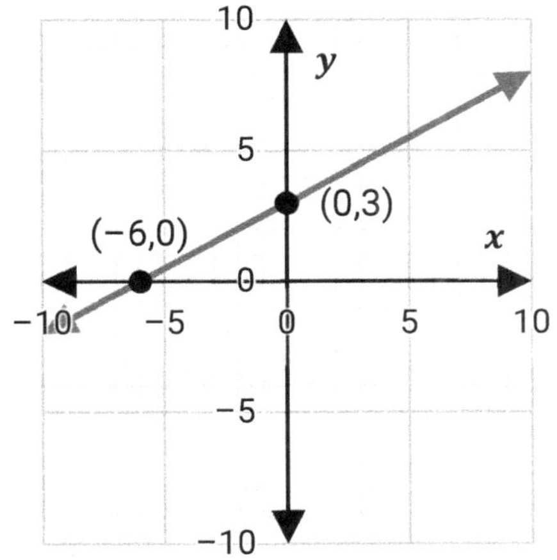

In many contexts it is standard to use the letter m for slope and thus to write the general form of a linear function as $f(x) = mx + b$, known as **slope-intercept form**.

The Squaring Function

The function $f(x) = x^2$ is the **squaring function**.

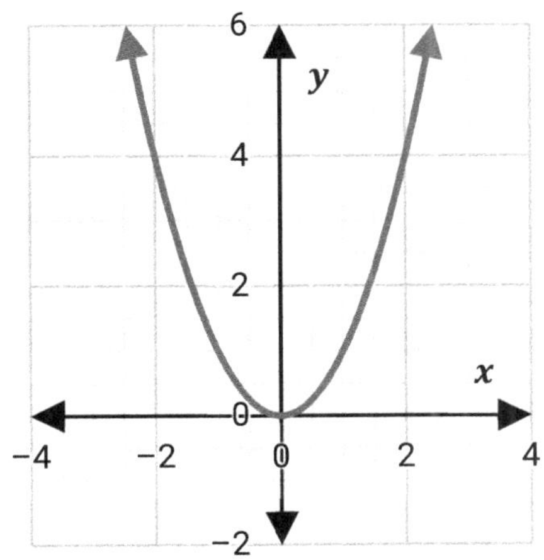

Its graph is U-shaped, opening upward as shown, a shape known as a **parabola**. It has a lowest point, its **vertex**, at the origin, which is also the location of its single x-intercept and single y-intercept. Thus, its **minimum** is $y = 0$, its domain is the set of all real numbers, and its range is the set of nonnegative real numbers (that is, $y \geq 0$). It is an even function and thus symmetric with respect to the y-axis (which we call the **axis of symmetry**), meaning that the left half of the graph is the mirror image of the right half, with the mirror standing on the y-axis.

QUADRATIC FUNCTIONS

A function of the form $f(x) = ax^2 + bx + c$, where a, b, and c are real numbers (with $a \neq 0$), is a **quadratic function** (the squaring function is a quadratic function with $a = 1$, $b = 0$, and $c = 0$). Its domain is the set of all real numbers, and its graph is a parabola. It is symmetric with respect to its axis of symmetry, the vertical line $x = -b/(2a)$. If $a > 0$, the parabola opens upward, so that its vertex is at its lowest point (its minimum) and its range consists of all real numbers greater than or equal to this minimum y-value. If $a < 0$, the parabola opens downward, so that its vertex is at its highest point (its maximum) and its range consists of all real numbers less than or equal to this maximum y-value. Its y-intercept is $y = c$ since $f(0) = c$, and it may have zero, one, or two x-intercepts. For example, the function $f(x) = x^2 - 6x + 5$ has $a = 1, b = -6$, and $c = 5$.

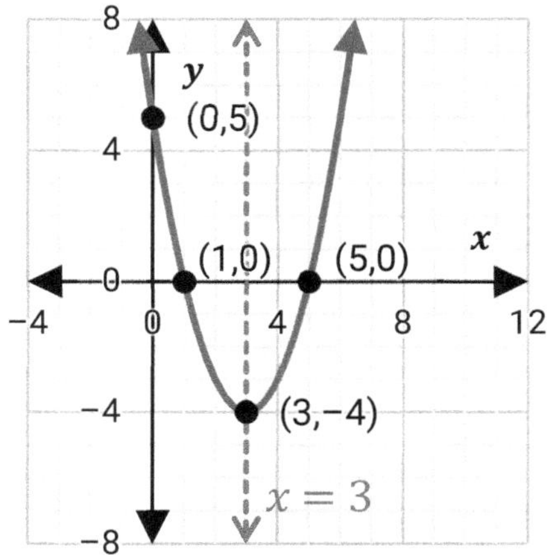

Its graph opens upward (because $a > 0$) and its axis of symmetry is the vertical line $x = 3$ (since $-b/(2a) = -(-6)/(2 \cdot 1) = 3$). Its y-intercept is at $y = 5$. It turns out to have its vertex at the point $(3, -4)$, making its minimum value $y = -4$. So, its domain is the set of all real numbers, and its range is $y \geq -4$. It also turns out to have two x-intercepts, at $x = 1$ and at $x = 5$ (since $f(1) = 0$ and $f(5) = 0$).

POLYNOMIAL FUNCTIONS

A function of the form $f(x) = a^n x^n + a^{n-1} x^{n-1} + \cdots + a_2 x^2 + a_1 x + a_0$, where n is a whole number and $a_0, a_1, a_2, \ldots a_{n-1}, a_n$ are real numbers, is a **polynomial function of degree *n***. Its domain is the set of all real numbers (it is complicated to describe its range in general), and its y-intercept is $y = a_0$ (since $f(0) = a_0$). Constant functions, linear functions, and quadratic functions are polynomial functions of degrees 0, 1, and 2, respectively. In general, a polynomial function of degree n has up to n zeros (x-intercepts) and up to $n - 1$ "bends." For example, the fourth degree polynomial function $f(x) = x^4 - 11x^3 + 41x^2 - 61x + 30$, whose graph appears here, has four x-intercepts (at $x = 1, x = 2, x = 3$, and $x = 5$) and three "bends," and its y-intercept (not visible on the graph) is at $y = 30$.

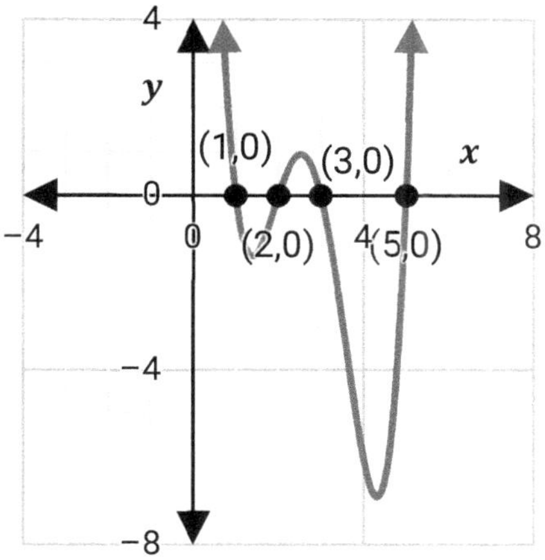

Rational Functions

A function of the form $f(x) = P(x)/Q(x)$, where P and Q are polynomials, is a rational function (we note that the word *rational* includes the root word *ratio*, indicating that a rational function is a ratio of polynomial functions). The domain of a rational function is all real numbers except the zeros of $Q(x)$ since division by zero is undefined (the range can be difficult to describe in general). Its y-intercept is $f(0)$, if this is defined; and its x-intercepts are the zeros of $P(x)$ that are in the domain of f, if there are any. A rational function may also have vertical asymptotes (vertical lines that the graph approaches without crossing) and a horizontal asymptote (a horizontal line that the curve approaches as x becomes very small or very large (toward the left and right edges of the graph). For example, the rational function $f(x) = (2x^2 + x - 1)/(x^2 + x - 2)$ has as its domain the set of all real numbers except $x = -2$ and $x = 1$ (since those numbers make the denominator zero).

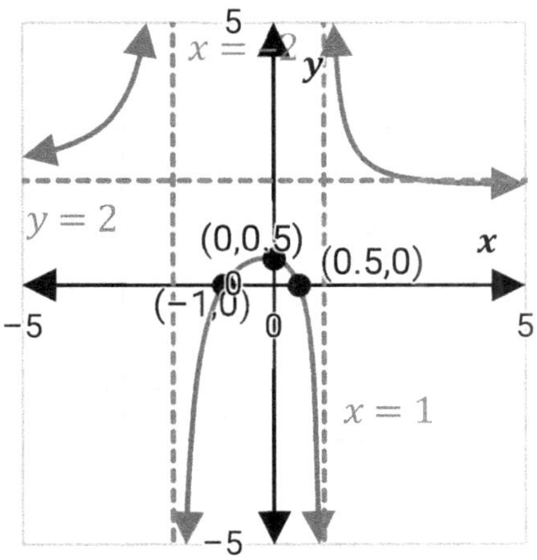

It has a y-intercept of $y = 1/2$ since $f(0) = (-1)/(-2) = 1/2$, and it has x-intercepts at $x = -1$ and at $x = 1/2$ since those numbers make the numerator zero. It has vertical asymptotes at $x = -2$ and $x = 1$ (not coincidentally, these are the numbers omitted from the domain) and a horizontal asymptote at $y = 2$. It is important to note that vertical asymptotes cannot be crossed in rational functions, but horizontal asymptotes can be crossed if the function tends near the asymptote at infinity and does not go past all possible turning points.

Review Video: Simplifying Rational Polynomial Functions
Visit mometrix.com/academy and enter code: 351038

Review Video: Horizontal Asymptotes
Visit mometrix.com/academy and enter code: 747796

THE SQUARE ROOT FUNCTION

The function $f(x) = \sqrt{x}$ is the square root function.

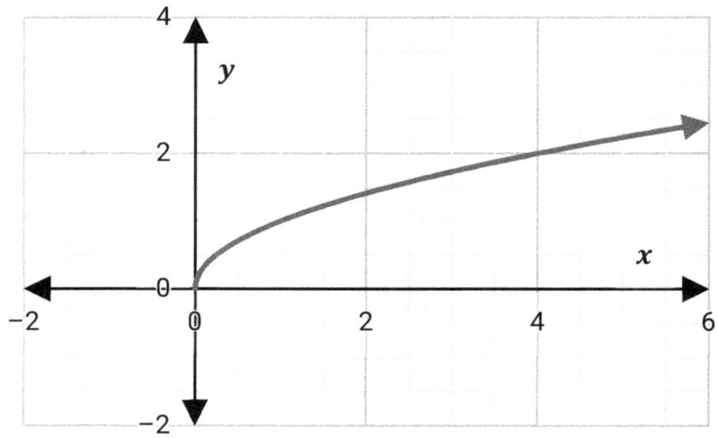

It is an increasing function, and its domain and range are both the set of all nonnegative real numbers. It has one x-intercept and one y-intercept, both appearing at the origin. Its graph is the upper half of a parabola opening to the right. The square root function is the inverse of the squaring function with domain restricted to the nonnegative real numbers (that is, $f(x) = x^2$ for $x \geq 0$).

PIECEWISE-DEFINED FUNCTIONS

As the name suggests, a **piecewise-defined function** (or, simply, a **piecewise function**) is a function defined by different rules on different pieces of the domain. We define such a function using the following form:

Function Name	Rule to Apply	Piece of the Domain on Which the Rule Applies
$f(x) =$	Rule 1,	First Piece of the Domain
	Rule 2,	Second Piece of the Domain
	Rule 3,	Third Piece of the Domain
	etc.,	etc.

The pieces of the domain should not overlap, and together they should cover the whole domain. For example, we might craft a piecewise-defined function by

$$f(x) = \begin{cases} x^2, & \text{if } x < 2 \\ 3x - 5, & \text{if } x \geq 2 \end{cases}$$

The two pieces of the domain—namely, $x < 2$ and $x \geq 2$—do not overlap, and together they include all real numbers. To evaluate the function for a particular argument x, we determine which piece of the domain includes x and then apply the corresponding rule. For instance, to find $f(4)$, we note that $4 \geq 2$; so, we apply the rule $3x - 5$ to get the value $f(4) = 3 \cdot 4 - 5 = 7$. Similarly, to find $f(-6)$, we note that $-6 < 2$; so, we apply the rule x^2 to get the value $f(-6) = (-6)^2 = 36$.

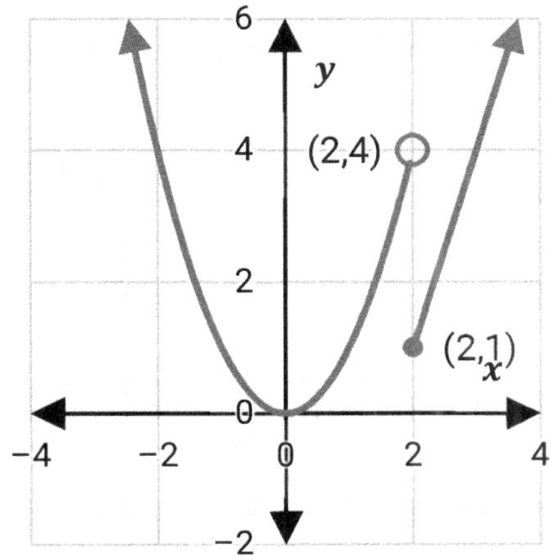

To graph this function, we sketch the graph of the parabola $y = x^2$ on the part of the plane where $x < 2$ and we sketch the line $y = 3x - 5$ on the part of the plane where $x \geq 2$. This produces a graph with a jump at $x = 2$ (a discontinuity—piecewise-defined functions are useful for producing graphs with discontinuities). We plot an open circle at the point (2,4), the end of the left part of the graph, to show that this point is not part of the graph. And we plot a solid dot at the point (2,1), the start of the right part of the graph, to show that this point *is* part of the graph.

Review Video: Piecewise Functions
Visit mometrix.com/academy and enter code: 707921

The Absolute Value Function

A particularly useful piecewise-defined function is the absolute value function. It is so important that instead of naming it $f(x)$ or $g(x)$, we denote it using the special notation $|x|$. Its definition is

$$|x| = \begin{cases} -x, & \text{if } x < 0 \\ x, & \text{if } x \geq 0 \end{cases}$$

For instance, $|8| = 8$ (since $8 \geq 0$) and $|-5| = -(-5) = 5$, since $-5 < 0$. So, the absolute value function acts like the identity function for nonnegative numbers (it leaves them unchanged), and it gives the opposite of negative numbers (it effectively strips off the minus sign). Thus, we can think of the absolute value of a real number as its distance from zero on the number line, without taking into consideration whether the number is larger than or smaller than zero. For instance, $|-3| = 3$ and $|3| = 3$, showing that both -3 and 3 are three units away from zero. The absolute value function is an even function with a V-shaped graph that looks like the line $y = x$ (the identity function) on the right "half" of the plane (for $x \geq 0$) and the line $y = -x$ on the left "half" of the plane (for $x < 0$).

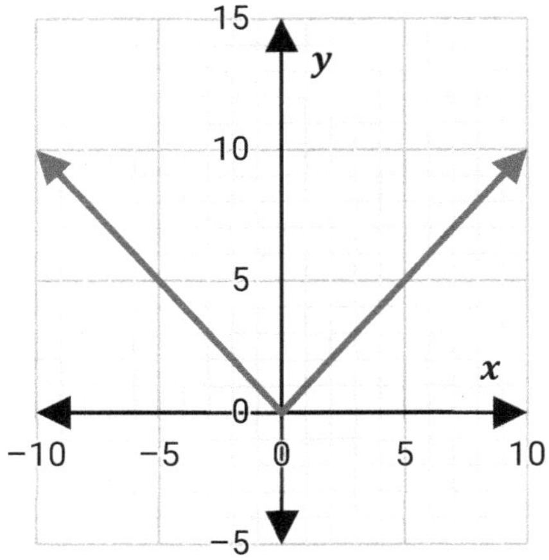

Working with Functions

Manipulation of Functions

Translation occurs when values are added to or subtracted from the x- or y-values. If a constant is added to the y-portion of each point, the graph shifts up. If a constant is subtracted from the y-portion of each point, the graph shifts down. This is represented by the expression $f(x) \pm k$, where k is a constant. If a constant is added to the x-portion of each point, the graph shifts left. If a constant is subtracted from the x-portion of each point, the graph shifts right. This is represented by the expression $f(x \pm k)$, where k is a constant.

Stretching, compression, and reflection occur when different parts of a function are multiplied by different groups of constants. If the function as a whole is multiplied by a real number constant greater than 1, $(k \times f(x))$, the graph is stretched vertically. If k in the previous equation is greater than zero but less than 1, the graph is compressed vertically. If k is less than zero, the graph is reflected about the x-axis, in addition to being either stretched or compressed vertically if k is less than or greater than −1, respectively. If instead, just the x-term is multiplied by a constant greater than 1 $(f(k \times x))$, the graph is compressed horizontally. If k in the previous equation is greater than zero but less than 1, the graph is stretched horizontally. If k is less than zero, the graph is reflected about the y-axis, in addition to being either stretched or compressed horizontally if k is greater than or less than −1, respectively.

Review Video: Manipulation of Functions
Visit mometrix.com/academy and enter code: 669117

Applying the Basic Operations to Functions

For each of the basic operations, we will use these functions as examples: $f(x) = x^2$ and $g(x) = x$.

To find the sum of two functions f and g, assuming the domains are compatible, simply add the two functions together: $(f + g)(x) = f(x) + g(x) = x^2 + x$.

To find the difference of two functions f and g, assuming the domains are compatible, simply subtract the second function from the first: $(f - g)(x) = f(x) - g(x) = x^2 - x$.

To find the product of two functions f and g, assuming the domains are compatible, multiply the two functions together: $(f \times g)(x) = f(x) \times g(x) = x^2 \times x = x^3$.

To find the quotient of two functions f and g, assuming the domains are compatible, divide the first function by the second: $\left(\frac{f}{g}\right)(x) = \frac{f(x)}{g(x)} = \frac{x^2}{x} = x\,; x \neq 0$.

The example given in each case is fairly simple, but on a given problem, if you are looking only for the value of the sum, difference, product, or quotient of two functions at a particular x-value, it may be simpler to solve the functions individually and then perform the given operation using those values.

The composite of two functions f and g, written as $(f \circ g)(x)$ simply means that the output of the second function is used as the input of the first. This can also be written as $f(g(x))$. In general, this can be solved by substituting $g(x)$ for all instances of x in $f(x)$ and simplifying. Using the example functions $f(x) = x^2 - x + 2$ and $g(x) = x + 1$, we can find that $(f \circ g)(x)$ or $f(g(x))$ is equal to $f(x + 1) = (x + 1)^2 - (x + 1) + 2$, which simplifies to $x^2 + x + 2$.

It is important to note that $(f \circ g)(x)$ is not necessarily the same as $(g \circ f)(x)$. The process is not always commutative like addition or multiplication expressions. It *can* be commutative, but most often this is not the case.

EVALUATING LINEAR FUNCTIONS

A **function** can be expressed as an equation that relates an input to an output where each input corresponds to exactly one output. The input of a function is defined by the x-variable, and the output is defined by the y-variable. For example, consider the function $y = 2x + 6$. The value of y, the output, is determined by the value of the x, the input. If the value of x is 3, the value of y is $y = 2(3) + 6 = 6 + 6 = 12$. This means that when $x = 3$, $y = 12$. This can be expressed as the ordered pair (3,12).

It is common for function equations to use the form $f(x) =$ instead of $y =$. However, $f(x)$ and y represent the same thing. We read $f(x)$ as "f of x." The expression "f of x" implies that the value of f depends on the value of x. The function used in the example above could be expressed as $y = 2x + 6$ or $f(x) = 2x + 6$. Both functions represent the same line when graphed.

Functions that are expressed in the form $f(x) =$ are evaluated in the same way the equations are evaluated in the form $y =$. For example, when evaluating the function $f(x) = 3x - 2$ for $f(6)$, substitute 6 in for x, and simplify. In this case, $f(x) = 3x - 2$ becomes $f(6) = 3(6) - 2 = 18 - 2 = 16$. When x is 6, $f(x)$ is 16.

Example: To find the value of $f(8)$, calculate as follows:

$$f(x) = 3x - 2$$
$$f(8) = 3(8) - 2$$
$$f(8) = 22$$

Review Video: Evaluating Functions
Visit mometrix.com/academy and enter code: 588515

ADVANCED FUNCTIONS

STEP FUNCTIONS

The double brackets indicate a step function. For a step function, the value inside the double brackets is rounded down to the nearest integer. The graph of the function $f_0(x) = [\![x]\!]$ appears on the left graph. In comparison $f(x) = 2\left[\!\left[\frac{1}{3}(x-1)\right]\!\right]$ is on the right graph. The coefficient of 2 shows that it's stretched vertically by a factor of 2 (so there's a vertical distance of 2 units between successive "steps"). The coefficient of $\frac{1}{3}$ in front of the x shows that it's stretched horizontally by a factor of 3 (so each "step" is three units long), and the $x-1$ shows that it's displaced one unit to the right.

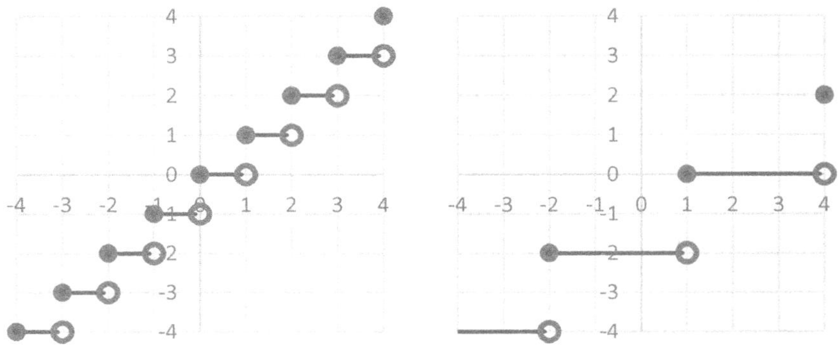

TRANSCENDENTAL FUNCTIONS

Roughly speaking, algebraic functions are functions defined by formulas involving numbers and variables combined by addition, subtraction, multiplication, division, powers (but not variable powers), and roots. **Transcendental functions** are functions that are not algebraic. A function that includes logarithms, trigonometric functions, or variables as exponents, is transcendental, not algebraic, even if the function also includes polynomials or roots.

EXPONENTIAL FUNCTIONS

Exponential functions are functions that have the form $y = b^x$, where base $b > 0$ and $b \neq 1$. The exponential function can also be written $f(x) = b^x$. The following properties apply to exponential expressions:

Property	Description
$a^x a^y = a^{x+y}$	The product of exponentials with the same base equals the base raised to the sum of the powers
$a^x/a^y = a^{x-y}$	The quotient of exponentials with the same base equals the base raised to the difference of the powers
$(a^x)^y = a^{xy}$	An exponential raised to a power equals the base raised to the product of the powers
$(ab)^x = a^x b^x$	Exponentiation distributes over multiplication
$(a/b)^x = a^x/b^x$	Exponentiation distributes over division

The graph of an example exponential function, $f(x) = 2^x$, is below:

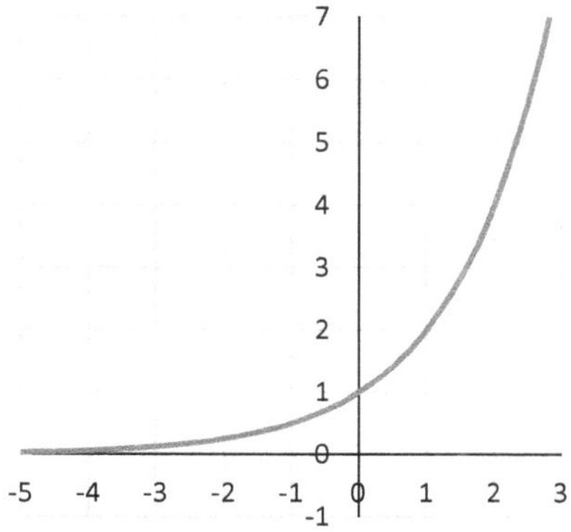

Note in the graph that the y-value approaches zero to the left and infinity to the right. One of the key features of an exponential function is that there will be one end that goes off to infinity and another that asymptotically approaches a lower bound. Common forms of exponential functions include the following:

Geometric sequences: The terms of a geometric sequence have the form $a_n = a_1 \times r^{n-1}$, where a_n is the value of the nth term, a_1 is the initial term, and r is the common ratio between succesive terms. Note that $a_1 \times r^{1-1} = a_1 \times r^0 = a_1 \times 1 = a_1$

Review Video: Geometric Sequences
Visit mometrix.com/academy and enter code: 140779

General exponential growth or decay: The general exponential growth or decay function is $f(t) = a(1 + r)^t$, where the value of $f(t)$ is the amount (of whatever quantity we are measuring) at time t, the constant a is the initial amount (the amount at time $t = 0$), and the constant r is the fixed rate of increase (if $r > 0$) or decrease (if $r < 0$) per unit time. For example, if we invest \$1000 at an interest rate of 6% interest compounded annually and we measure time in years, then our investment will grow according to an exponential model with $a = \$1000$ and $r = 6\% = 0.06$. Thus, after t years the value of our investment will be $f(t) = \$1000(1 + 0.06)^t$. After 5 years, for instance, the investment will grow to $f(5) = \$1000(1 + 0.06)^5 \approx \$1338.23$.

Compound interest: We can modify the general exponential growth function above slightly to get a formula for interest compounded n times per year. Replacing the constant a with P (for principal, the initial amount we invest), the total value of our investment after t years is $f(t) = P\left(1 + \frac{r}{n}\right)^{nt}$, where r is the nominal annual interest rate. Thus, if we change our \$1000 investment above by having the interest rate be 6% compounded *semiannually* (two times per year—so $n = 2$), the value of our investment after t years is $f(t) = \$1000\left(1 + \frac{0.06}{2}\right)^{2t} = \$1000(1 + 0.03)^{2t}$. After 5 years, for instance, the investment will grow to $f(5) = \$1000(1 + 0.03)^{10} \approx \$1343.92$. This is slightly higher than the previous investment because the interest compounds twice as often.

Review Video: Compound Interest Formula
Visit mometrix.com/academy and enter code: 100366

Review Video: Interest Functions
Visit mometrix.com/academy and enter code: 559176

Population growth and continuously compounded interest: When a quantity grows (or decays) continuously at a rate that stays constant relative to the size the quantity has already attained, then we can model its size at time t with the function $f(t) = ae^{rt}$, where a is the initial amount, r is the relative growth rate (which we often call simply the **growth rate**), and e is the irrational constant known as Euler's number (approximately 2.718; scientific calculators have a button for finding e^x, which is known as the **natural exponential function**). For instance, under some circumstances, if an initial population (of people, plants, or animals, for example) of size a grows at constant relative rate r, then the size of the population at time t is $f(t) = ae^{rt}$. Similarly, if we invest principal P (instead of a) for t years at nominal annual interest rate r compounded *continuously*, then after t years the value of the investment is $f(t) = Pe^{rt}$.

For example, suppose the initial population of a town is $a = 1{,}200$ people and the relative annual growth rate is $r = 5\% = 0.05$. Then the population of the town after t years is $f(t) = ae^{rt} = 1200e^{0.05t}$. After 10 years, for instance, the town population is $f(10) = 1200e^{0.05\cdot 10} = 1200e^{0.5} \approx$ 1978 people.

Review Video: Population Growth
Visit mometrix.com/academy and enter code: 109278

Logarithmic Functions

The **logarithmic function base b** is the function $y = \log_b x$ or $f(x) = \log_b x$, where the base b may be any positive number except one. The most common bases for logarithms are base 10 (the **common logarithm**) and base e (the **natural logarithm**). We usually write the common logarithm as $y = \log x$ (that is, $\log x$, with no base listed, means $\log_{10} x$) and the natural logarithm as $y = \ln x$ (that is, $\ln x$ means $\log_e x$).

Exponential functions and logarithmic functions with the same base are inverse functions. That is, if $f(x) = b^x$, then $f^{-1}(x) = \log_b x$. This means that the two equations $y = b^x$ (exponential form) and $x = \log_b y$ (logarithmic form) express the same relationship between the quantities x and y. We often solve problems involving logarithms by rewriting them in exponential form, and vice versa. Also, because of this inverse relationship, logarithms and exponentials cancel each other. That is, $\log_b b^x = x$ and $b^{\log_b x} = x$.

The following properties apply to logarithmic expressions:

Property	Description
$\log_b 1 = 0$	The log of 1 is equal to 0 for any base
$\log_b b = 1$	The log of the base is equal to 1
$\log_b b^p = p$	The log of the base raised to a power is equal to that power
$\log_b MN = \log_b M + \log_b N$	The log of a product is the sum of the log of each factor
$\log_b \frac{M}{N} = \log_b M - \log_b N$	The log of a quotient is equal to the log of the dividend minus the log of the divisor
$\log_b M^p = p \log_b M$	The log of a value raised to a power is equal to the power times the log of the value

Logarithms are helpful in solving equations in which the variable appears in an exponent. For instance, consider the example above in which we model the population of a town by the function $f(t) = 1200e^{0.05t}$. Suppose we want to know how long it will take the population of the town to double to 2400 people from its original value of 1200. We find this by solving the equation $1200e^{0.05t} = 2400$. First, we divide both sides of the equation by 1200 to get $e^{0.05t} = 2$. Then we take natural logarithms of both sides to get $\ln e^{0.05t} = \ln 2$, which simplifies (because natural logarithms and natural exponentials are inverse functions) to $0.05t = \ln 2$. Finally, we divide both sides by 0.05 to get $t = (\ln 2)/0.05$, which we evaluate with a calculator to get $t \approx 13.9$ years.

The graph of an example logarithmic function, $f(x) = \log_2(x + 2)$, is below:

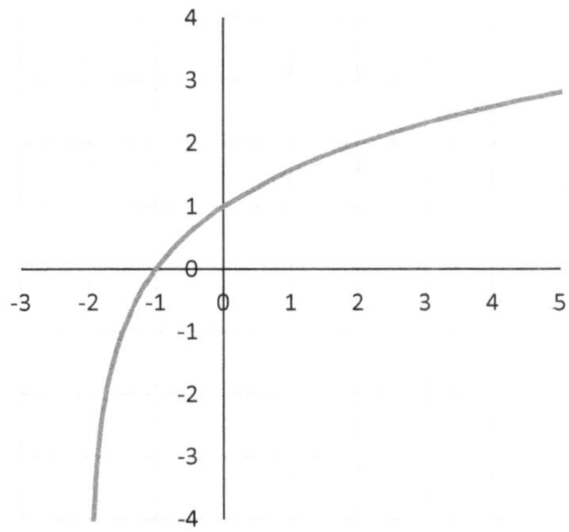

Review Video: Logarithmic Function
Visit mometrix.com/academy and enter code: 658985

Matrices

Matrix Basics

A **matrix** (plural: matrices) is a rectangular array of numbers or variables, often called **elements**, which are arranged in columns and rows. A matrix is generally represented by a capital letter, with its elements represented by the corresponding lowercase letter with two subscripts indicating the row and column of the element. For example, n_{ab} represents the element in row a column b of matrix N.

$$N = \begin{bmatrix} n_{11} & n_{12} & n_{13} \\ n_{21} & n_{22} & n_{23} \end{bmatrix}$$

A matrix can be described in terms of the number of rows and columns it contains in the format $a \times b$, where a is the number of rows and b is the number of columns. The matrix shown above is a 2×3 matrix. Any $a \times b$ matrix where $a = b$ is a square matrix. A **vector** is a matrix that has exactly one column (**column vector**) or exactly one row (**row vector**).

The **main diagonal** of a matrix is the set of elements on the diagonal from the top left to the bottom right of a matrix. Because of the way it is defined, only square matrices will have a main diagonal. For the matrix shown below, the main diagonal consists of the elements $n_{11}, n_{22}, n_{33}, n_{44}$.

$$\begin{bmatrix} n_{11} & n_{12} & n_{13} & n_{14} \\ n_{21} & n_{22} & n_{23} & n_{24} \\ n_{31} & n_{32} & n_{33} & n_{34} \\ n_{41} & n_{42} & n_{43} & n_{44} \end{bmatrix}$$

A 3×4 matrix such as the one shown below would not have a main diagonal because there is no straight line of elements between the top left corner and the bottom right corner that joins the elements.

$$\begin{bmatrix} n_{11} & n_{12} & n_{13} & n_{14} \\ n_{21} & n_{22} & n_{23} & n_{24} \\ n_{31} & n_{32} & n_{33} & n_{34} \end{bmatrix}$$

A **diagonal matrix** is a square matrix that has a zero for every element in the matrix except the elements on the main diagonal. All the elements on the main diagonal must be nonzero numbers.

$$\begin{bmatrix} n_{11} & 0 & 0 & 0 \\ 0 & n_{22} & 0 & 0 \\ 0 & 0 & n_{33} & 0 \\ 0 & 0 & 0 & n_{44} \end{bmatrix}$$

If every element on the main diagonal of a diagonal matrix is equal to one, the matrix is called an **identity matrix**. The identity matrix is often represented by the letter I.

$$I = \begin{bmatrix} 1 & 0 & 0 & 0 \\ 0 & 1 & 0 & 0 \\ 0 & 0 & 1 & 0 \\ 0 & 0 & 0 & 1 \end{bmatrix}$$

A **zero matrix** is a matrix that has zero as the value for every element in the matrix.

$$\begin{bmatrix} 0 & 0 & 0 & 0 \\ 0 & 0 & 0 & 0 \\ 0 & 0 & 0 & 0 \\ 0 & 0 & 0 & 0 \end{bmatrix}$$

The zero matrix is the *identity for matrix addition*. Do not confuse the zero matrix with the identity matrix.

The **negative of a matrix** is also known as the additive inverse of a matrix. If matrix N is the given matrix, then matrix $-N$ is its negative. This means that every element n_{ab} is equal to $-n_{ab}$ in the negative. To find the negative of a given matrix, change the sign of every element in the matrix and keep all elements in their original corresponding positions in the matrix.

If two matrices have the same order and all corresponding elements in the two matrices are the same, then the two matrices are **equal matrices**.

A matrix N may be **transposed** to matrix N^T by changing all rows into columns and changing all columns into rows. The easiest way to accomplish this is to swap the positions of the row and column notations for each element. For example, suppose the element in the second row of the third column of matrix N is $n_{23} = 6$. In the transposed matrix N^T, the transposed element would be $n_{32} = 6$, and it would be placed in the third row of the second column.

$$N = \begin{bmatrix} 1 & 2 & 3 \\ 4 & 5 & 6 \end{bmatrix}; \; N^T = \begin{bmatrix} 1 & 4 \\ 2 & 5 \\ 3 & 6 \end{bmatrix}$$

To quickly transpose a matrix by hand, begin with the first column and rewrite a new matrix with those same elements in the same order in the first row. Write the elements from the second column of the original matrix in the second row of the transposed matrix. Continue this process until all columns have been completed. If the original matrix is identical to the transposed matrix, the matrices are symmetric.

Basic Operations with Matrices

There are two categories of basic operations with regard to matrices: operations between a matrix and a scalar, and operations between two matrices.

Scalar Operations

A scalar being added to a matrix is treated as though it were being added to each element of the matrix. The same is true for subtraction, multiplication and division:

$$M + \mathrm{k} = \begin{bmatrix} m_{11} + k & m_{12} + k \\ m_{21} + k & m_{22} + k \end{bmatrix}$$

$$M - k = \begin{bmatrix} m_{11} - k & m_{12} - k \\ m_{21} - k & m_{22} - k \end{bmatrix}$$

$$M \times \mathrm{k} = \begin{bmatrix} m_{11} \times k & m_{12} \times k \\ m_{21} \times k & m_{22} \times k \end{bmatrix}$$

$$M \div k = \begin{bmatrix} m_{11} \div k & m_{12} \div k \\ m_{21} \div k & m_{22} \div k \end{bmatrix}$$

Matrix Addition and Subtraction

All four of the basic operations can be used with operations between matrices (although division is usually discarded in favor of multiplication by the inverse), but there are restrictions on the situations in which they can be used. Matrices that meet all the qualifications for a given operation are called **conformable matrices**. However, conformability is specific to the operation; two matrices that are conformable for addition are not necessarily conformable for multiplication.

For two matrices to be conformable for addition or subtraction, they must be of the same dimension; otherwise, the operation is not defined. If matrix M is a 3×2 matrix and matrix N is a 2×3 matrix, the operations $M + N$ and $M - N$ are meaningless. If matrices M and N are the same size, the operation is as simple as adding or subtracting all of the corresponding elements:

$$\begin{bmatrix} m_{11} & m_{12} \\ m_{21} & m_{22} \end{bmatrix} + \begin{bmatrix} n_{11} & n_{12} \\ n_{21} & n_{22} \end{bmatrix} = \begin{bmatrix} m_{11} + n_{11} & m_{12} + n_{12} \\ m_{21} + n_{21} & m_{22} + n_{22} \end{bmatrix}$$

$$\begin{bmatrix} m_{11} & m_{12} \\ m_{21} & m_{22} \end{bmatrix} - \begin{bmatrix} n_{11} & n_{12} \\ n_{21} & n_{22} \end{bmatrix} = \begin{bmatrix} m_{11} - n_{11} & m_{12} - n_{12} \\ m_{21} - n_{21} & m_{22} - n_{22} \end{bmatrix}$$

Algebra and Functions

The result of addition or subtraction is a matrix of the same dimension as the two original matrices involved in the operation.

MATRIX MULTIPLICATION

The first thing it is necessary to understand about matrix multiplication is that it is not commutative. In scalar multiplication, the operation is commutative, meaning that $a \times b = b \times a$. For matrix multiplication, this is not the case: $A \times B \neq B \times A$. The terminology must be specific when describing matrix multiplication. The operation $A \times B$ can be described as A multiplied (or **post-multiplied**) by B, or B **pre-multiplied** by A.

For two matrices to be conformable for multiplication, they need not be of the same dimension, but specific dimensions must correspond. Taking the example of two matrices M and N to be multiplied $M \times N$, matrix M must have the same number of columns as matrix N has rows. Put another way, if matrix M has the dimensions $a \times b$ and matrix N has the dimensions $c \times d$, b must equal c if the two matrices are to be conformable for this multiplication. The matrix that results from the multiplication will have the dimensions $a \times d$. If a and d are both equal to 1, the product is simply a scalar. Square matrices of the same dimensions are always conformable for multiplication, and their product is always a matrix of the same size.

The simplest type of matrix multiplication is a 1×2 matrix (a row vector) times a 2×1 matrix (a column vector). These will multiply in the following way:

$$\begin{bmatrix} m_{11} & m_{12} \end{bmatrix} \times \begin{bmatrix} n_{11} \\ n_{21} \end{bmatrix} = m_{11}n_{11} + m_{12}n_{21}$$

The two matrices are conformable for multiplication because matrix M has the same number of columns as matrix N has rows. Because the other dimensions are both 1, the result is a scalar. Expanding our matrices to 1×3 and 3×1, the process is the same:

$$\begin{bmatrix} m_{11} & m_{12} & m_{13} \end{bmatrix} \times \begin{bmatrix} n_{11} \\ n_{21} \\ n_{31} \end{bmatrix} = m_{11}n_{11} + m_{12}n_{21} + m_{13}n_{31}$$

Once again, the result is a scalar. This type of basic matrix multiplication is the building block for the multiplication of larger matrices.

To multiply larger matrices, treat each **row from the first matrix** and each **column from the second matrix** as individual vectors and follow the pattern for multiplying vectors. The scalar value found from multiplying the first-row vector by the first column vector is placed in the first row, first column of the new matrix. The scalar value found from multiplying the second-row vector by the first column vector is placed in the second row, first column of the new matrix. Continue this pattern until each row of the first matrix has been multiplied by each column of the second matrix.

Below is an example of the multiplication of a 3×2 matrix and a 2×3 matrix.

$$\begin{bmatrix} m_{11} & m_{12} \\ m_{21} & m_{22} \\ m_{31} & m_{32} \end{bmatrix} \times \begin{bmatrix} n_{11} & n_{12} & n_{13} \\ n_{21} & n_{22} & n_{23} \end{bmatrix} = \begin{bmatrix} m_{11}n_{11} + m_{12}n_{21} & m_{11}n_{12} + m_{12}n_{22} & m_{11}n_{13} + m_{12}n_{23} \\ m_{21}n_{11} + m_{22}n_{21} & m_{21}n_{12} + m_{22}n_{22} & m_{21}n_{13} + m_{22}n_{23} \\ m_{31}n_{11} + m_{32}n_{21} & m_{31}n_{12} + m_{32}n_{22} & m_{31}n_{13} + m_{32}n_{23} \end{bmatrix}$$

This process starts by taking the first column of the second matrix and running it through each row of the first matrix. Removing all but the first M row and first N column, we would see only the following:

$$[m_{11} \;\; m_{12}] \times \begin{bmatrix} n_{11} \\ n_{21} \end{bmatrix} = m_{11}n_{11} + m_{12}n_{21}$$

The first product would then be $m_{11}n_{11} + m_{12}n_{21}$. This process will be continued for each column of the N matrix to find the first full row of the product matrix, as shown below.

$$[m_{11}n_{11} + m_{12}n_{21} \quad m_{11}n_{12} + m_{12}n_{22} \quad m_{11}n_{13} + m_{12}n_{23}]$$

After completing the first row, the next step would be to simply move to the second row of the M matrix and repeat the process until all of the rows have been finished. The result is a 3×3 matrix.

$$\begin{bmatrix} m_{11} & m_{12} \\ m_{21} & m_{22} \\ m_{31} & m_{32} \end{bmatrix} \times \begin{bmatrix} n_{11} & n_{12} & n_{13} \\ n_{21} & n_{22} & n_{23} \end{bmatrix} = \begin{bmatrix} m_{11}n_{11} + m_{12}n_{21} & m_{11}n_{12} + m_{12}n_{22} & m_{11}n_{13} + m_{12}n_{23} \\ m_{21}n_{11} + m_{22}n_{21} & m_{21}n_{12} + m_{22}n_{22} & m_{21}n_{13} + m_{22}n_{23} \\ m_{31}n_{11} + m_{32}n_{21} & m_{31}n_{12} + m_{32}n_{22} & m_{31}n_{13} + m_{32}n_{23} \end{bmatrix}$$

If the operation were done in reverse $(N \times M)$, the result would be a 2×2 matrix.

$$\begin{bmatrix} n_{11} & n_{12} & n_{13} \\ n_{21} & n_{22} & n_{23} \end{bmatrix} \times \begin{bmatrix} m_{11} & m_{12} \\ m_{21} & m_{22} \\ m_{31} & m_{32} \end{bmatrix} = \begin{bmatrix} m_{11}n_{11} + m_{21}n_{12} + m_{31}n_{13} & m_{12}n_{11} + m_{22}n_{12} + m_{32}n_{13} \\ m_{11}n_{21} + m_{21}n_{22} + m_{31}n_{23} & m_{12}n_{21} + m_{22}n_{22} + m_{32}n_{23} \end{bmatrix}$$

Review Video: Matrices: The Basics
Visit mometrix.com/academy and enter code: 516658

ADVANCED MATRICES

ELEMENTARY ROW OPERATIONS

Elementary row operations include multiplying a row by a non-zero scalar, adding scalar multiples of two rows, and switching rows. These operations can be done using matrix multiplication with specialized transformation matrices. Elementary row operations are left multiplied, $A' = MA$.To do the same type of operations on the columns, then the transformation matrix is right multiplied, $A' = AM$. **Row switching** is achieved by swapping the corresponding rows in the identity matrix. For example, consider switching row 2 and row 3 in a 3×3 matrix:

$$M_{R_2 \leftrightarrow R_3} = \begin{bmatrix} 1 & 0 & 0 \\ 0 & 0 & 1 \\ 0 & 1 & 0 \end{bmatrix}$$

The transformation matrix for **row multiplication** is also based on the identity matrix with the scalar multiplication factor in place of the 1 in the corresponding row. Multiplying row 1 by –4 in a 3×3 matrix:

$$M_{-4R_1 \to R_1} = \begin{bmatrix} -4 & 0 & 0 \\ 0 & 1 & 0 \\ 0 & 0 & 1 \end{bmatrix}$$

The transformation matrix for **row addition** consists of the identity matrix with a 1 in the element corresponding to the two rows being added in the column where you want the result to go. Adding row 2 to row 1 in a 3×3 matrix:

$$M_{R_1+R_2 \to R_1} = \begin{bmatrix} 1 & 1 & 0 \\ 0 & 1 & 0 \\ 0 & 0 & 1 \end{bmatrix}$$

Review Video: Matrices: Elementary Row Operations
Visit mometrix.com/academy and enter code: 493170

Determinants and Inverses of Matrices

The **determinant** of a matrix is a scalar value that is calculated by taking into account all the elements of a square matrix. A determinant only exists for square matrices. Finding the determinant of a 2×2 matrix is as simple as remembering a simple equation. For a 2×2 matrix $M = \begin{bmatrix} m_{11} & m_{12} \\ m_{21} & m_{22} \end{bmatrix}$, the determinant is obtained by the equation $|M| = m_{11}m_{22} - m_{12}m_{21}$. Anything larger than 2×2 requires multiple steps. Take matrix $N = \begin{bmatrix} a & b & c \\ d & e & f \\ g & h & j \end{bmatrix}$. The determinant of N is calculated as $|N| = a\begin{vmatrix} e & f \\ h & j \end{vmatrix} - b\begin{vmatrix} d & f \\ g & j \end{vmatrix} + c\begin{vmatrix} d & e \\ g & h \end{vmatrix}$ or $|N| = a(ej - fh) - b(dj - fg) + c(dh - eg)$.

There is a shortcut for 3×3 matrices: add the products of each unique set of elements diagonally left-to-right and subtract the products of each unique set of elements diagonally right-to-left. In matrix N, the left-to-right diagonal elements are (a, e, j), (b, f, g), and (c, d, h). The right-to-left diagonal elements are (a, f, h), (b, d, j), and (c, e, g). $\det(N) = aej + bfg + cdh - afh - bdj - ceg$.

Calculating the determinants of matrices larger than 3×3 is rarely, if ever, done by hand.

The **inverse** of a matrix M is the matrix that, when multiplied by matrix M, yields a product that is the identity matrix. Multiplication of matrices will be explained in greater detail shortly. Not all matrices have inverses. Only a square matrix whose determinant is not zero has an inverse. If a matrix has an inverse, that inverse is unique to that matrix. For any matrix M that has an inverse, the inverse is represented by the symbol M^{-1}. To calculate the inverse of a 2×2 square matrix, use the following pattern:

$$M = \begin{bmatrix} m_{11} & m_{12} \\ m_{21} & m_{22} \end{bmatrix};\ M^{-1} = \begin{bmatrix} \frac{m_{22}}{|M|} & \frac{-m_{12}}{|M|} \\ \frac{-m_{21}}{|M|} & \frac{m_{11}}{|M|} \end{bmatrix}$$

Another way to find the inverse of a matrix by hand is use an augmented matrix and elementary row operations. An **augmented matrix** is formed by appending the entries from one matrix onto the end of another. For example, given a 2×2 invertible matrix $N = \begin{bmatrix} a & b \\ c & d \end{bmatrix}$, you can find the inverse N^{-1} by creating an augmented matrix by appending a 2×2 identity matrix: $\left[\begin{array}{cc|cc} a & b & 1 & 0 \\ c & d & 0 & 1 \end{array}\right]$. To find the inverse of the original 2×2 matrix, perform elementary row operations to convert the original matrix on the left to an identity matrix: $\left[\begin{array}{cc|cc} 1 & 0 & e & f \\ 0 & 1 & g & h \end{array}\right]$. For instance, the first step might be to multiply the second row by $\frac{b}{d}$ and then subtract it from the first row to make its second column a zero. The end result is that the 2×2 section on the right will become the inverse of the original matrix: $N^{-1} = \begin{bmatrix} e & f \\ g & h \end{bmatrix}$.

> **Review Video:** Transposition, Determinants, and Augmentation of Matrices
> Visit mometrix.com/academy and enter code: 521365

Reduced Row-Echelon Forms

When a system of equations has a solution, finding the transformation of the augmented matrix will result in one of three reduced row-echelon forms. Only one of these forms will give a unique solution to the system of equations, however. The following examples show the solutions indicated by particular results:

$\begin{bmatrix} 1 & 0 & 0 & x_0 \\ 0 & 1 & 0 & y_0 \\ 0 & 0 & 1 & z_0 \end{bmatrix}$ gives the unique solution $x = x_0;\ y = y_0; z = z_0$

$\begin{bmatrix} 1 & 0 & k_1 & x_0 \\ 0 & 1 & k_2 & y_0 \\ 0 & 0 & 0 & 0 \end{bmatrix}$ gives a non-unique solution $x = x_0 - k_1 z; y = y_0 - k_2 z$

$\begin{bmatrix} 1 & j_1 & k_1 & x_0 \\ 0 & 0 & 0 & 0 \\ 0 & 0 & 0 & 0 \end{bmatrix}$ gives a non-unique solution $x = x_0 - j_1 y - k_1 z$

Geometric Transformations

The four *geometric transformations* are **translations**, **reflections**, **rotations**, and **dilations**. When geometric transformations are expressed as matrices, the process of performing the transformations is simplified. For calculations of the geometric transformations of a planar figure, make a $2 \times n$ matrix, where n is the number of vertices in the planar figure. Each column represents the rectangular coordinates of one vertex of the figure, with the top row containing the values of the x-coordinates and the bottom row containing the values of the y-coordinates. For example, given a planar triangular figure with coordinates (x_1, y_1), (x_2, y_2), and (x_3, y_3), the corresponding matrix is $\begin{bmatrix} x_1 & x_2 & x_3 \\ y_1 & y_2 & y_3 \end{bmatrix}$. You can then perform the necessary transformations on this matrix to determine the coordinates of the resulting figure.

> **Review Video:** Matrices: Geometric Transformations
> Visit mometrix.com/academy and enter code: 612781

Translation

A **translation** moves a figure along the x-axis, the y-axis, or both axes without changing the size or shape of the figure. To calculate the new coordinates of a planar figure following a translation, set up a matrix of the coordinates and a matrix of the translation values and add the two matrices.

$$\begin{bmatrix} h & h & h \\ v & v & v \end{bmatrix} + \begin{bmatrix} x_1 & x_2 & x_3 \\ y_1 & y_2 & y_3 \end{bmatrix} = \begin{bmatrix} h + x_1 & h + x_2 & h + x_3 \\ v + y_1 & v + y_2 & v + y_3 \end{bmatrix}$$

where h is the number of units the figure is moved along the x-axis (horizontally) and v is the number of units the figure is moved along the y-axis (vertically).

Algebra and Functions

Reflection

To find the **reflection** of a planar figure over the x-axis, set up a matrix of the coordinates of the vertices and pre-multiply the matrix by the 2×2 matrix $\begin{bmatrix} 1 & 0 \\ 0 & -1 \end{bmatrix}$ so that $\begin{bmatrix} 1 & 0 \\ 0 & -1 \end{bmatrix}\begin{bmatrix} x_1 & x_2 & x_3 \\ y_1 & y_2 & y_3 \end{bmatrix} = \begin{bmatrix} x_1 & x_2 & x_3 \\ -y_1 & -y_2 & -y_3 \end{bmatrix}$. To find the reflection of a planar figure over the y-axis, set up a matrix of the coordinates of the vertices and pre-multiply the matrix by the 2×2 matrix $\begin{bmatrix} -1 & 0 \\ 0 & 1 \end{bmatrix}$ so that $\begin{bmatrix} -1 & 0 \\ 0 & 1 \end{bmatrix}\begin{bmatrix} x_1 & x_2 & x_3 \\ y_1 & y_2 & y_3 \end{bmatrix} = \begin{bmatrix} -x_1 & -x_2 & -x_3 \\ y_1 & y_2 & y_3 \end{bmatrix}$. To find the reflection of a planar figure over the line $y = x$, set up a matrix of the coordinates of the vertices and pre-multiply the matrix by the 2×2 matrix $\begin{bmatrix} 0 & 1 \\ 1 & 0 \end{bmatrix}$ so that $\begin{bmatrix} 0 & 1 \\ 1 & 0 \end{bmatrix}\begin{bmatrix} x_1 & x_2 & x_3 \\ y_1 & y_2 & y_3 \end{bmatrix} = \begin{bmatrix} y_1 & y_2 & y_3 \\ x_1 & x_2 & x_3 \end{bmatrix}$. Remember that the order of multiplication is important when multiplying matrices. The commutative property does not apply.

Rotation

To find the coordinates of the figure formed by rotating a planar figure about the origin θ degrees in a counterclockwise direction, set up a matrix of the coordinates of the vertices and pre-multiply the matrix by the 2×2 matrix $\begin{bmatrix} \cos\theta & -\sin\theta \\ \sin\theta & \cos\theta \end{bmatrix}$. For example, if you want to rotate a figure 90° clockwise around the origin, you would have to convert the degree measure to 270° counterclockwise and solve the 2×2 matrix you have set as the pre-multiplier: $\begin{bmatrix} \cos 270° & \sin 270° \\ -\sin 270° & \cos 270° \end{bmatrix} = \begin{bmatrix} 0 & -1 \\ 1 & 0 \end{bmatrix}$. Use this as the pre-multiplier for the matrix $\begin{bmatrix} x_1 & x_2 & x_3 \\ y_1 & y_2 & y_3 \end{bmatrix}$ and solve to find the new coordinates.

Dilation

To find the **dilation** of a planar figure by a scale factor of k, set up a matrix of the coordinates of the vertices of the planar figure and pre-multiply the matrix by the 2×2 matrix $\begin{bmatrix} k & 0 \\ 0 & k \end{bmatrix}$ so that $\begin{bmatrix} k & 0 \\ 0 & k \end{bmatrix}\begin{bmatrix} x_1 & x_2 & x_3 \\ y_1 & y_2 & y_3 \end{bmatrix} = \begin{bmatrix} kx_1 & kx_2 & kx_3 \\ ky_1 & ky_2 & ky_3 \end{bmatrix}$. This is effectively the same as multiplying the matrix by the scalar k, but the matrix equation would still be necessary if the figure were being dilated by different factors in vertical and horizontal directions. The scale factor k will be greater than 1 if the figure is being enlarged, and between 0 and 1 if the figure is being shrunk. Again, remember that when multiplying matrices, the order of the matrices is important. The commutative property does not apply, and the matrix with the coordinates of the figure must be the second matrix.

Matrix Systems

Solving Systems of Equations

Matrices can be used to represent the coefficients of a system of linear equations and can be very useful in solving those systems. Take for instance three equations with three variables where all a, b, c, and d are known constants:

$$a_1x + b_1y + c_1z = d_1$$
$$a_2x + b_2y + c_2z = d_2$$
$$a_3x + b_3y + c_3z = d_3$$

To solve this system, define three matrices:

$$A = \begin{bmatrix} a_1 & b_1 & c_1 \\ a_2 & b_2 & c_2 \\ a_3 & b_3 & c_3 \end{bmatrix};\ D = \begin{bmatrix} d_1 \\ d_2 \\ d_3 \end{bmatrix};\ X = \begin{bmatrix} x \\ y \\ z \end{bmatrix}$$

The three equations in our system can be fully represented by a single matrix equation:

$$AX = D$$

We know that the identity matrix times X is equal to X, and we know that any matrix multiplied by its inverse is equal to the identity matrix.

$$A^{-1}AX = IX = X;\ \text{thus}\ X = A^{-1}D$$

Our goal then is to find the inverse of A, or A^{-1}. Once we have that, we can pre-multiply matrix D by A^{-1} (post-multiplying here is an undefined operation) to find matrix X.

Systems of equations can also be solved using the transformation of an augmented matrix in a process similar to that for finding a matrix inverse. Begin by arranging each equation of the system in the following format:

$$\begin{gathered} a_1x + b_1y + c_1z = d_1 \\ a_2x + b_2y + c_2z = d_2 \\ a_3x + b_3y + c_3z = d_3 \end{gathered}$$

Define matrices A and D and combine them into augmented matrix A_a:

$$A = \begin{bmatrix} a_1 & b_1 & c_1 \\ a_2 & b_2 & c_2 \\ a_3 & b_3 & c_3 \end{bmatrix};\ D = \begin{bmatrix} d_1 \\ d_2 \\ d_3 \end{bmatrix};\ A_a = \begin{bmatrix} a_1 & b_1 & c_1 & d_1 \\ a_2 & b_2 & c_2 & d_2 \\ a_3 & b_3 & c_3 & d_3 \end{bmatrix}$$

To solve the augmented matrix and the system of equations, use elementary row operations to form an identity matrix in the first 3×3 section. When this is complete, the values in the last column are the solutions to the system of equations:

$$\begin{bmatrix} 1 & 0 & 0 & x \\ 0 & 1 & 0 & y \\ 0 & 0 & 1 & z \end{bmatrix}$$

If an identity matrix is not possible, the system of equations has no unique solution. Sometimes only a partial solution will be possible. The following are partial solutions you may find:

$\begin{bmatrix} 1 & 0 & k_1 & x_0 \\ 0 & 1 & k_2 & y_0 \\ 0 & 0 & 0 & 0 \end{bmatrix}$ gives the non-unique solution $x = x_0 - k_1z;\ y = y_0 - k_2z$

$\begin{bmatrix} 1 & j_1 & k_1 & x_0 \\ 0 & 0 & 0 & 0 \\ 0 & 0 & 0 & 0 \end{bmatrix}$ gives the non-unique solution $x = x_0 - j_1y - k_1z$

Algebra and Functions

This process can be used to solve systems of equations with any number of variables, but three is the upper limit for practical purposes. Anything more ought to be done with a graphing calculator.

Review Video: Matrices: Data Systems
Visit mometrix.com/academy and enter code: 579763

Sequences and Series

Sequences

A **sequence** is an ordered set of numbers that continues in a defined pattern. The function that defines a sequence has a domain composed of the set of positive integers. Each member of the sequence is an element, or individual term. Each element is identified by the notation a_n, where a is the term of the sequence, and n is the integer identifying which term in the sequence a is.

There are two different ways to represent a sequence that contains the element a_n. The first is the simple notation $\{a_n\}$. The second is the expanded notation of a sequence: $a_1, a_2, a_3, \dots a_n, \dots$. Notice that the expanded form does not end with the nth term. There is no indication that the nth term is the last term in the sequence, only that the nth term is an element of the sequence.

Arithmetic Sequences

An **arithmetic sequence**, or arithmetic progression, is a special kind of sequence in which a specific quantity, called the common difference, is added to each term to make the next term. The common difference may be positive or negative. The general form of an arithmetic sequence containing n terms is $a_1, a_1 + d, a_1 + 2d, \dots, a_1 + (n-1)d$, where d is the common difference. The general formula for any term of an arithmetic sequence is $a_n = a_1 + (n-1)d$, where a_n is the term you are looking for and d is the common difference. To find the sum of the first n terms of an arithmetic sequence, use the formula $s_n = \frac{n}{2}(a_1 + a_n)$.

Review Video: Arithmetic Sequence
Visit mometrix.com/academy and enter code: 676885

Monotonic Sequences

A **monotonic sequence** is a sequence that is either nonincreasing or nondecreasing. A **nonincreasing** sequence is one whose terms either get progressively smaller in value or remain the same. Such a sequence is always bounded above, that is, all elements of the sequence must be less than some real number. A **nondecreasing** sequence is one whose terms either get progressively larger in value or remain the same. Such a sequence is always bounded below, that is, all elements of the sequence must be greater than some real number.

Recursive Sequences

When one element of a sequence is defined in terms of a previous element or elements of the sequence, the sequence is a **recursive sequence**. For example, given the recursive definition $a_1 = 1; a_2 = 1; a_n = a_{n-1} + a_{n-2}$ for all $n > 2$, you get the sequence 1,1,2,3,5,8, … . This is known as the Fibonacci sequence: a continuing sequence of numbers in which each number (after a_2) is the sum of the two previous numbers. The Fibonacci sequence can be defined as starting with either 1,1 or 0,1. Both definitions are considered correct in mathematics. Make sure you know which definition you are working with when dealing with Fibonacci numbers.

Sometimes in a recursive sequence, the terms can be found using a general formula that does not involve the previous terms of the sequence. Such a formula is called a **closed-form** expression for a recursive definition—an alternate formula that will generate the same sequence of numbers. However, not all sequences based on recursive definitions will have a closed-form expression. Some sequences will require the use of the recursive definition.

The Golden Ratio and the Fibonacci Sequence

The golden ratio is approximately 1.6180339887 and is often represented by the Greek letter phi, Φ. The exact value of Φ is $\frac{(1+\sqrt{5})}{2}$ and it is one of the solutions to $x - \frac{1}{x} = 1$. The golden ratio can be found using the Fibonacci sequence, since the ratio of a term to the previous term approaches Φ as the sequence approaches infinity:

n	a_n	a_{n-1}	$\frac{a_n}{a_{n-1}}$
3	2	1	2
4	3	2	1.5
5	5	3	$1.\overline{6}$
6	8	5	1.6
7	13	8	1.625
8	21	13	$1.\overline{615384}$
9	34	21	$1.\overline{619047}$
⋮	⋮	⋮	⋮
20	6,765	4,181	1.618033963 …

Geometric Sequences

A geometric sequence is a sequence in which each term is multiplied by a constant number (called the common ratio) to get the next term. Essentially, it's the same concept as an arithmetic sequence, but with multiplication instead of addition.

Consider the following example of a geometric sequence: Andy opens a savings account with \$10. During each subsequent week, he plans to double the amount from the previous week.

Sequence: $10, 20, 40, 80, 160, \ldots$

Function: $a_n = 10 \times 2^{n-1}$

This is a geometric sequence with a common ratio of 2. All geometric sequences represent exponential functions. The nth term in any geometric sequence is $a_n = a_1 \times r^{n-1}$, where a_n represents the value of the nth term, a_1 is the initial term, r is the common ratio, and n is the number of terms. Thus, substituting the initial value of 10 and common ratio of 2 gives the function $a_n = 10 \times 2^{n-1}$.

> **Review Video: Geometric Sequences**
> Visit mometrix.com/academy and enter code: 140779

Advanced Sequences and Series

Limit of a Sequence

Some sequences will have a **limit**—a value the sequence approaches, or sometimes even reaches, but never passes. A sequence with a limit is called a **convergent** sequence because all the values of the sequence seemingly converge at that point. Sequences that do not converge at a particular limit are **divergent** sequences. The easiest way to determine whether a sequence converges or diverges is to find the limit of the sequence. If the limit is a real number, the sequence is convergent. If the limit is infinity, the sequence is divergent.

Remember the following rules for finding limits:

- $\lim_{n\to\infty} k = k$, for all real numbers k
- $\lim_{n\to\infty} \frac{1}{n} = 0$
- $\lim_{n\to\infty} n = \infty$
- $\lim_{n\to\infty} \frac{k}{n^p} = 0$, for all real numbers k and positive rational numbers p
- The limit of the sum of two sequences is equal to the sum of the limits of the two sequences: $\lim_{n\to\infty}(a_n + b_n) = \lim_{n\to\infty} a_n + \lim_{n\to\infty} b_n$
- The limit of the difference between two sequences is equal to the difference between the limits of the two sequences: $\lim_{n\to\infty}(a_n - b_n) = \lim_{n\to\infty} a_n - \lim_{n\to\infty} b_n$
- The limit of the product of two sequences is equal to the product of the limits of the two sequences: $\lim_{n\to\infty}(a_n \times b_n) = \lim_{n\to\infty} a_n \times \lim_{n\to\infty} b_n$
- The limit of the quotient of two sequences is equal to the quotient of the limits of the two sequences, with some exceptions: $\lim_{n\to\infty}\left(\frac{a_n}{b_n}\right) = \frac{\lim_{n\to\infty} a_n}{\lim_{n\to\infty} b_n}$. In the quotient formula, it is important that $b_n \neq 0$ and that $\lim_{n\to\infty} b_n \neq 0$.
- The limit of a sequence multiplied by a scalar is equal to the scalar multiplied by the limit of the sequence: $\lim_{n\to\infty} ka_n = k \lim_{n\to\infty} a_n$, where k is any real number

Review Video: Limit of a Sequence
Visit mometrix.com/academy and enter code: 847732

Infinite Series

Both arithmetic and geometric sequences have formulas to find the sum of the first n terms in the sequence, assuming you know what the first term is. The sum of all the terms in a sequence is called a series. An **infinite series** is an infinite sum. In other words, it is what you get by adding up all the terms in an infinite sequence: $\sum_{n=1}^{\infty} a_n = a_1 + a_2 + a_3 + \cdots + a_n + \cdots$. This notation can be shortened to $\sum_{n=1}^{\infty} a_n$ or $\sum a_n$.

While we can't add up an infinite list of numbers one at a time, we can still determine the infinite sum. As we add the terms in a series, we can imagine an infinite sequence of partial sums, where the first partial sum is the first element of the series, the second partial sum is the sum of the first two elements of the series, and the nth partial sum is the sum of the first n elements of the series.

Every infinite sequence of partial sums (infinite series) either converges or diverges. As with the test for convergence in a sequence, finding the limit of the sequence of partial sums will indicate whether it is a converging series or a diverging series. If there exists a real number S such that $\lim_{n\to\infty} S_n = S$, where S_n is the sequence of partial sums, then the series converges. If the limit equals infinity, then the series diverges. If $\lim_{n\to\infty} S_n = S$ and S is a real number, then S is also the convergence value of the series.

To find the sum as n approaches infinity for the sum of two convergent series, find the sum as n approaches infinity for each individual series and add the results.

$$\sum_{n=1}^{\infty}(a_n + b_n) = \sum_{n=1}^{\infty} a_n + \sum_{n=1}^{\infty} b_n$$

The same idea works for subtraction.

$$\sum_{n=1}^{\infty}(a_n - b_n) = \sum_{n=1}^{\infty} a_n - \sum_{n=1}^{\infty} b_n$$

To find the sum as n approaches infinity for the product of a constant (also called a scalar) and a convergent series, find the sum as n approaches infinity for the series and multiply the result by the scalar.

$$\sum_{n=1}^{\infty} ka_n = k\sum_{n=1}^{\infty} a_n$$

Review Video: Infinite Series
Visit mometrix.com/academy and enter code: 271404

The nth **term test for divergence** means taking the limit of a sequence a_n as n goes to infinity $\left(\lim_{n\to\infty} a_n\right)$ and checking whether the limit is zero. If the limit is not zero, then the series $\sum a_n$ is a diverging series. This test only works to prove divergence, however. If the limit is zero, the test is inconclusive, meaning the series could be either convergent or divergent.

Review Video: Nth Term Test for Divergence
Visit mometrix.com/academy and enter code: 346400

Permutations and Combinations

Permutations and Combinations

Factorials

The **factorial** is a function that can be performed on any **non-negative integer**. It is represented by the ! sign written after the integer on which it is being performed. The factorial of an integer is the product of all positive integers less than or equal to the number. For example, 4! (read "4 factorial") is calculated as $4 \times 3 \times 2 \times 1 = 24$.

Since 0 is not itself a positive integer, nor does it have any positive integers less than it, 0! cannot be calculated using this method. Instead, 0! is defined by convention to equal 1. This makes sense if you consider the pattern of descending factorials:

$$5! = 120$$
$$4! = \frac{5!}{5} = \frac{120}{5} = 24$$
$$3! = \frac{4!}{4} = \frac{24}{4} = 6$$
$$2! = \frac{3!}{3} = \frac{6}{3} = 2$$
$$1! = \frac{2!}{2} = \frac{2}{2} = 1$$
$$0! = \frac{1!}{1} = \frac{1}{1} = 1$$

PERMUTATIONS

For any given set of data, the individual elements in the set may be arranged in different groups containing different numbers of elements arranged in different orders. For example, given the set of integers from one to three, inclusive, the elements of the set are 1, 2, and 3: written as $\{1, 2, 3\}$. They may be arranged as follows: 1, 2, 3, 12, 21, 13, 31, 23, 32, 123, 132, 213, 231, 312, and 321. These ordered sequences of elements from the given set of data are called **permutations**. It is important to note that in permutations, the order of the elements in the sequence is important. The sequence 123 is not the same as the sequence 213. Also, no element in the given set may be used more times as an element in a permutation than it appears as an element in the original set. For example, 223 is not a permutation in the above example because the number 2 only appears one time in the given set.

To find the number of permutations of r items from a set of n items, use the formula ${}_nP_r = \frac{n!}{(n-r)!}$. When using this formula, each element of r must be unique. Also, this assumes that different arrangements of the same set of elements yields different outcomes. For example, 123 is not the same as 321; order is important.

A special case arises while finding the number of possible permutations of n items from a set of n items. Because $n = r$, the equation for the number of permutations becomes simply $P = n!$ The same result is true for $r = n - 1$. Both of these cases are a result of the fact that 0! and 1! are both equal to 1.

If a set contains one or more groups of **indistinguishable or interchangeable elements** (e.g., the set $\{1, 2, 3, 3\}$, which has a group of two indistinguishable 3's), there is a different formula for finding distinct permutations of all n elements. Use the formula $P = \frac{n!}{m_1!m_2!...m_k!}$, where P is the number of permutations, n is the total number of elements in the set, and m_1 through m_k are the number of identical elements in each group (e.g., for the set $\{1, 1, 2, 2, 2, 3, 3\}$, $m_1 = 2$, $m_2 = 3$, and $m_3 = 2$). It is important to note that each repeated number is counted as its own element for the purpose of defining n (e.g., for the set $\{1, 1, 2, 2, 2, 3, 3\}$, $n = 7$, not 3).

To find the number of possible permutations of **any number of elements** in a set of unique elements, you must apply the permutation formulas multiple times. For example, to find the total number of possible permutations of the set $\{1, 2, 3\}$ first apply the permutation formula for situations where $n = r$ as follows: $P = n! = 3! = 6$. This gives the number of permutations of the three elements when all three elements are used. To find the number of permutations when only two of the three elements are used, use the formula ${}_nP_r = \frac{n!}{(n-r)!}$, where n is 3 and r is 2.

$$ {}_nP_r = \frac{n!}{(n-r)!} \Rightarrow {}_3P_2 = \frac{3!}{(3-2)!} = \frac{6}{1} = 6 $$

To find the number of permutations when one element is used, use the formula ${}_nP_r = \frac{n!}{(n-r)!}$, where n is 3 and r is 1.

$$ {}_nP_r = \frac{n!}{(n-r)!} \Rightarrow {}_3P_1 = \frac{3!}{(3-1)!} = \frac{3!}{2!} = \frac{6}{2} = 3 $$

Find the sum of the three formulas: $6 + 6 + 3 = 15$ total possible permutations.

Alternatively, the general formula for total possible permutations can be written as follows:

$$ P_T = \sum_{i=1}^{n} \frac{n!}{(i-1)!} $$

Combinations

Combinations are essentially defined as permutations where the order in which the elements appear does not matter. Going back to the earlier example of the set $\{1, 2, 3\}$, the possible combinations that can be made from that set are 1, 2, 3, 12, 13, 23, and 123.

In a set containing n elements, the number of combinations of r items from the set can be found using the formula ${}_nC_r = \frac{n!}{r!(n-r)!}$. Notice the similarity to the formula for permutations. In effect, you are dividing the number of permutations by $r!$ to get the number of combinations, and the formula may be written ${}_nC_r = \frac{{}_nP_r}{r!}$. When finding the number of combinations, it is important to remember that the elements in the set must be unique (i.e., there must not be any duplicate items), and that no item may be used more than once in any given sequence.

Algebra and Functions

Trigonometry

Trigonometric Functions

The basic trigonometric functions are sine (abbreviated 'sin'), cosine (abbreviated 'cos'), and tangent (abbreviated 'tan'). The simplest way to think of them is as describing the ratio of the side lengths of a right triangle in relation to the angles of the triangle.

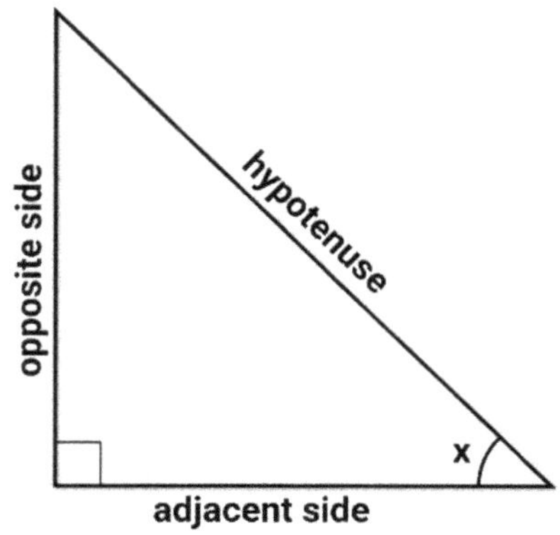

$$\sin x = \frac{\text{opposite side}}{\text{hypotenuse}}$$
$$\cos x = \frac{\text{adjacent side}}{\text{hypotenuse}}$$
$$\tan x = \frac{\text{opposite side}}{\text{adjacent side}}$$

Applying Trigonometric Functions

Suppose we have a point (x, y) at distance r from the origin such that the line segment between this point and the origin forms an angle of measure θ with the positive x-axis (We allow θ to be any real number, with the angle measured counterclockwise from the x-axis if $\theta > 0$ and clockwise if $\theta < 0$). Then we can define the trigonometric functions sine, cosine, and tangent by the formulas $\sin\theta = y/r$, $\cos\theta = x/r$, and $\tan\theta = y/x$ (and their reciprocal functions cosecant, secant, and cotangent accordingly). These functions are periodic because increasing the angle θ by 2π radians (that is, 360°) turns the line segment all the way around the origin so that it ends at the same point (x, y) again.

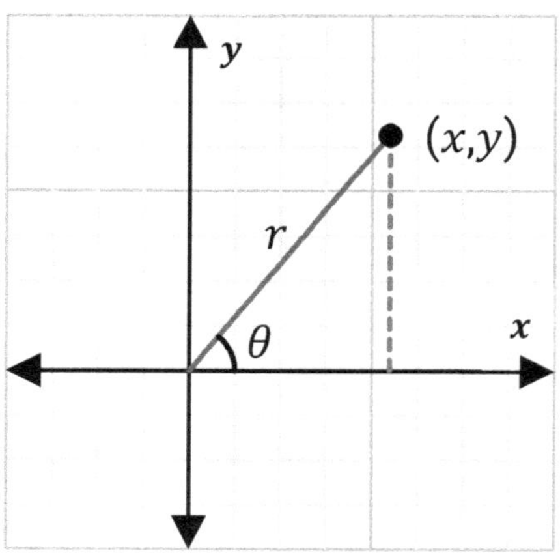

If the angle in the above diagram is acute, we can think of the line segment as the hypotenuse of a right triangle so that we get the familiar formulas $\sin\theta = (\text{opposite leg})/\text{hypotenuse}$, $\cos\theta = (\text{adjacent leg})/\text{hypotenuse}$, and $\tan\theta = (\text{opposite leg})/(\text{adjacent leg})$. In this form, the trigonometric functions are useful for finding missing sides of triangles. For example, suppose that standing 30 feet from the base of a tree, we have to look up at a 58° angle to see the top of the tree (see diagram). If we let x be the height of the tree, then we get the equation $\tan 58° = x/(30\text{ feet})$.

Multiplying both sides of the equation by 30 gives us the height of the tree, namely, $x = 30 \tan 58° \approx 48$ feet.

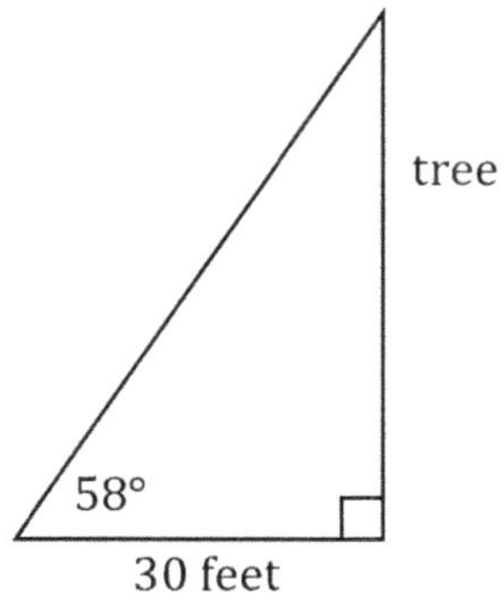

Graphing Trigonometric Functions

Sine

To graph the sine function, we let x (rather than θ) be the angle measure (usually in radians) and y be the value of the sine. That is, we graph the equation $y = \sin x$. From the definition of sine above we can show that $\sin 0 = 0$, $\sin(\pi/2) = 1$, $\sin \pi = 0$, $\sin(3\pi/2) = -1$, and $\sin 2\pi = 0$. Thus, the graph of the sine function begins at the origin, peaks at the point $(\pi/2, 1)$, crosses the x-axis at $(\pi, 0)$, has its lowest point at $(3\pi/2, -1)$, and returns to the x-axis at $(2\pi, 0)$. Because the sine function is periodic with period 2π radians (or 360°), this curve represents one complete cycle of the graph. The remainder of the graph of $y = \sin x$ consists of infinitely many repetitions of this cycle, producing the familiar "sine wave."

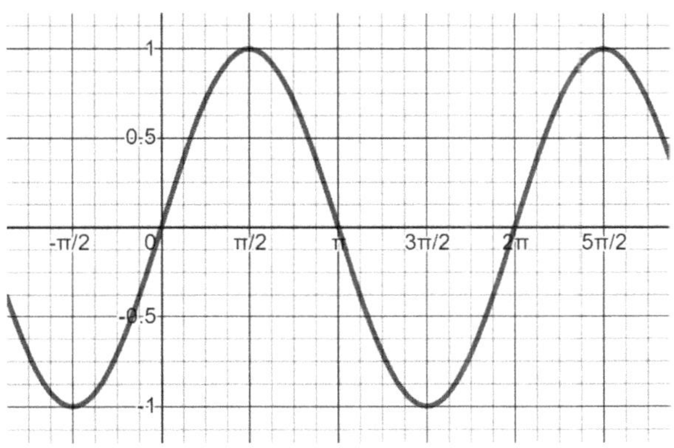

Review Video: Sine
Visit mometrix.com/academy and enter code: 193339

By suitable choice of the parameters A, B, C, and D, we can construct a "custom" sine function $y = A\sin(Bx + C) + D$ with any desired amplitude (the vertical distance the curve extends above and below its midline), period (distance between successive peaks), phase shift (horizontal shift), and midline (the horizontal line halfway between the peaks and troughs).

COSINE

Similarly, the graph of the cosine function is the graph of the equation $y = \cos x$. Proceeding as above, we get a graph with the same shape as that of the sine function but shifted by $\pi/2$ to the left.

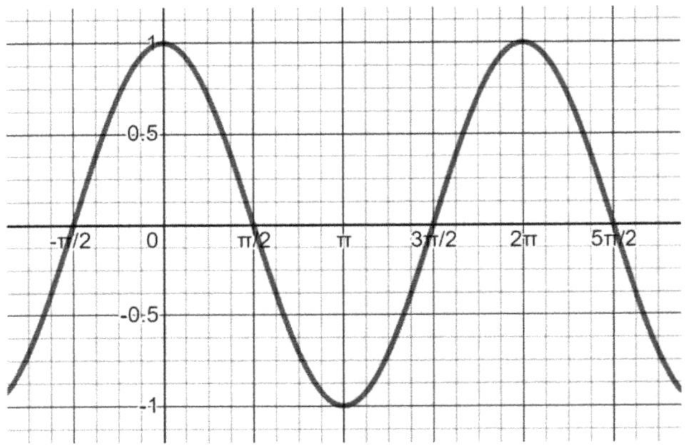

Review Video: Cosine
Visit mometrix.com/academy and enter code: 361120

As for the sine function, we can construct custom cosine graphs with desired amplitude, period, phase shift, and midline by choosing suitable values of the parameters in the equation $y = A\cos(Bx + C) + D$.

TANGENT

The graph of the tangent function (that is, the graph of the equation $y = \tan x$) turns out to be periodic with period π radians (or 180°). It consists of an infinite sequence of smooth curves crossing the x-axis at every multiple of π radians (180°) and having as asymptotes the vertical lines $x = \frac{k\pi}{2}$ radians ($x = k \times 90°$), for every odd integer k (which also means that the domain of the tangent function excludes these x-values).

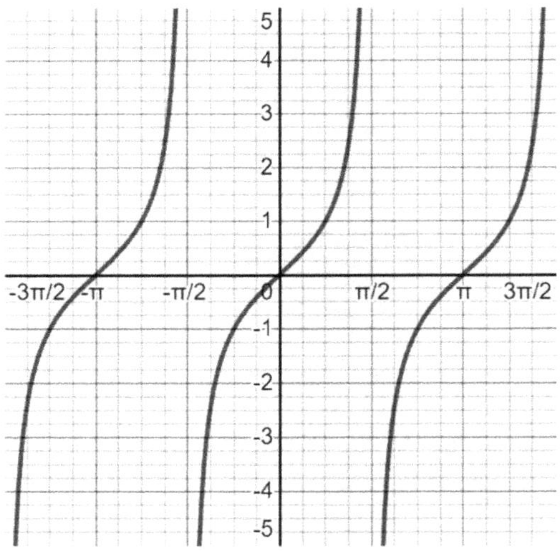

Review Video: Tangent
Visit mometrix.com/academy and enter code: 947639

Complex Numbers

Complex numbers consist of a real component and an imaginary component. Complex numbers are expressed in the form $a + bi$ with real component a and imaginary component bi. The imaginary unit i is equal to $\sqrt{-1}$. That means $i^2 = -1$. The imaginary unit provides a way to find the square root of a negative number. For example, $\sqrt{-25}$ is $5i$. You should expect questions asking you to add, subtract, multiply, divide, and simplify complex numbers. You may see a question that says, "Add $3 + 2i$ and $5 - 7i$" or "Subtract $4 + i\sqrt{5}$ from $2 + i\sqrt{5}$." Or you may see a question that says, "Multiply $6 + 2i$ by $8 - 4i$" or "Divide $1 - 3i$ by $9 - 7i$."

Operations on Complex Numbers

Operations with complex numbers resemble operations with variables in algebra. When adding or subtracting complex numbers, you can only combine like terms—real terms with real terms and imaginary terms with imaginary terms. For example, if you are asked to simplify the expression $-2 + 4i - (-3 + 7i) - 5i$, you should first remove the parentheses to yield $-2 + 4i + 3 - 7i - 5i$. Combining like terms yields $1 - 8i$. One interesting aspect of imaginary numbers is that if i has an exponent greater than 1, it can be simplified. Example: $i^2 = -1$, $i^3 = -i$, and $i^4 = 1$. When multiplying complex numbers, remember to simplify each i with an exponent greater than 1. For example, you might see a question that says, "Simplify $(2 - i)(3 + 2i)$." You need to distribute and multiply to get $6 + 4i - 3i - 2i^2$. This is further simplified to $6 + i - 2(-1)$, or $8 + i$.

Simplifying Expressions with Complex Denominators

If an expression contains an i in the denominator, it must be simplified. Remember, roots cannot be left in the denominator of a fraction. Since i is equivalent to $\sqrt{-1}$, i cannot be left in the denominator of a fraction. You must rationalize the denominator of a fraction that contains a complex denominator by multiplying the numerator and denominator by the conjugate of the denominator. The conjugate of the complex number $a + bi$ is $a - bi$. You can simplify $\frac{2}{5i}$ by simply multiplying $\frac{2}{5i} \times \frac{i}{i}$, which yields $-\frac{2}{5}i$. And you can simplify $\frac{5+3i}{2-4i}$ by multiplying $\frac{5+3i}{2-4i} \times \frac{2+4i}{2+4i}$. This yields $\frac{10+20i+6i-12}{4-8i+8i+16}$ which simplifies to $\frac{-2+26i}{20}$ or $\frac{-1+13i}{10}$, which can also be written as $-\frac{1}{10} + \frac{13}{10}i$.

The Unit Circle

Degrees, Radians, and the Unit Circle

It is important to understand the deep connection between trigonometry and circles. Specifically, the two main units, **degrees** (°) and **radians** (rad), that are used to measure angles are related this way: 360° in one full circle and 2π radians in one full circle: $(360° = 2\pi \text{ rad})$. The conversion factor relating the two is often stated as $\frac{180°}{\pi}$. For example, to convert $\frac{3\pi}{2}$ radians to degrees, multiply by the conversion factor: $\frac{3\pi}{2} \times \frac{180°}{\pi} = 270°$. As another example, to convert 60° to radians, divide by the conversion factor or multiply by the reciprocal: $60° \times \frac{\pi}{180°} = \frac{\pi}{3}$ radians.

Recall that the standard equation for a circle is $(x - h)^2 + (y - k)^2 = r^2$. A **unit circle** is a circle with a radius of 1 $(r = 1)$ that has its center at the origin $(h = 0, k = 0)$. Thus, the equation for the unit circle simplifies from the standard equation down to $x^2 + y^2 = 1$.

Standard position is the position of an angle of measure θ whose vertex is at the origin, the initial side crosses the unit circle at the point $(1, 0)$, and the terminal side crosses the unit circle at some other point (a, b). In the standard position, $\sin\theta = b$, $\cos\theta = a$, and $\tan\theta = \frac{b}{a}$.

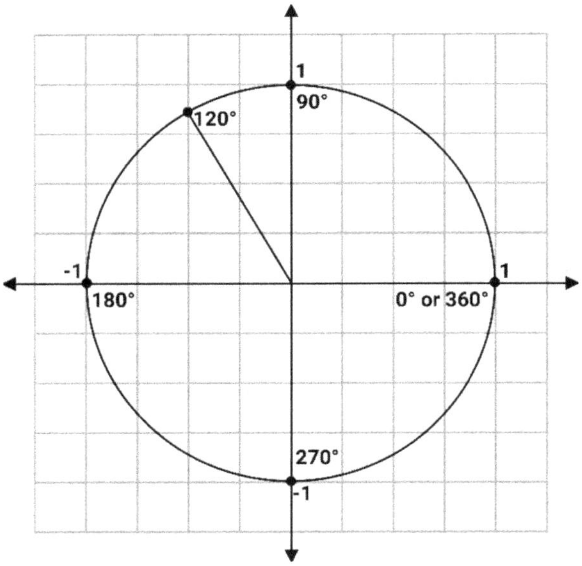

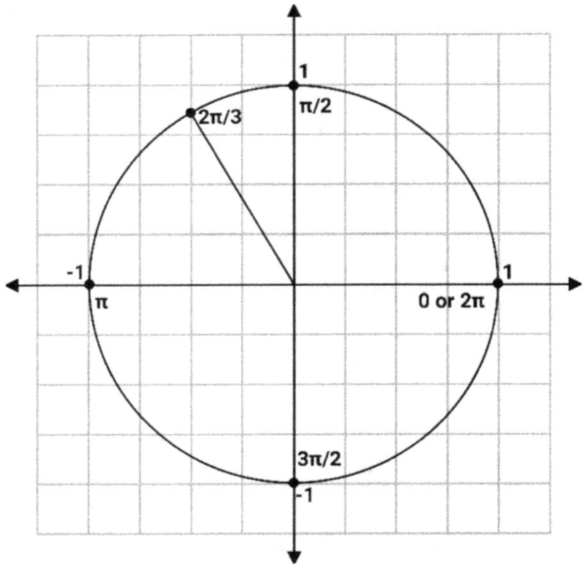

Review Video: Unit Circle
Visit mometrix.com/academy and enter code: 333922

TABLE OF COMMONLY ENCOUNTERED ANGLES

$0° = 0$ radians, $30° = \frac{\pi}{6}$ radians, $45° = \frac{\pi}{4}$ radians, $60° = \frac{\pi}{3}$ radians, and $90° = \frac{\pi}{2}$ radians

$\sin 0° = 0$	$\cos 0° = 1$	$\tan 0° = 0$
$\sin 30° = \frac{1}{2}$	$\cos 30° = \frac{\sqrt{3}}{2}$	$\tan 30° = \frac{\sqrt{3}}{3}$
$\sin 45° = \frac{\sqrt{2}}{2}$	$\cos 45° = \frac{\sqrt{2}}{2}$	$\tan 45° = 1$
$\sin 60° = \frac{\sqrt{3}}{2}$	$\cos 60° = \frac{1}{2}$	$\tan 60° = \sqrt{3}$
$\sin 90° = 1$	$\cos 90° = 0$	$\tan 90° = \text{undefined}$
$\csc 0° = \text{undefined}$	$\sec 0° = 1$	$\cot 0° = \text{undefined}$
$\csc 30° = 2$	$\sec 30° = \frac{2\sqrt{3}}{3}$	$\cot 30° = \sqrt{3}$
$\csc 45° = \sqrt{2}$	$\sec 45° = \sqrt{2}$	$\cot 45° = 1$
$\csc 60° = \frac{2\sqrt{3}}{3}$	$\sec 60° = 2$	$\cot 60° = \frac{\sqrt{3}}{3}$
$\csc 90° = 1$	$\sec 90° = \text{undefined}$	$\cot 90° = 0$

The values in the upper half of this table are values you should have memorized or be able to find quickly and those in the lower half can easily be determined as the reciprocal of the corresponding function.

Defined and Reciprocal Functions

The tangent function is defined as the ratio of the sine to the cosine: $\tan x = \frac{\sin x}{\cos x}$.

To take the reciprocal of a number means to place that number as the denominator of a fraction with a numerator of 1. The reciprocal functions are thus defined quite simply.

Cosecant	$\csc x$	$\frac{1}{\sin x}$
Secant	$\sec x$	$\frac{1}{\cos x}$
Cotangent	$\cot x$	$\frac{1}{\tan x}$

It is important to know these reciprocal functions, but they are not as commonly used as the three basic functions.

Review Video: Defined and Reciprocal Functions
Visit mometrix.com/academy and enter code: 996431

Inverse Trigonometric Functions

Inverse Functions

Each of the trigonometric functions accepts an angular measure, either degrees or radians, and gives a numerical value as the output. The inverse functions do the opposite; they accept a numerical value and give an angular measure as the output.

The inverse of sine, or arcsine, commonly written as either $\sin^{-1} x$ or $\arcsin x$, gives the angle whose sine is x. Similarly:

The inverse of $\cos x$ is written as $\cos^{-1} x$ or $\arccos x$ and means the angle whose cosine is x.
The inverse of $\tan x$ is written as $\tan^{-1} x$ or $\arctan x$ and means the angle whose tangent is x.
The inverse of $\csc x$ is written as $\csc^{-1} x$ or $\text{arccsc}\, x$ and means the angle whose cosecant is x.
The inverse of $\sec x$ is written as $\sec^{-1} x$ or $\text{arcsec}\, x$ and means the angle whose secant is x.

The inverse of $\cot x$ is written as $\cot^{-1} x$ or $\text{arccot}\, x$ and means the angle whose cotangent is x.

Review Video: Inverse Trig Functions
Visit mometrix.com/academy and enter code: 156054

Important Note About Solving Trigonometric Equations

When solving for an angle with a known trigonometric value, you must consider the sign and include all angles with that value. Your calculator will probably only give one value as an answer, typically in the following ranges:

- For $\sin^{-1} x$, $\left[-\frac{\pi}{2}, \frac{\pi}{2}\right]$ or $[-90°, 90°]$
- For $\cos^{-1} x$, $[0, \pi]$ or $[0°, 180°]$
- For $\tan^{-1} x$, $\left[-\frac{\pi}{2}, \frac{\pi}{2}\right]$ or $[-90°, 90°]$

It is important to determine if there is another angle in a different quadrant that also satisfies the problem. To do this, find the other quadrant(s) with the same sign for that trigonometric function and find the angle that has the same reference angle. Then check whether this angle is also a solution.

- In the first quadrant, all six trigonometric functions are positive.
- In the second quadrant, sin and csc are positive.
- In the third quadrant, tan and cot are positive.
- In the fourth quadrant, cos and sec are positive.

If you remember the phrase, "ALL Students Take Classes," you will be able to remember the sign of each trigonometric function in each quadrant. ALL represents all the signs in the first quadrant. The "S" in "Students" represents the sine function and its reciprocal in the second quadrant. The "T" in "Take" represents the tangent function and its reciprocal in the third quadrant. The "C" in "Classes" represents the cosine function and its reciprocal.

Trigonometric Identities

Sum and Difference

To find the sine, cosine, or tangent of the sum or difference of two angles, use one of the following formulas where α and β are two angles with known sine, cosine, or tangent values as needed:

$$\sin(\alpha \pm \beta) = \sin\alpha\cos\beta \pm \cos\alpha\sin\beta$$
$$\cos(\alpha \pm \beta) = \cos\alpha\cos\beta \mp \sin\alpha\sin\beta$$
$$\tan(\alpha \pm \beta) = \frac{\tan\alpha \pm \tan\beta}{1 \mp \tan\alpha\tan\beta}$$

HALF ANGLE

To find the sine or cosine of half of a known angle, use the following formulas where θ is an angle with a known exact cosine value:

$$\sin\left(\frac{\theta}{2}\right) = \pm\sqrt{\frac{(1 - cos\theta)}{2}}$$

$$\cos\left(\frac{\theta}{2}\right) = \pm\sqrt{\frac{(1 + \cos\theta)}{2}}$$

To determine the sign of the answer, you must recognize which quadrant the given angle is in and apply the correct sign for the trigonometric function you are using. If you need to find an expression for the exact sine or cosine of an angle that you do not know, such as sine 22.5°, you can rewrite the given angle as a half angle, such as $\sin\left(\frac{45°}{2}\right)$, and use the formula above:

$$\sin\left(\frac{45°}{2}\right) = \pm\sqrt{\frac{(1 - \cos(45°))}{2}} = \pm\sqrt{\frac{\left(1 - \frac{\sqrt{2}}{2}\right)}{2}} = \pm\sqrt{\frac{(2 - \sqrt{2})}{4}} = \pm\frac{1}{2}\sqrt{(2 - \sqrt{2})}$$

To find the tangent or cotangent of half of a known angle, use the following formulas where θ is an angle with known exact sine and cosine values:

$$\tan\frac{\theta}{2} = \frac{sin\theta}{1 + \cos\theta}$$

$$\cot\frac{\theta}{2} = \frac{\sin\theta}{1 - \cos\theta}$$

These formulas will work for finding the tangent or cotangent of half of any angle unless the cosine of θ happens to make the denominator of the identity equal to 0.

The Pythagorean theorem states that $a^2 + b^2 = c^2$ for all right triangles. The trigonometric identity that derives from this principle is stated in this way: $\sin^2\theta + \cos^2\theta = 1$.

Dividing each term by either $\sin^2\theta$ or $\cos^2\theta$ yields two other identities, respectively:

$$1 + \cot^2\theta = \csc^2\theta$$

$$\tan^2\theta + 1 = \sec^2\theta$$

Review Video: Sum and Difference Trigonometric Identities
Visit mometrix.com/academy and enter code: 468838

Double Angles

In each case, use one of the double angle formulas. To find the sine or cosine of twice a known angle, use one of the following formulas:

$$\sin(2\theta) = 2\sin\theta\cos\theta$$

$$\begin{aligned}\cos(2\theta) &= \cos^2\theta - \sin^2\theta \\ &= 2\cos^2\theta - 1 \\ &= 1 - 2\sin^2\theta\end{aligned}$$

To find the tangent or cotangent of twice a known angle, use the formulas where θ is an angle with known exact sine, cosine, tangent, and cotangent values:

$$\tan(2\theta) = \frac{2\tan\theta}{1 - \tan^2\theta}$$

$$\cot(2\theta) = \frac{\cot\theta - \tan\theta}{2}$$

Products

To find the product of the sines and cosines of two different angles, use one of the following formulas where α and β are two unique angles:

$$\sin\alpha\sin\beta = \frac{1}{2}[\cos(\alpha - \beta) - \cos(\alpha + \beta)]$$

$$\cos\alpha\cos\beta = \frac{1}{2}[\cos(\alpha + \beta) + \cos(\alpha - \beta)]$$

$$\sin\alpha\cos\beta = \frac{1}{2}[\sin(\alpha + \beta) + \sin(\alpha - \beta)]$$

$$\cos\alpha\sin\beta = \frac{1}{2}[\sin(\alpha + \beta) - \sin(\alpha - \beta)]$$

Review Video: Half-Angle, Double Angle, and Product Trig Identities
Visit mometrix.com/academy and enter code: 274252

Complementary

The trigonometric cofunction identities use the trigonometric relationships of complementary angles (angles whose sum is 90°). These are:

$$\cos x = \sin(90° - x)$$

$$\csc x = \sec(90° - x)$$

$$\cot x = \tan(90° - x)$$

Domain, Range, and Asymptotes in Trigonometry

The domain is the set of all possible real number values of x on the graph of a trigonometric function. Some graphs will impose limits on the values of x.

The range is the set of all possible real number values of y on the graph of a trigonometric function. Some graphs will impose limits on the values of y.

Asymptotes are lines that the graph of a trigonometric function approaches but never reaches. Asymptotes exist for values of x in the graphs of the tangent, cotangent, secant, and cosecant. The sine and cosine graphs do not have any asymptotes.

DOMAIN, RANGE, AND ASYMPTOTES OF THE SIX TRIGONOMETRIC FUNCTIONS

The domain, range, and asymptotes for each of the trigonometric functions are as follows:

- In the **sine** function, the domain is all real numbers, the range is $-1 \leq y \leq 1$, and there are no asymptotes.
- In the **cosine** function, the domain is all real numbers, the range is $-1 \leq y \leq 1$, and there are no asymptotes.
- In the **tangent** function, the domain is $x \in \mathbb{R}; x \neq \frac{\pi}{2} + k\pi$, the range is all real numbers, and the asymptotes are the lines $x = \frac{\pi}{2} + k\pi$.
- In the **cosecant** function, the domain is $x \in \mathbb{R}; x \neq k\pi$, the range is $(-\infty, -1]$ and $[1, \infty)$, and the asymptotes are the lines $x = k\pi$.
- In the **secant** function, the domain is $x \in \mathbb{R}; x \neq \frac{\pi}{2} + k\pi$, the range is $(-\infty, 1]$ and $[1, \infty)$, and the asymptotes are the lines $x = \frac{\pi}{2} + k\pi$.
- In the **cotangent** function, the domain is $x \in \mathbb{R}; x \neq k\pi$, the range is all real numbers, and the asymptotes are the lines $x = k\pi$.

In each of the above cases, k represents any integer.

RECTANGULAR AND POLAR COORDINATES

Rectangular coordinates are those that lie on the square grids of the Cartesian plane. They should be quite familiar to you. The polar coordinate system is based on a circular graph, rather than the square grid of the Cartesian system. Points in the polar coordinate system are in the format (r, θ), where r is the distance from the origin (think radius of the circle) and θ is the smallest positive angle (moving counterclockwise around the circle) made with the positive horizontal axis.

To convert a point from rectangular (x, y) format to polar (r, θ) format, use the formula (x, y) to $(r, \theta) \Rightarrow r = \sqrt{x^2 + y^2}; \theta = \arctan\frac{y}{x}$ when $x \neq 0$.

If x is positive, use the positive square root value for r. If x is negative, use the negative square root value for r. If $x = 0$, use the following rules:

- If $y = 0$, then $\theta = 0$.
- If $y > 0$, then $\theta = \frac{\pi}{2}$.
- If $y < 0$, then $\theta = \frac{3\pi}{2}$.

To convert a point from polar (r, θ) format to rectangular (x, y) format, use the formula (r, θ) to $(x, y) \Rightarrow x = r\cos\theta\,; y = r\sin\theta$.

DE MOIVRE'S THEOREM

De Moivre's theorem is used to find the powers of complex numbers (numbers that contain the imaginary number i) written in polar form. Given a trigonometric expression that contains i, such

Algebra and Functions

as $z = r\cos x + ir\sin x$, where r is a real number and x is an angle measurement in polar form, use the formula $z^n = r^n(\cos nx + i\sin nx)$, where r and n are real numbers, x is the angle measure in polar form, and i is the imaginary number $i = \sqrt{-1}$. The expression $\cos x + i\sin x$ can be written cis x, making the formula appear in the format $z^n = r^n \text{ cis } nx$.

Note that De Moivre's theorem is only for angles in polar form. If you are given an angle in degrees, you must convert to polar form before using the formula.

Chapter Quiz

Ready to see how well you retained what you just read? Scan the QR code to go directly to the chapter quiz interface for this study guide. If you're using a computer, simply visit the online resources page at **mometrix.com/resources719/priimath5161** and click the Chapter Quizzes link.

Calculus

Calculus, also called analysis, is the branch of mathematics that studies the length, area, and volume of objects, and the rate of change of quantities (which can be expressed as slopes of curves). The two principal branches of calculus are differential and integral. **Differential calculus** is based on derivatives and takes the form,

$$\frac{d}{dx}f(x)$$

Integral calculus is based on integrals and takes the form,

$$\int f(x)dx$$

Some of the basic ideas of calculus were utilized as far back in history as Archimedes. However, its modern forms were developed by Newton and Leibniz.

LIMITS

The **limit of a function** is represented by the notation $\lim_{x \to a} f(x)$. It is read as "the limit of f of x as x approaches a." In many cases, $\lim_{x \to a} f(x)$ will simply be equal to $f(a)$, but not always. Limits are important because some functions are not defined or are not easy to evaluate at certain values of x.

The limit at the point is said to exist only if the limit is the same when approached from the right side as from the left: $\lim_{x \to a^+} f(x) = \lim_{x \to a^-} f(x)$). Notice the symbol by the a in each case. When x approaches a from the right, it approaches from the positive end of the number line. When x approaches a from the left, it approaches from the negative end of the number line.

If the limit as x approaches a differs depending on the direction from which it approaches, then the limit does not exist at a. In other words, if $\lim_{x \to a^+} f(x)$ does not equal $\lim_{x \to a^-} f(x)$, then the limit does not exist at a. The limit also does not exist if either of the one-sided limits does not exist.

Situations in which the limit does not exist include a function that jumps from one value to another at a, one that oscillates between two different values as x approaches a, or one that increases or decreases without bounds as x approaches a. If the limit you calculate has a value of $\frac{c}{0}$, where c is any constant, this means the function goes to infinity and the limit does not exist.

It is possible for two functions that do not have limits to be multiplied to get a new function that does have a limit. Just because two functions do not have limits, do not assume that the product will not have a limit.

Direct Substitution

The first thing to try when looking for a limit is direct substitution. To find the limit of a function $\lim_{x \to a} f(x)$ by direct substitution, substitute the value of a for x in the function and solve. The following patterns apply to finding the limit of a function by direct substitution:

$$\lim_{x \to a} b = b, \text{ where } b \text{ is any real number}$$
$$\lim_{x \to a} x = a$$
$$\lim_{x \to a} x^n = a^n, \text{ where } n \text{ is any positive integer}$$
$$\lim_{x \to a} \sqrt{x} = \sqrt{a}; a > 0$$
$$\lim_{x \to a} \sqrt[n]{x} = \sqrt[n]{a}, \text{ where } n \text{ is a positive integer and } a > 0 \text{ for all even values of } n$$
$$\lim_{x \to a} \frac{1}{x} = \frac{1}{a}; a \neq 0$$

You can also use substitution for finding the limit of a trigonometric function, a polynomial function, or a rational function. Be sure that in manipulating an expression to find a limit that you do not divide by terms equal to zero.

In finding the limit of a composite function, begin by finding the limit of the innermost function. For example, to find $\lim_{x \to a} f(g(x))$, first find the value of $\lim_{x \to a} g(x)$. Then substitute this value for x in $f(x)$ and solve. The result is the limit of the original problem.

Limits and Operations

When finding the limit of the sum or difference of two functions, find the limit of each individual function and then add or subtract the results. Example:

$$\lim_{x \to a}[f(x) \pm g(x)] = \lim_{x \to a} f(x) \pm \lim_{x \to a} g(x)$$

To find the limit of the product or quotient of two functions, find the limit of each individual function and then multiply or divide the results. Example:

$$\lim_{x \to a}[f(x) \times g(x)] = \lim_{x \to a} f(x) \times \lim_{x \to a} g(x)$$

$$\lim_{x \to a} \frac{f(x)}{g(x)} = \frac{\lim_{x \to a} f(x)}{\lim_{x \to a} g(x)}, \text{ where } g(x) \neq 0$$

$$\lim_{x \to a} g(x) \neq 0$$

When finding the quotient of the limits of two functions, make sure the denominator is not equal to zero. If it is, use differentiation or L'Hôpital's rule to find the limit.

To find the limit of a power of a function or a root of a function, find the limit of the function and then raise the limit to the original power or take the root of the limit. Example:

$$\lim_{x\to a}[f(x)]^n = \left[\lim_{x\to a} f(x)\right]^n$$

$$\lim_{x\to a} \sqrt[n]{f(x)} = \sqrt[n]{\lim_{x\to a} f(x)}, \text{ where } n \text{ is a positive integer}$$

$$\lim_{x\to a} f(x) > 0 \text{ for all even values of } n$$

To find the limit of a function multiplied by a scalar, find the limit of the function and multiply the result by the scalar. Example:

$$\lim_{x\to a} kf(x) = k \lim_{x\to a} f(x), \text{ where } k \text{ is a real number.}$$

Review Video: Limits
Visit mometrix.com/academy and enter code: 554961

L'HÔPITAL'S RULE

Sometimes solving $\lim_{x\to a}\frac{f(x)}{g(x)}$ by the direct substitution method will result in the numerator and denominator both being equal to zero, or both being equal to infinity. This outcome is called an indeterminate form. The limit cannot be directly found by substitution in these cases. L'Hôpital's rule is a useful method for finding the limit of a problem in the indeterminate form. L'Hôpital's rule allows you to find the limit using derivatives. Assuming both the numerator and denominator are differentiable, and that both are equal to zero when the direct substitution method is used, take the derivative of both the numerator and the denominator and then use the direct substitution method. For example, if $\lim_{x\to a}\frac{f(x)}{g(x)} = \frac{0}{0}$, take the derivatives of $f(x)$ and $g(x)$ and then find $\lim_{x\to a}\frac{f'(x)}{g'(x)}$. If $g'(x) \neq 0$, then you have found the limit of the original function. If $g'(x) = 0$ and $f'(x) = 0$, L'Hôpital's rule may be applied to the function $\frac{f'(x)}{g'(x)}$, and so on until either a limit is found, or it can be determined that the limit does not exist.

Review Video: L'Hopital's Rule
Visit mometrix.com/academy and enter code: 624400

SQUEEZE THEOREM

The squeeze theorem is known by many names, including the sandwich theorem, the sandwich rule, the squeeze lemma, the squeezing theorem, and the pinching theorem. No matter what you call it, the principle is the same. To prove the limit of a difficult function exists, find the limits of two functions, one on either side of the unknown, that are easy to compute. If the limits of these functions are equal, then that is also the limit of the unknown function. In mathematical terms, the theorem is:

If $g(x) \leq f(x) \leq h(x)$ for all values of x where $f(x)$ is the function with the unknown limit, and if $\lim_{x \to a} g(x) = \lim_{x \to a} h(x)$, then this limit is also equal to $\lim_{x \to a} f(x)$.

To find the limit of an expression containing an absolute value sign, take the absolute value of the limit. If $\lim_{n \to \infty} a_n = L$, where L is the numerical value for the limit, then $\lim_{n \to \infty} |a_n| = |L|$. Also, if $\lim_{n \to \infty} |a_n| = 0$, then $\lim_{n \to \infty} a_n = 0$. The trick comes when you are asked to find the limit as n approaches from the left. Whenever the limit is being approached from the left, it is being approached from the negative end of the domain. The absolute value sign makes everything in the equation positive, essentially eliminating the negative side of the domain. In this case, rewrite the equation without the absolute value signs and add a negative sign in front of the expression. Example:

$$\lim_{n \to 0^-} |x| \text{ becomes } \lim_{n \to 0^-} (-x)$$

Review Video: Squeeze Theorem
Visit mometrix.com/academy and enter code: 383104

DERIVATIVES

The derivative of a function is a measure of how much that function is changing at a specific point, and is the slope of a line tangent to a curve at the specific point. The derivative of a function $f(x)$ is written $f'(x)$, and read, "f prime of x." Other notations for the derivative include $D_x f(x)$, y', $D_x y$, $\frac{dy}{dx}$, and $\frac{d}{dx} f(x)$. The definition of the derivative of a function is $f'(x) = \lim_{h \to 0} \frac{f(x+h) - f(x)}{h}$. However, this formula is rarely used.

There is a simpler method you can use to find the derivative of a polynomial. Given a function $f(x) = a_n x^n + a_{n-1} x^{n-1} + a_{n-2} x^{n-2} + \cdots + a_1 x + a_0$, multiply each exponent by its corresponding coefficient to get the new coefficient and reduce the value of the exponent by one. Coefficients with no variable are dropped. This gives $f'(x) = n a_n x^{n-1} + (n-1) a_{n-1} x^{n-2} + \cdots + a_1$, a pattern that can be repeated for each successive derivative.

Review Video: Definition of a Derivative
Visit mometrix.com/academy and enter code: 787269

Differentiable functions are functions that have a derivative. Some basic rules for finding derivatives of functions are:

$$f(x) = c \Rightarrow f'(x) = 0; \text{ where } c \text{ is a constant}$$
$$f(x) = x \Rightarrow f'(x) = 1$$
$$f(x) = x^n \Rightarrow f'(x) = n x^{n-1}; \text{ where } n \text{ is a real number}$$

$$\left(cf(x)\right)' = cf'(x);\ \text{where } c \text{ is a constant}$$
$$(f+g)'(x) = f'(x) + g'(x)$$
$$(fg)'(x) = f(x)g'(x) + f'(x)g(x)$$
$$\left(\frac{f}{g}\right)'(x) = \frac{f'(x)g(x) - f(x)g'(x)}{[g(x)]^2}$$
$$(f \circ g)'(x) = f'\left(g(x)\right) \times g'(x)$$

This last formula is also known as the **chain rule**. If you are finding the derivative of a polynomial that is raised to a power, let the polynomial be represented by $g(x)$ and use the chain rule. The chain rule is one of the most important concepts to grasp in the early stages of learning calculus. Many other rules and shortcuts are based upon the chain rule.

Review Video: Derivative Properties and Formulas
Visit mometrix.com/academy and enter code: 735227

Review Video: Product and Quotient Rules in Calculus
Visit mometrix.com/academy and enter code: 649197

Review Video: The Chain Rule - An Integral Part of Calculus
Visit mometrix.com/academy and enter code: 938732

DIFFERENCE QUOTIENT AND DERIVATIVE

A secant is a line that connects two points on a curve. The **difference quotient** gives the slope of an arbitrary secant line that connects the point $\left(x, f(x)\right)$ with a nearby point $\left(x+h, f(x+h)\right)$ on the graph of the function f. The difference quotient is the same formula that is always used to determine a slope—the change in y divided by the change in x. It is written as $\frac{f(x+h)-f(x)}{h}$.

A tangent is a line that touches a curve at one point. The tangent and the curve have the same slope at the point where they touch. The derivative is the function that gives the slope of both the tangent and the curve of the function at that point. The derivative is written as the limit of the difference quotient, or:

$$\lim_{h \to 0} \frac{f(x+h) - f(x)}{h}$$

If the function is f, the derivative is denoted as $f'(x)$, and it is the slope of the function f at point $\left(x, f(x)\right)$. It is expressed as:

$$f'(x) = \lim_{h \to 0} \frac{f(x+h) - f(x)}{h}$$

IMPLICIT FUNCTIONS

An **implicit function** is one where it is impossible, or very difficult, to express one variable in terms of another by normal algebraic methods. This would include functions that have both variables raised to a power greater than 1, functions that have two variables multiplied by each other, or a combination of the two. To differentiate such a function with respect to x, take the derivative of each term that contains a variable, either x or y. When differentiating a term with y, use the chain rule, first taking the derivative with respect to y, and then multiplying by $\frac{dy}{dx}$. If a term contains both

Calculus

x and y, you will have to use the product rule as well as the chain rule. Once the derivative of each individual term has been found, use the rules of algebra to solve for $\frac{dy}{dx}$ to get the final answer.

Review Video: Implicit Differentiation
Visit mometrix.com/academy and enter code: 102151

Derivatives of Trigonometric Functions

Trigonometric functions are any functions that include one of the six trigonometric expressions. The following rules for derivatives apply for all trigonometric differentiation:

$$\frac{d}{dx}(\sin x) = \cos x, \quad \frac{d}{dx}(\cos x) = -\sin x, \quad \frac{d}{dx}(\tan x) = \sec^2 x$$

For functions that are a combination of trigonometric and algebraic expressions, use the chain rule:

$$\frac{d}{dx}(\sin u) = \cos u \frac{du}{dx} \qquad \frac{d}{dx}(\sec u) = \tan u \sec u \frac{du}{dx}$$
$$\frac{d}{dx}(\cos u) = -\sin u \frac{du}{dx} \qquad \frac{d}{dx}(\csc u) = -\csc u \cot u \frac{du}{dx}$$
$$\frac{d}{dx}(\tan u) = \sec^2 u \frac{du}{dx} \qquad \frac{d}{dx}(\cot u) = -\csc^2 u \frac{du}{dx}$$

Functions involving the inverses of the trigonometric functions can also be differentiated.

$$\frac{d}{dx}(\sin^{-1} u) = \frac{1}{\sqrt{1-u^2}}\frac{du}{dx} \qquad \frac{d}{dx}(\csc^{-1} u) = \frac{-1}{|u|\sqrt{u^2-1}}\frac{du}{dx}$$
$$\frac{d}{dx}(\cos^{-1} u) = \frac{-1}{\sqrt{1-u^2}}\frac{du}{dx} \qquad \frac{d}{dx}(\sec^{-1} u) = \frac{1}{|u|\sqrt{u^2-1}}\frac{du}{dx}$$
$$\frac{d}{dx}(\tan^{-1} u) = \frac{1}{1+u^2}\frac{du}{dx} \qquad \frac{d}{dx}(\cot^{-1} u) = \frac{-1}{1+u^2}\frac{du}{dx}$$

In each of the above expressions, u represents a differentiable function. Also, the value of u must be such that the radicand, if applicable, is a positive number. Remember the expression $\frac{du}{dx}$ means to take the derivative of the function u with respect to the variable x.

Review Video: Derivatives of Trigonometry Functions
Visit mometrix.com/academy and enter code: 132724

Derivatives of Exponential and Logarithmic Functions

Exponential functions are in the form e^x, which has itself as its derivative: $\frac{d}{dx} e^x = e^x$. For functions that have a function as the exponent rather than just an x, use the formula $\frac{d}{dx} e^u = e^u \frac{du}{dx}$. The inverse of the exponential function is the natural logarithm. To find the derivative of the natural logarithm, use the formula $\frac{d}{dx}\ln u = \frac{1}{u}\frac{du}{dx}$.

If you are trying to solve an expression with a variable in the exponent, use the formula $a^x = e^{x \ln a}$, where a is a positive real number and x is any real number. To find the derivative of a function in this format, use the formula $\frac{d}{dx}a^x = a^x \ln a$. If the exponent is a function rather than a single variable x, use the formula $\frac{d}{dx}a^u = a^u \ln a \frac{du}{dx}$. If you are trying to solve an expression involving a logarithm, use the formula $\frac{d}{dx}(\log_a x) = \frac{1}{x \ln a}$ or $\frac{d}{dx}(\log_a |u|) = \frac{1}{u \ln a}\frac{du}{dx}; u \neq 0$.

Review Video: Derivatives of Exponential and Logarithmic Functions
Visit mometrix.com/academy and enter code: 594367

CONTINUITY

A function can be either continuous or discontinuous. A conceptual way to describe continuity is this: A function is continuous if its graph can be traced with a pen without lifting the pen from the page. In other words, there are no breaks or gaps in the graph of the function. However, this is only a description, not a technical definition. A function is continuous at the point $x = a$ if the three following conditions are met:

1. $f(a)$ is defined
2. $\lim_{x \to a} f(x)$ exists
3. $\lim_{x \to a} f(x) = f(a)$

If any of these conditions are not met, the function is discontinuous at the point $x = a$.

A function can be continuous at a point, continuous over an interval, or continuous everywhere. The above rules define continuity at a point. A function that is continuous over an interval $[a, b]$ is continuous at the points a and b and at every point between them. A function that is continuous everywhere is continuous for every real number, that is, for all points in its domain.

DISCONTINUITY

Discontinuous functions are categorized according to the type or cause of discontinuity. Three examples are point, infinite, and jump discontinuity. A function with a point discontinuity has one value of x for which it is not continuous. A function with infinite discontinuity has a vertical asymptote at $x = a$ and $f(a)$ is undefined. It is said to have an infinite discontinuity at $x = a$. A function with jump discontinuity has one-sided limits from the left and from the right, but they are not equal to one another, that is, $\lim_{x \to a^-} f(x) \neq \lim_{x \to a^+} f(x)$. It is said to have a jump discontinuity at $x = a$.

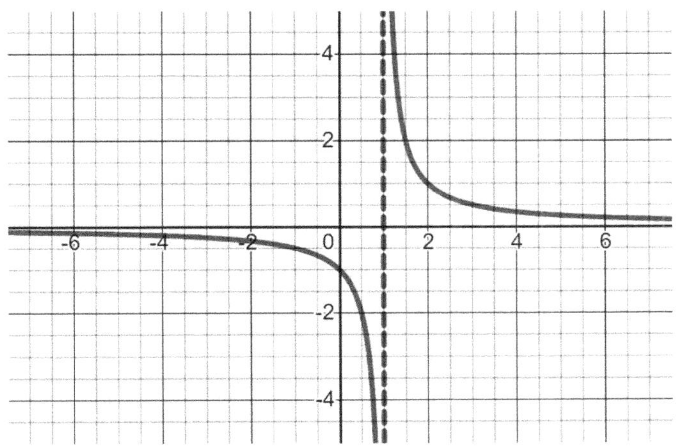

Calculus

The function $f(x) = \sqrt{x}$ is not differentiable because its domain is $[0, \infty)$ and it has a discontinuity at $x = 0$. Therefore, a tangent could not be drawn at that point.

DIFFERENTIABILITY

A function is said to be differentiable at point $x = a$ if it has a derivative at that point, that is, if $f'(a)$ exists. For a function to be differentiable, it must be continuous because the slope cannot be defined at a point of discontinuity. Furthermore, for a function to be differentiable, its graph must not have any sharp turn for which it is impossible to draw a tangent line. The sine function is an example of a differentiable function. It is continuous, and a tangent line can be drawn anywhere along its graph.

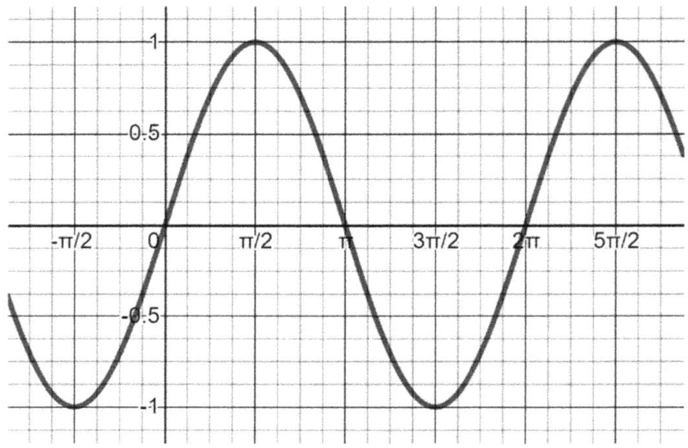

The absolute value function, $f(x) = |x|$, is an example of a function that is not differentiable:

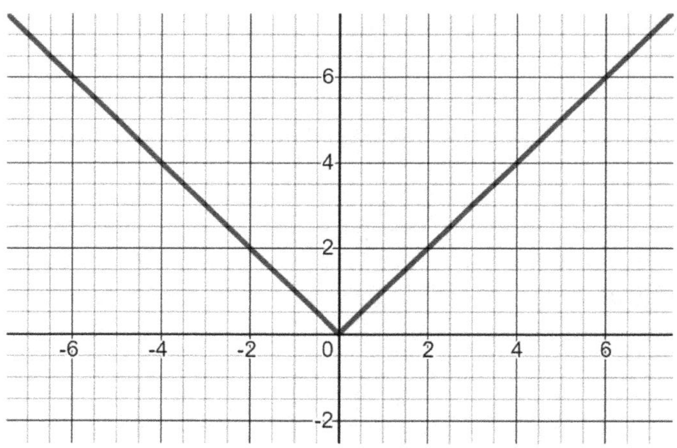

It is continuous, but it has a sharp turn at $x = 0$ which prohibits the drawing of a tangent at that point. All differentiable functions are continuous, but not all continuous functions are differentiable, as the absolute value function demonstrates.

The function $f(x) = \sqrt{x}$ is not differentiable because its domain is $[0, \infty)$ and it has a discontinuity at $x = 0$. Therefore, a tangent could not be drawn at that point.

APPROXIMATING A DERIVATIVE FROM A TABLE OF VALUES

The derivative of a function at a particular point is equal to the slope of the graph of the function at that point. For a nonlinear function, it can be thought of as the limit of the slope of a line drawn

between two other points on the function as those points become closer to the point in question. Such a line drawn through two points on the function is called a **secant** of the function.

This definition of the derivative in terms of the secant allows us to approximate the derivative of a function at a point from a table of values: we take the slope of the line through the points on either side. That is, if the point lies between (x_1, y_1) and (x_2, y_2), the slope of the secant—the approximate derivative—is $\frac{y_2-y_1}{x_2-x_1}$. (This is also equal to the average slope over the interval $[x_1, x_2]$.)

For example, consider the function represented by the following table:

x	0	2	4	6	8	10
y	1	5	8	9	7	4

Suppose we want to know the derivative of the function when $x = 3$. This lies between the points $(2, 5)$ and $(4, 8)$; the approximate derivative is $\frac{8-5}{4-2} = \frac{3}{2}$.

Position, Velocity, and Acceleration

Velocity is a specific type of rate of change. It refers to the rate of change of the position of an object with relation to a reference frame. **Acceleration** is the rate of change of velocity.

Average velocity over a period of time is found using the formula $\bar{v} = \frac{s(t_2)-s(t_1)}{t_2-t_1}$, where t_1 and t_2 are specific points in time and $s(t_1)$ and $s(t_2)$ are the distances traveled at those points in time.

Instantaneous velocity at a specific time, t, is found using the limit $v = \lim_{h\to 0} \frac{s(t+h)-s(t)}{h}$, or $v = s'(t)$.

Remember that velocity at a given point is found using the first derivative, and acceleration at a given point is found using the second derivative. Therefore, the formula for acceleration at a given point in time is found using the formula $a(t) = v'(t) = s''(t)$, where a is acceleration, v is velocity, and s is displacement.

Review Video: Position, Velocity, and Acceleration
Visit mometrix.com/academy and enter code: 714040

Using First and Second Derivatives

The **first derivative** of a function is equal to the **rate of change** of the function. The sign of the rate of change shows whether the value of the function is **increasing** or **decreasing**. A positive rate of change—and therefore a positive first derivative—represents that the function is increasing at that point. A negative rate of change represents that the function is decreasing. If the rate of change is zero, the function is not changing, i.e., it is constant.

For example, consider the function $f(x) = x^3 - 6x^2 - 15x + 12$. The derivative of this function is $f'(x) = 3x^2 - 12x - 15 = 3(x^2 - 4x - 5) = 3(x - 5)(x + 1)$. This derivative is a quadratic function with zeroes at $x = 5$ and $x = -1$; by plugging in points in each interval we can find that $f'(x)$ is positive when $x < -1$ and when $x > 5$ and negative when $-1 < x < 5$. Thus $f(x)$ is increasing in the interval $(-\infty, -1) \cup (5, \infty)$ and decreasing in the interval $(-1, 5)$.

Review Video: Rate of Change
Visit mometrix.com/academy and enter code: 100367

Calculus

Extrema

The **maximum** and **minimum** values of a function are collectively called the **extrema** of the function. Both maxima and minima can be local, also known as relative, or absolute. A local maximum or minimum refers to the value of a function near a certain value of x. An absolute maximum or minimum refers to the value of a function on a given interval.

The local maximum of a function is the largest value that the function attains near a certain value of x. For example, function f has a local maximum at $x = b$ if $f(b)$ is the largest value that f attains as it approaches b.

Conversely, the local minimum is the smallest value that the function attains near a certain value of x. In other words, function f has a local minimum at $x = b$ if $f(b)$ is the smallest value that f attains as it approaches b.

The absolute maximum of a function is the largest value of the function over a certain interval. The function f has an absolute maximum at $x = b$ if $f(b) \geq f(x)$ for all x in the domain of f.

The absolute minimum of a function is the smallest value of the function over a certain interval. The function f has an absolute minimum at $x = b$ if $f(b) \leq f(x)$ for all x in the domain of f.

Critical Points

Remember Rolle's theorem, which states that if two points have the same value in the range that there must be a point between them where the slope of the graph is zero. This point is located at a peak or valley on the graph. A **peak** is a maximum point, and a **valley** is a minimum point. The relative minimum is the lowest point on a graph for a given section of the graph. It may or may not be the same as the absolute minimum, which is the lowest point on the entire graph. The relative maximum is the highest point on one section of the graph. Again, it may or may not be the same as the absolute maximum. A relative extremum (plural extrema) is a relative minimum or relative maximum point on a graph.

A **critical point** is a point $(x, f(x))$ that is part of the domain of a function, such that either $f'(x) = 0$ or $f'(x)$ does not exist. If either of these conditions is true, then x is either an inflection point or a point at which the slope of the curve changes sign. If the slope changes sign, then a relative minimum or maximum occurs.

In graphing an equation with relative extrema, use a sign diagram to approximate the shape of the graph. Once you have determined the relative extrema, calculate the sign of a point on either side of each critical point. This will give a general shape of the graph, and you will know whether each critical point is a relative minimum, a relative maximum, or a point of inflection.

First Derivative Test

Remember that critical points occur where the slope of the curve is 0. Also remember that the **first derivative** of a function gives the slope of the curve at a particular point on the curve. Because of this property of the first derivative, the first derivative test can be used to determine if a critical point is a minimum or maximum. If $f'(x)$ is negative at a point to the left of a critical number and $f'(x)$ is positive at a point to the right of a critical number, then the critical number is a relative minimum. If $f'(x)$ is positive to the left of a critical number and $f'(x)$ is negative to the right of a critical number, then the critical number is a relative maximum. If $f'(x)$ has the same sign on both sides, then the critical number is a point of inflection.

Review Video: First Derivative Test
Visit mometrix.com/academy and enter code: 205981

Second Derivative Test

The **second derivative**, designated by $f''(x)$, is helpful in determining whether the relative extrema of a function are relative maximums or relative minimums. If the second derivative at the critical point is greater than zero, the critical point is a relative minimum. If the second derivative at the critical point is less than zero, the critical point is a relative maximum. If the second derivative at the critical point is equal to zero, you must use the first derivative test to determine whether the point is a relative minimum or a relative maximum.

There are a couple of ways to determine the concavity of the graph of a function. To test a portion of the graph that contains a point with domain p, find the second derivative of the function and evaluate it for p. If $f''(p) > 0$, then the graph is concave upward at that point. If $f''(p) < 0$, then the graph is concave downward at that point.

The **point of inflection** on the graph of a function is the point at which the concavity changes from concave downward to concave upward or from concave upward to concave downward. The easiest way to find the points of inflection is to find the second derivative of the function and then solve the equation $f''(x) = 0$. Remember that if $f''(p) > 0$, the graph is concave upward, and if $f''(p) < 0$, the graph is concave downward. Logically, the concavity changes at the point when $f''(p) = 0$:

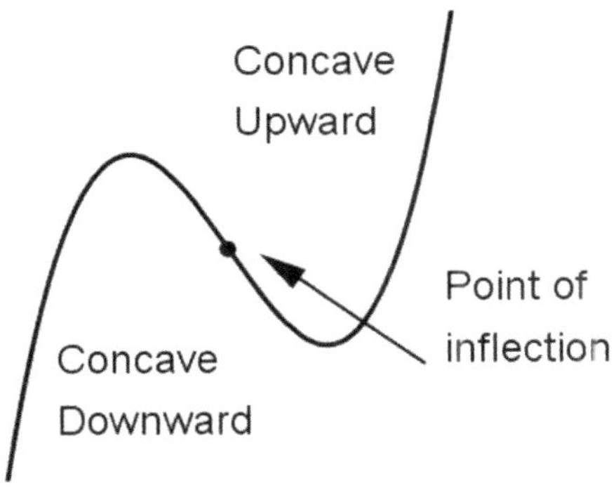

It is important to note in situations where $f'(p) = f''(p) = 0$ it is necessary to use the first derivative test to determine if p is an inflection point.

The derivative tests that have been discussed thus far can help you get a rough picture of what the graph of an unfamiliar function looks like. Begin by solving the equation $f(x) = 0$ to find all the zeros of the function, if they exist. Plot these points on the graph. Then, find the first derivative of the function and solve the equation $f'(x) = 0$ to find the critical points. Remember the numbers obtained here are the x portions of the coordinates. Substitute these values for x in the original function and solve for y to get the full coordinates of the points. Plot these points on the graph. Take the second derivative of the function and solve the equation $f''(x) = 0$ to find the points of inflection. Substitute in the original function to get the coordinates and graph these points. Test points on both sides of the critical points to test for concavity and draw the curve.

Review Video: Second Derivative Test
Visit mometrix.com/academy and enter code: 655614

DERIVATIVE PROBLEMS

A derivative represents the rate of change of a function; thus, derivatives are a useful tool for solving any problem that involves finding the rate at which a function is changing. In its simplest form, such a problem might provide a formula for a quantity as a function of time and ask for its rate of change at a particular time.

If the temperature in a chamber in degrees Celsius is equal to $T(t) = 20 + e^{-\left(\frac{t}{2}\right)}$, where t is the time in seconds, then the derivative of the function represents the rate of change of the temperature over time. The rate of change is equal to $\frac{dT}{dt} = \frac{d}{dt}\left(20 + e^{-\left(\frac{t}{2}\right)}\right) = -\frac{1}{2}e^{-\left(\frac{t}{2}\right)}$, and the initial rate of change is $T'(0) = -\frac{1}{2}e^{-\left(\frac{0}{2}\right)} = -\frac{1}{2}\frac{°C}{s}$.

Suppose we are told that the net profit that a small company makes when it produces and sells x units of a product is equal to $P(x) = 200x - 20{,}000$. The derivative of this function would be the *additional profit for each additional unit sold*, a quantity known as the marginal profit. The marginal profit in this case is $P'(x) = 200$.

SOLVING RELATED RATES PROBLEMS

A **related rate problem** is one in which one variable has a relation with another variable, and the rate of change of one of the variables is known. With that information, the rate of change of the other variable can be determined. The first step in solving related rates problems is defining the known rate of change. Then, determine the relationship between the two variables, then the derivatives (the rates of change), and finally substitute the problem's specific values. Consider the following example:

The side of a cube is increasing at a rate of 2 feet per second. Determine the rate at which the volume of the cube is increasing when the side of the cube is 4 feet long.

For the problem in question, the known rate of change can be expressed as $s'(t) = \frac{ds}{dt} = 2\frac{\text{ft}}{\text{s}}$, where s is the length of the side and t is the elapsed time in seconds. The relationship between the two variables of the cube is $v = s^3$, where v is the volume of the cube and s is the length of the side. The unknown rate of change to determine is the volume. As both v and s change with time, $v = s^3$ becomes $v(t) = [s(t)]^3$

Now, the chain rule is applied to differentiate both sides of the equation with respect to t.

$$d\frac{v(t)}{dt} = \frac{d[s(t)]^3}{dt}; \quad \frac{dv}{dt} = \frac{(3[s(t)]^2)ds}{dt}$$

Finally, the specific value of $s = 4$ feet is substituted, and the equation is evaluated.

$$\begin{aligned} \frac{dv}{dt} &= \frac{(3[s(t)]^2)ds}{dt} \\ &= 3(4)^2 \times 2 = 96\frac{\text{ft}^3}{\text{s}} \\ &= 96 \text{ cubic}\frac{\text{ft}}{\text{s}} \end{aligned}$$

Therefore, when a side of the cube is 4 feet long, the volume of the cube is increasing at a rate of 96 cubic $\frac{\text{feet}}{\text{second}}$.

Review Video: Solving Related Rates Problems
Visit mometrix.com/academy and enter code: 321959

Solving Optimization Problems

An **optimization problem** is a problem in which we are asked to find the value of a variable that maximizes or minimizes a particular value. Because the minimum or maximum occurs at a critical point, and because the critical point occurs when the derivative of the function is zero, we can solve an optimization problem by setting the derivative of the function to zero and solving for the desired variable.

For example, suppose a farmer has 720 m of fencing, and wants to use it to fence in a 2 by 3 block of identical rectangular pens. What dimensions of the pens will maximize their area?

We can draw a diagram:

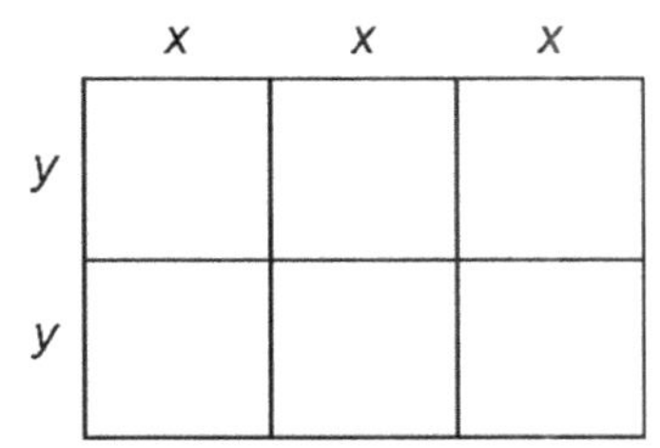

We want to maximize the area of each pen, $A(x, y) = xy$. However, we have the additional constraint that the farmer has only 720 m of fencing. In terms of x and y, we can count the number of segments of each length, 9 for x and 8 for y, so the total amount of fencing required will be $9x + 8y$. Our constraint becomes $9x + 8y = 720$; solving for y yields $y = -\frac{9}{8}x + 90$. We can substitute that into the area equation to get $A(x) = x\left(-\frac{9}{8}x + 90\right) = -\frac{9}{8}x^2 + 90x$. Taking the derivative yields $A'(x) = -\frac{9}{4}x + 90$; setting that equal to zero and solving for x yields $x = 40$. $y = -\frac{9}{8}(40) + 90 = 45$; thus, the maximum dimensions of the pen are 40 by 45 meters.

Review Video: Solving Optimization Problems
Visit mometrix.com/academy and enter code: 628734

Characteristics of Functions (Using Calculus)

Rolle's theorem states that if a differentiable function has two different values in the domain that correspond to a single value in the range, then the function must have a point between them where the slope of the tangent to the graph is zero. This point will be a maximum or a minimum value of the function between those two points. The maximum or minimum point is the point at which $f'(c) = 0$, where c is within the appropriate interval of the function's domain. The following graph shows a function with one maximum in the second quadrant and one minimum in the fourth quadrant.

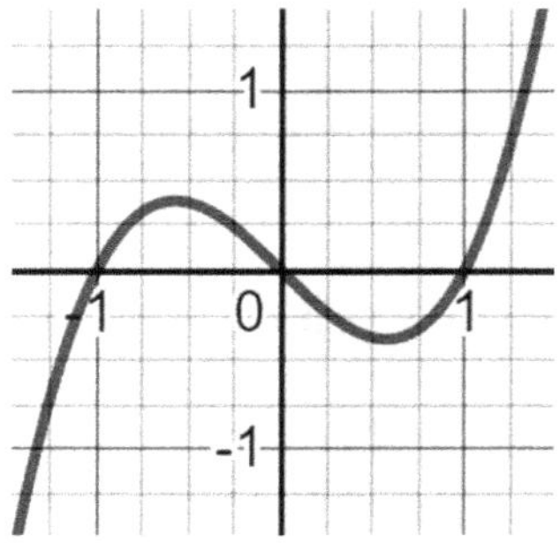

MEAN VALUE THEOREM

According to the **mean value theorem**, between any two points on a curve, there exists a tangent to the curve whose slope is parallel to the chord formed by joining those two points. Remember the formula for slope: $m = \frac{\Delta y}{\Delta x}$. In a function, $f(x)$ represents the value for y. Therefore, if you have two points on a curve, m and n, the corresponding points are $(m, f(m))$ and $(n, f(n))$. Assuming $m < n$, the formula for the slope of the chord joining those two points is $\frac{f(n)-f(m)}{n-m}$. This must also be the slope of a line parallel to the chord, since parallel lines have equal slopes. Therefore, there must be a value p between m and n such that $f'(p) = \frac{f(n)-f(m)}{n-m}$.

For a function to have continuity, its graph must be an unbroken curve. That is, it is a function that can be graphed without having to lift the pencil to move it to a different point. To say a function is continuous at point p, you must show the function satisfies three requirements. First, $f(p)$ must exist. If you evaluate the function at p, it must yield a real number. Second, there must exist a relationship such that $\lim_{x \to p} f(x) = f(p)$. Finally, the following relationship must be true:

$$\lim_{x \to p^+} F(x) = \lim_{x \to p^-} F(x) = F(p)$$

If all three of these requirements are met, a function is considered continuous at p. If any one of them is not true, the function is not continuous at p.

Review Video: Mean Value Theorem
Visit mometrix.com/academy and enter code: 633482

TANGENTS

Tangents are lines that touch a curve in exactly one point and have the same slope as the curve at that point. To find the slope of a curve at a given point and the slope of its tangent line at that point, find the derivative of the function of the curve. If the slope is undefined, the tangent is a vertical line. If the slope is zero, the tangent is a horizontal line.

A line that is normal to a curve at a given point is perpendicular to the tangent at that point. Assuming $f'(x) \neq 0$, the equation for the normal line at point (a, b) is: $y - b = -\frac{1}{f'(a)}(x - a)$. The easiest way to find the slope of the normal is to take the negative reciprocal of the slope of the tangent. If the slope of the tangent is zero, the slope of the normal is undefined. If the slope of the tangent is undefined, the slope of the normal is zero.

ANTIDERIVATIVES (INTEGRALS)

The antiderivative of a function is the function whose first derivative is the original function. Antiderivatives are typically represented by capital letters, while their first derivatives are represented by lower case letters. For example, if $F' = f$, then F is the antiderivative of f. Antiderivatives are also known as indefinite integrals. When taking the derivative of a function, any constant terms in the function are eliminated because their derivative is 0. To account for this possibility, when you take the indefinite integral of a function, you must add an unknown constant C to the end of the function. Because there is no way to know what the value of the original constant was when looking just at the first derivative, the integral is indefinite.

To find the indefinite integral, reverse the process of differentiation. Below are the formulas for constants and powers of x.

$$\int 0\, dx = C$$

$$\int k\, dx = kx + C$$

$$\int x^n\, dx = \frac{x^{n+1}}{n+1} + C, \text{where } n \neq -1$$

Recall that in the differentiation of powers of x, you multiplied the coefficient of the term by the exponent of the variable and then reduced the exponent by one. In integration, the process is reversed: add one to the value of the exponent, and then divide the coefficient of the term by this number to get the integral. Because you do not know the value of any constant term that might have been in the original function, add C to the end of the function once you have completed this process for each term.

Review Video: Indefinite Integrals
Visit mometrix.com/academy and enter code: 541913

Finding the integral of a function is the opposite of finding the derivative of the function. Where possible, you can use the trigonometric or logarithmic differentiation formulas in reverse, and add C to the end to compensate for the unknown term. In instances where a negative sign appears in the differentiation formula, move the negative sign to the opposite side (multiply both sides by -1) to reverse for the integration formula. You should end up with the following formulas:

$$\int \cos x\, dx = \sin x + C$$

$$\int \sec x \tan x\, dx = \sec x + C$$

$$\int \sin x\, dx = -\cos x + C$$

$$\int \csc x \cot x\, dx = -\csc x + C$$

$$\int \sec^2 x\, dx = \tan x + C$$

$$\int \csc^2 x\, dx = -\cot x + C$$

$$\int \frac{1}{x}\, dx = \ln|x| + C$$

$$\int e^x\, dx = e^x + C$$

Calculus

Integration by substitution is the integration version of the chain rule for differentiation. The formula for integration by substitution is given by the equation

$$\int f(g(x))g'(x)dx = \int f(u)du\,; \quad u = g(x) \text{ and } du = g'(x)dx$$

When a function is in a format that is difficult or impossible to integrate using traditional integration methods and formulas due to multiple functions being combined, use the formula shown above to convert the function to a simpler format that can be integrated directly.

Integration by parts is the integration version of the product rule for differentiation. Whenever you are asked to find the integral of the product of two different functions or parts, integration by parts can make the process simpler. Recall for differentiation $(fg)'(x) = f(x)g'(x) + g(x)f'(x)$. This can also be written $\frac{d}{dx}(u \times v) = u\frac{dv}{dx} + v\frac{du}{dx}$, where $u = f(x)$ and $v = g(x)$. Rearranging to integral form gives the formula:

$$\int u\,dv = uv - \int v\,du$$

$$\int f(x)g'(x)\,dx = f(x)g(x) - \int f'(x)g(x)\,dx$$

When using integration by parts, the key is selecting the best functions to substitute for u and v so that you make the integral easier to solve and not harder.

While the indefinite integral has an undefined constant added at the end, the definite integral can be calculated as an exact real number. To find the definite integral of a function over a closed interval, use the formula $\int_n^m f(x)\,dx = F(m) - F(n)$ where F is the integral of f. Because you have been given the boundaries of n and m, no undefined constant C is needed.

Review Video: Integration by Parts
Visit mometrix.com/academy and enter code: 459972

Review Video: Integration by Substitution
Visit mometrix.com/academy and enter code: 740649

First Fundamental Theorem of Calculus

The **first fundamental theorem of calculus** shows that the process of indefinite integration can be reversed by finding the first derivative of the resulting function. It also gives the relationship between differentiation and integration over a closed interval of the function. For example, assuming a function is continuous over the interval $[m, n]$, you can find the definite integral by using the formula

$$\int_m^n f(x)\,dx = F(n) - F(m)$$

To find the **average value** of the function over the given interval, use the formula:

$$\frac{1}{n-m}\int_m^n f(x)\,dx$$

Review Video: First Fundamental Theorem of Calculus
Visit mometrix.com/academy and enter code: 248431

SECOND FUNDAMENTAL THEOREM OF CALCULUS

The **second fundamental theorem of calculus** is related to the first. This theorem states that, assuming the function is continuous over the interval you are considering, taking the derivative of the integral of a function will yield the original function. The general format for this theorem for any point having a domain value equal to c in the given interval is:

$$\frac{d}{dx}\int_c^x f(t)\,dt = f(x)$$

For each of the following **properties of integrals** of function f, the variables m, n, and p represent values in the domain of the given interval of $f(x)$. The function is assumed to be integrable across all relevant intervals.

Swapping the limits of integration:	$\int_m^n f(x)\,dx = -\int_n^m f(x)\,dx$
Function multiplied by a constant:	$\int_m^n kf(x)dx = k\int_m^n f(x)\,dx$
Separating the integral into parts:	$\int_m^n f(x)\,dx = \int_m^p f(x)\,dx + \int_p^n f(x)\,dx$
If the limits of integration are equivalent:	$\int_n^n f(x)\,dx = 0$
If $f(x)$ is an even function and the limits of integration are symmetric:	$\int_{-m}^m f(x)\,dx = 2\int_0^m f(x)\,dx$
If $f(x)$ is an odd function and the limits of integration are symmetric:	$\int_{-m}^m f(x)\,dx = 0$

Review Video: Second Fundamental Theorem of Calculus
Visit mometrix.com/academy and enter code: 524689

MATCHING FUNCTIONS TO DERIVATIVES OR ACCUMULATIONS

DERIVATIVES

We can use what we know about the meaning of a derivative to match the graph of a function with a graph of its derivative. For one thing, we know that where the function has a critical point, the derivative is zero. Therefore, at every x value at which the graph of a function has a maximum or minimum, the derivative must cross the x-*axis*—and conversely, everywhere the graph of the derivative crosses the x axis, the function must have a critical point: either a maximum, a minimum, or an inflection point. If this is still not enough to identify the correct match, we can also use the fact that the sign of the derivative corresponds to whether the function is increasing or decreasing:

everywhere the graph of the derivative is above the x-axis, the function must be increasing (its slope is positive), and everywhere the graph of the derivative is below the x-axis, the function must be decreasing (its slope is negative).

For example, below are graphs of the function and its derivative. The maxima and minima of the function (left) are circled, and the zeroes of the derivative (right) are circled.

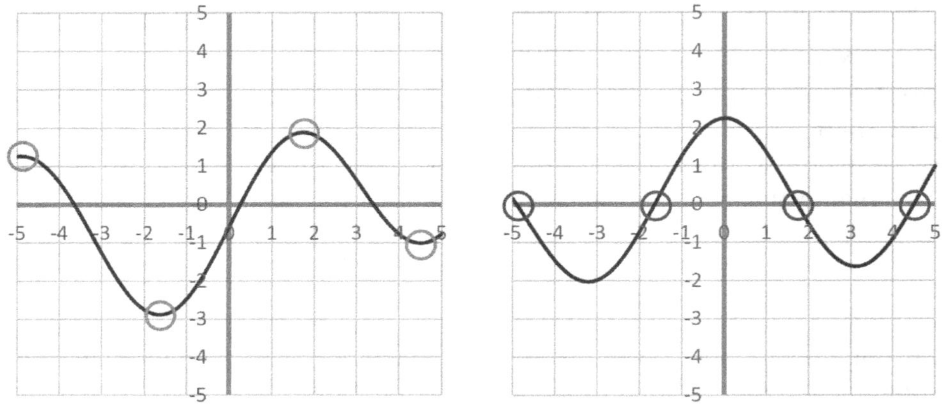

Accumulations

The **accumulation** of a function is another name for its antiderivative, or integral. We can use the relationship between a function and its antiderivative to match the corresponding graphs. For example, we know that where the graph of the function is above the x-axis, the function is positive, thus the accumulation must be increasing (its slope is positive); where the graph of the function is below the x-axis, the accumulation must be decreasing (its slope is negative). It follows that where the function changes from positive to negative—where the graph crosses the x-axis with a negative slope—, its accumulation changes from increasing to decreasing—so the accumulation has a local maximum. Where the function changes from negative to positive—where its graph crosses the x-axis with a positive slope—, the accumulation has a local minimum.

For example, below are graphs of a function and its accumulation. The points on the function (left) where the graph crosses the x-axis are circled; the local minima and maxima of the accumulation (right) are circled.

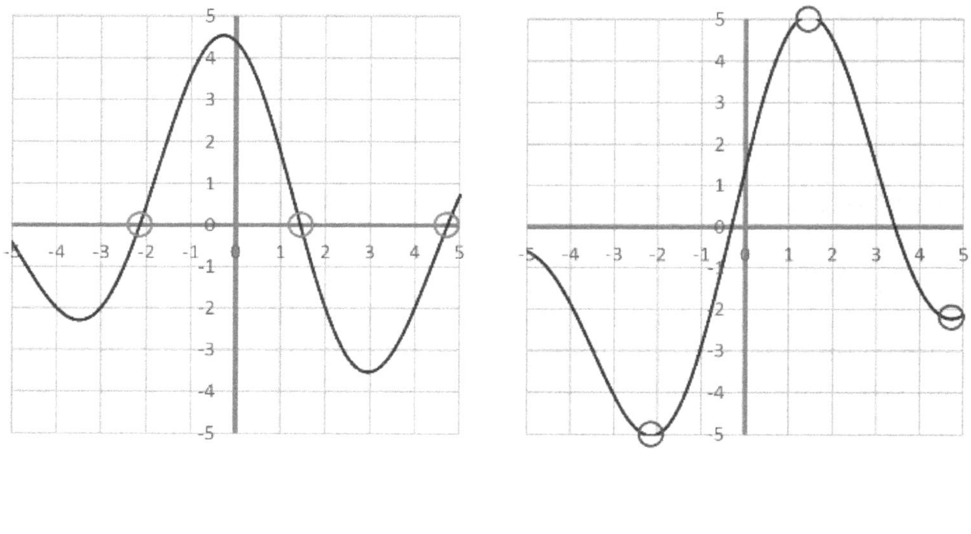

RIEMANN SUMS

A **Riemann sum** is a sum used to approximate the definite integral of a function over a particular interval by dividing the area under the function into vertical rectangular strips and adding the areas of the strips. The height of each strip is equal to the value of the function at some point within the interval covered by the strip. Formally, if we divide the interval over which we are finding the area into n intervals bounded by the $n + 1$ points $\{x_i\}$ (where x_0 and x_n are the left and right bounds of the interval), then the Riemann sum is $\sum_{i=1}^{n} f(x_i^*)\Delta x_i$, where $\Delta x_i = x_i - x_{i-1}$ and x_i^* is some point in the interval $[x_{i-1}, x_i]$. In principle, any point in the interval can be chosen, but common choices include the left endpoint of the interval (yielding the **left Riemann sum**), the right endpoint (yielding the **right Riemann sum**), and the midpoint of the interval (the basis of the **midpoint rule**). Usually, it is convenient to set all the intervals to the same width, although the definition of the Riemann sum does not require this.

The following graphic shows the rectangular strips used for one possible Riemann sum of a particular function:

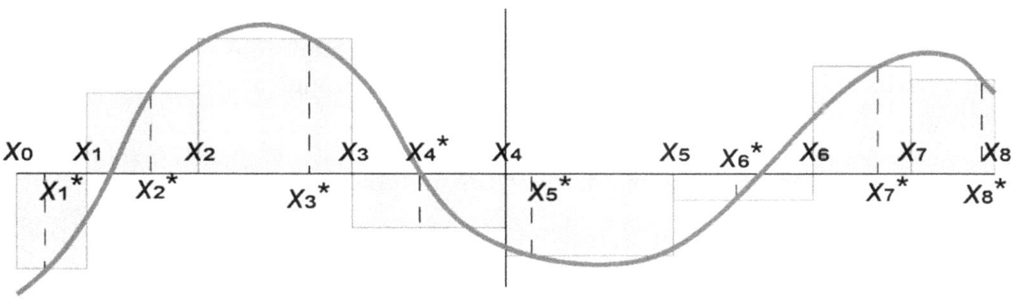

LEFT AND RIGHT RIEMANN SUMS

A **Riemann sum** is an approximation to the definite integral of a function over a particular interval performed by dividing it into smaller intervals and summing the products of the width of each interval and the value of the function evaluated at some point within the interval. The **left Riemann sum** is a Riemann sum in which the function is evaluated at the left endpoint of each interval. In the **right Riemann sum**, the function is evaluated at the right endpoint of each interval.

Review Video: Left and Right Riemann Sums
Visit mometrix.com/academy and enter code: 998275

When the function is increasing, the left Riemann sum will always underestimate the function. This is because we are evaluating the function at the minimum point within each interval; the integral of the function in the interval will be larger than the estimate. Conversely, the right Riemann sum is evaluating the function at the maximum point within each interval, thus it will always overestimate

Calculus

the function. Consider the following diagrams, in which the area under the same increasing function is shown approximated by a left Riemann sum and a right Riemann sum:

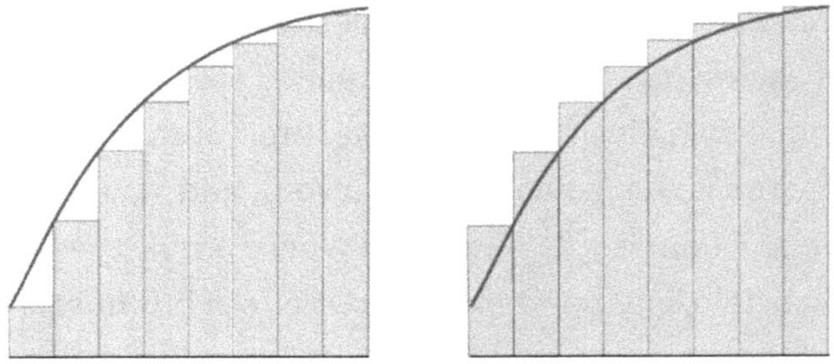

For a decreasing function these considerations are reversed: a left Riemann sum will overestimate the integral, and a right Riemann sum will underestimate it.

MIDPOINT RULE

The **midpoint rule** is a way of approximating the definite integral of a function over an interval by dividing the interval into smaller sub-intervals, multiplying the width of each sub-interval by the value of the function at the midpoint of the sub-interval, and then summing these products. This is a special case of the Riemann sum, specifying the midpoint of the interval as the point at which the function is to be evaluated. The approximation found using the midpoint rule is usually more accurate than that found using the left or right Riemann sum, though as the number of intervals becomes very large the difference becomes negligible.

For example, suppose we are asked to estimate by the midpoint rule the integral of $f(x) = \frac{1}{x}$ in the interval [2, 4]. We can divide this interval into four intervals of width $\frac{1}{2}$: [2, 2.5], [2.5, 3], [3, 3.5], and [3.5, 4]. (The more intervals, the more accurate the estimate, but we'll use a small number of intervals in this example to keep it simple.) The midpoint rule then gives an estimate of $\frac{1}{2}(f(2.25)) + \frac{1}{2}(f(2.75)) + \frac{1}{2}(f(3.25)) + \frac{1}{2}(f(3.75)) = \frac{1}{2}\left(\frac{4}{9}\right) + \frac{1}{2}\left(\frac{4}{11}\right) + \frac{1}{2}\left(\frac{4}{13}\right) + \frac{1}{2}\left(\frac{4}{15}\right) \approx 0.691$, not far from the actual value of $\int_2^4 \frac{1}{x} dx = [\ln x]_2^4 \approx 0.693$.

Review Video: Midpoint Rule
Visit mometrix.com/academy and enter code: 790070

TRAPEZOID RULE

The **trapezoid rule** is a method of approximating the definite integral of a function by dividing the area under the function into a series of trapezoidal strips, the upper corners of the trapezoid touching the function, and adding the areas of the strips. The following diagram shows the use of the trapezoid rule to estimate the integral of the function $y = 2^x$ in the interval $[0, 3]$:

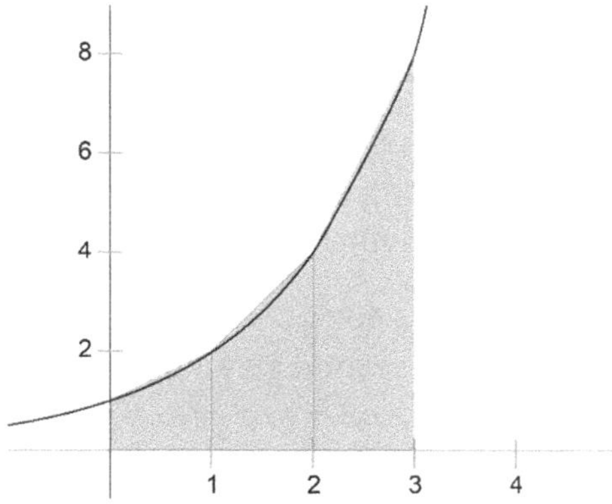

Mathematically, if we define the endpoints of the n subdivisions to be $\{x_{i-1}, x_i\}$: $(1 \leq i \leq n)$, where x_0 and x_n are the endpoints of the entire interval over which we are estimating the integral, then the result of the application of the trapezoid rule is equal to $\sum_{i=1}^{n}\left(\frac{f(x_{i-1})+f(x_i)}{2}\right)(x_i - x_{i-1})$. For the example shown above, that yields $\left(\frac{2^1+2^0}{2}\right)(1-0) + \left(\frac{2^2+2^1}{2}\right)(2-1) + \left(\frac{2^3+2^2}{2}\right)(3-2) = \frac{21}{2}$, or 10.5—not far from the actual value of $\int_0^3 2^x dx = \int_0^3 e^{x \ln 2}\, dx = \left[\frac{2^x}{\ln 2}\right]_0^3 \approx 10.1$. (Of course, we could have achieved more accuracy by using smaller subdivisions.)

The trapezoid rule is related to the Riemann sum, but usually gives more accurate results than the left or right Riemann sum for the same number of intervals. In fact, it isn't hard to prove that the answer given by the trapezoid rule is equal to the average of the left and right Riemann sums using the same partition.

Review Video: Trapezoid Rule
Visit mometrix.com/academy and enter code: 170482

Calculus

LIMIT OF RIEMANN SUMS

As the number of sub-intervals becomes larger, and the width of each sub-interval becomes smaller, the approximation becomes increasingly accurate, and at the limit as the number of sub-intervals approaches infinity and their width approaches zero, the value becomes exact. In fact, the definite integral is often defined as a limit of Riemann sums.

It's possible to find the definite integral by this method. Suppose we want to find the integral of $f(x) = x^2$ over the interval $[0, 2]$. We'll divide this interval into n sub-intervals of equal width and evaluate the function at the right endpoint of each sub-interval. (This choice is arbitrary; at the limit the answer would be the same if we chose the left endpoint, or any other point within the interval.) Our Riemann sum becomes $\sum_{i=1}^{n}\frac{2}{n}\left(\frac{2}{n}i\right)^2 = \frac{8}{n^3}\sum_{i=1}^{n} i^2$, where $\sum_{i=1}^{n} i^2 = \frac{1}{6}n(n+1)(2n+1)$, thus this becomes $\frac{8}{n^3}\cdot\frac{1}{6}n(n+1)(2n+1) = \frac{4}{3}\left(1+\frac{1}{n}\right)\left(2+\frac{1}{n}\right)$. At the limit as $n \to \infty$, this becomes $\frac{4}{3}(1)(2) = \frac{8}{3}$. This is the same result as we get by integrating directly: $\int_0^2 x^2 dx = \left[\frac{1}{3}x^3\right]_0^2 = \frac{1}{3}(2)^3 - \frac{1}{3}(0)^3 = \frac{8}{3}$.

Uses for Integration

Calculating Distances

When given the velocity of an object over time, it's possible to find a distance by integration. The velocity is the rate of change of the position; therefore, the displacement is the accumulation of the velocity: that is, the integral of the velocity is the displacement. However, if asked to find the total distance traveled (as opposed to the displacement), it's important to take the sign into account: we must integrate not just the velocity, but the absolute value of the velocity, which essentially means integrating separately over each interval in which the velocity has a different sign.

For example, suppose we're asked to find the total distance traveled from $t = 0$ to $t = 8$ by an object moving with a velocity in meters per second given by the equation $v(t) = 2\sqrt{t} - t$. This function is zero when $2\sqrt{t} - t = 0 \Rightarrow \sqrt{t}(2 - \sqrt{t}) = 0 \Rightarrow t = 0$ or 4. $v(t)$ is positive when $0 < t < 4$ and negative when $t > 4$. Thus, the distance travelled is

$$\begin{aligned}
\int_0^8 |v(t)|dt &= \int_0^8 \left|2\sqrt{t} - t\right|dt \\
&= \int_0^4 (2\sqrt{t} - t)dt - \int_4^8 (2\sqrt{t} - t)dt \\
&= \left[\frac{4}{3}t^{\frac{3}{2}} - \frac{1}{2}t^2\right]_0^4 - \left[\frac{4}{3}t^{\frac{3}{2}} - \frac{1}{2}t^2\right]_4^8 \\
&= \left(\left[\frac{4}{3}(4)^{\frac{3}{2}} - \frac{1}{2}4^2\right] - \left[\frac{4}{3}(0)^{\frac{3}{2}} - \frac{1}{2}(0)^2\right]\right) - \left(\left[\frac{4}{3}(8)^{\frac{3}{2}} - \frac{1}{2}(8)^2\right] - \left[\frac{4}{3}(4)^{\frac{3}{2}} - \frac{1}{2}(4)^2\right]\right) \\
&= \left(\left[\frac{32}{3} - 8\right] - 0\right) - \left(\left[\frac{64}{3}\sqrt{2} - 32\right] - \left[\frac{32}{3} - 8\right]\right) \\
&= \frac{8}{3} - \frac{64}{3}\sqrt{2} + \frac{96}{3} + \frac{8}{3} \\
&= \frac{8}{3}(1 - 8\sqrt{2} + 12 + 1) \\
&= \frac{8}{3}(14 - 8\sqrt{2}) \approx 7.16 \text{ meters}
\end{aligned}$$

Review Video: Calculating Distances Using Integration
Visit mometrix.com/academy and enter code: 866719

Calculating Areas

One way to calculate the area of an irregular shape is to find a formula for the width of the shape along the x direction as a function of the y coordinate, and then integrate over y, or vice versa. What this amounts to is dividing the area into thin strips and adding the areas of the strips—and then taking the limit as the width of the strips approaches zero.

For example, suppose we want to find the area enclosed by the functions $y_1 = x^2$ and $y_2 = (2 - x^2)$. The height of this enclosure is equal to $y_2 - y_1 = 2 - 2x^2$; we can find the area by integrating this height over x. The two shapes intersect at the points $(1, 1)$ and $(-1, 1)$, thus our limits of integration are -1 and 1. Thus the area can be found as:

$$\int_{-1}^{1}(2-2x^2)dx = \left[2x-\frac{2}{3}x^3\right]_{-1}^{1} = \left(2(1)-\frac{2}{3}(1)\right)-\left(2(-1)-\frac{2}{3}(-1)\right)=\frac{8}{3}$$

Review Video: Calculating Areas Using Integration
Visit mometrix.com/academy and enter code: 118949

Calculating Volumes

One way to calculate the volume of a three-dimensional shape is to find a formula for its cross-sectional area perpendicular to some axis and then integrate over that axis. Effectively, this divides the shape into thin, flat slices and adds the volumes of the slices—and then takes the limit as the thickness of the slices approaches zero.

For example, suppose we want to find the volume of the ellipsoid $4x^2 + 4y^2 + z^2 = 36$. If we take a cross-section perpendicular to the z-axis, this has the formula $4x^2 + 4y^2 = 36 - z^2$, or $x^2 + y^2 = 9 - \frac{z^2}{4}$; this is the formula of a circle with a radius of $\sqrt{9 - \frac{z^2}{4}}$, and thus has an area of $\pi\left(9 - \frac{z^2}{4}\right)$. To find the volume, we integrate this formula over z. The maximum and minimum values of z occur when $x = y = 0$, and then $z^2 = 36$, thus $z = \pm 6$; these are our limits of integration. Thus, the volume is:

$$\begin{aligned}\int_{-6}^{6}\pi\left(9-\frac{z^2}{4}\right)dz &= \pi\left[9z-\frac{z^3}{12}\right]_{-6}^{6}\\ &= \pi\left[\left(9(6)-\frac{6^3}{12}\right)-\left(9(-6)-\frac{(-6)^3}{12}\right)\right]\\ &= \pi[(54-18)-(-54+18)]\\ &= 72\pi \approx 226.2\end{aligned}$$

Review Video: Calculating Volume Using Integration
Visit mometrix.com/academy and enter code: 100341

Calculus

Geometry

Transform passive reading into active learning! After immersing yourself in this chapter, put your comprehension to the test by taking a quiz. The insights you gained will stay with you longer this way. Scan the QR code to go directly to the chapter quiz interface for this study guide. If you're using a computer, simply visit the online resources page at **mometrix.com/resources719/priimath5161** and click the Chapter Quizzes link.

Measurement

Rounding and Estimation

Rounding is reducing the digits in a number while still trying to keep the value similar. The result will be less accurate but in a simpler form and easier to use. Whole numbers can be rounded to the nearest ten, hundred, or thousand, for instance.

To round a number, we make it a little smaller (rounding down) or a little larger (rounding up) to get a number that ends in zeros. We specify the number of zeros by naming the last place that we will not "zero out." For example, to round 8,327 to the nearest hundred, we round down to 8,300, zeroing out every digit to the right of the hundreds place. To round 4,728 to the nearest thousand, we round up to 5,000, increasing the thousands digit by one (to make the number larger) and zeroing out every digit to the right of the thousands place.

We decide whether to round down or up by looking at the first digit we are going to zero out. If it is less than 5 (namely, 0, 1, 2, 3, or 4) we round down. If it is greater than or equal to 5 (namely, 5, 6, 7, 8, or 9) we round up by adding 1 to the place we are rounding to. So, rounding 8,327 to the nearest hundred, we round down to 8,300 because the tens digit, 2, is less than 5. And rounding 4,728 to the nearest thousand, we round up to 5,000, increasing the thousands digit by 1, because the hundreds digit, 7, is greater than or equal to 5.

This even works with decimals. For example, rounding 39.7426 to the nearest tenth, we round down to 39.7000 (or simply 39.7) because the hundredths digit, 4, is less than 5. And rounding 0.019823 to the nearest thousandth, we round up to 0.020000 (or simply 0.02) by increasing the thousandths digit by 1, because the ten-thousandths digit, 8, is greater than or equal to 5.

When you are asked to estimate the solution to a problem, you will need to provide only an approximate figure or **estimation** for your answer. In this situation, you will need to round each number in the calculation to the level indicated (nearest hundred, nearest thousand, etc.) or to a level that makes sense for the numbers involved. When estimating a sum **all numbers must be rounded to the same level**. You cannot round one number to the nearest thousand while rounding another to the nearest hundred.

For instance, suppose you are considering buying four pieces of equipment for your home office. Their prices are \$485, \$1,217, \$750, and \$643. To estimate their total cost, you might round each price to the nearest hundred and add the rounded figures, getting an estimate of \$500 + \$1,200 +

$\$800 + \$600 = \$3{,}100$. By estimating instead of making an exact calculation, you give up a little accuracy to get a simpler calculation.

Review Video: Rounding and Estimation
Visit mometrix.com/academy and enter code: 126243

SCIENTIFIC NOTATION

Scientific notation is a way of writing large numbers in a shorter form. The form $a \times 10^n$ is used in scientific notation, where a is greater than or equal to 1 but less than 10, and n is the number of places the decimal must move to get from the original number to a. Example: The number 230,400,000 is cumbersome to write. To write the value in scientific notation, place a decimal point between the first and second numbers, and include all digits through the last non-zero digit ($a = 2.304$). To find the appropriate power of 10, count the number of places the decimal point had to move ($n = 8$). The number is positive if the decimal moved to the left, and negative if it moved to the right. We can then write 230,400,000 as 2.304×10^8. If we look instead at the number 0.00002304, we have the same value for a, but this time the decimal moved 5 places to the right ($n = -5$). Thus, 0.00002304 can be written as 2.304×10^{-5}. Using this notation makes it simple to compare very large or very small numbers. By comparing exponents, it is easy to see that 3.28×10^4 is smaller than 1.51×10^5, because 4 is less than 5.

Review Video: Scientific Notation
Visit mometrix.com/academy and enter code: 976454

METRIC AND CUSTOMARY MEASUREMENTS

METRIC MEASUREMENT PREFIXES

Giga-	One billion	1 *giga*watt is one billion watts
Mega-	One million	1 *mega*hertz is one million hertz
Kilo-	One thousand	1 *kilo*gram is one thousand grams
Deci-	One-tenth	1 *deci*meter is one-tenth of a meter
Centi-	One-hundredth	1 *centi*meter is one-hundredth of a meter
Milli-	One-thousandth	1 *milli*liter is one-thousandth of a liter
Micro-	One-millionth	1 *micro*gram is one-millionth of a gram

Review Video: How the Metric System Works
Visit mometrix.com/academy and enter code: 163709

MEASUREMENT CONVERSION

When converting between units, the goal is to maintain the same meaning but change the way it is displayed. In order to go from a larger unit to a smaller unit, multiply the number of the known amount by the equivalent amount. When going from a smaller unit to a larger unit, divide the number of the known amount by the equivalent amount.

For complicated conversions, it may be helpful to set up conversion fractions. In these fractions, one fraction is the **conversion factor**. The other fraction has the unknown amount in the numerator. So, the known value is placed in the denominator. Sometimes, the second fraction has the known value from the problem in the numerator and the unknown in the denominator. Multiply the two fractions to get the converted measurement. Note that since the numerator and the denominator of the factor are equivalent, the value of the fraction is 1. That is why we can say that the result in the new units is equal to the result in the old units even though they have different numbers.

It can often be necessary to chain known conversion factors together. As an example, consider converting 512 square inches to square meters. We know that there are 2.54 centimeters in an inch and 100 centimeters in a meter, and we know we will need to square each of these factors to achieve the conversion we are looking for.

$$\frac{512\text{ in}^2}{1} \times \left(\frac{2.54\text{ cm}}{1\text{ in}}\right)^2 \times \left(\frac{1\text{ m}}{100\text{ cm}}\right)^2 = \frac{512\ \cancel{\text{in}^2}}{1} \times \left(\frac{6.4516\ \cancel{\text{cm}^2}}{1\ \cancel{\text{in}^2}}\right) \times \left(\frac{1\text{ m}^2}{10{,}000\ \cancel{\text{cm}^2}}\right) = 0.330\text{ m}^2$$

Review Video: Measurement Conversions
Visit mometrix.com/academy and enter code: 316703

Review Video: Converting Kilograms to Pounds
Visit mometrix.com/academy and enter code: 241463

Common Units and Equivalents

Metric Equivalents

1000 μg (microgram)	1 mg
1000 mg (milligram)	1 g
1000 g (gram)	1 kg
1000 kg (kilogram)	1 metric ton
1000 mL (milliliter)	1 L
1000 μm (micrometer)	1 mm
1000 mm (millimeter)	1 m
100 cm (centimeter)	1 m
1000 m (meter)	1 km

Distance and Area Measurement

Unit	Abbreviation	US equivalent	Metric equivalent
Inch	in	1 inch	2.54 centimeters
Foot	ft	12 inches	0.305 meters
Yard	yd	3 feet	0.914 meters
Mile	mi	5280 feet	1.609 kilometers
Acre	ac	4840 square yards	0.405 hectares
Square Mile	sq. mi. or mi.^2	640 acres	2.590 square kilometers

Capacity Measurements

Unit	Abbreviation	US equivalent	Metric equivalent
Fluid Ounce	fl oz	8 fluid drams	29.573 milliliters
Cup	c	8 fluid ounces	0.237 liter
Pint	pt.	16 fluid ounces	0.473 liter
Quart	qt.	2 pints	0.946 liter
Gallon	gal.	4 quarts	3.785 liters
Teaspoon	t or tsp.	1 fluid dram	5 milliliters
Tablespoon	T or tbsp.	4 fluid drams	15 or 16 milliliters
Cubic Centimeter	cc or cm^3	0.271 drams	1 milliliter

Weight Measurements

Unit	Abbreviation	US equivalent	Metric equivalent
Ounce	oz	16 drams	28.35 grams
Pound	lb	16 ounces	453.6 grams
Ton	tn.	2,000 pounds	907.2 kilograms

Volume and Weight Measurement Clarifications

Always be careful when using ounces and fluid ounces. They are not equivalent.

1 pint = 16 fluid ounces	1 fluid ounce ≠ 1 ounce
1 pound = 16 ounces	1 pint ≠ 1 pound

Having one pint of something does not mean you have one pound of it. In the same way, just because something weighs one pound does not mean that its volume is one pint.

In the United States, the word "ton" by itself refers to a short ton or a net ton. Do not confuse this with a long ton (also called a gross ton) or a metric ton (also spelled *tonne*), which have different measurement equivalents.

$$1 \text{ US ton} = 2000 \text{ pounds} \quad \neq \quad 1 \text{ metric ton} = 1000 \text{ kilograms}$$

Precision, Accuracy, and Error

Measurements of physical quantities (e.g., length, area, volume, weight, mass, and time) in the real world are never perfect. Measurements miss the true value, and repeated measurements produce different values. For this reason, fields that depend on good measurement have technical terms, precision and accuracy, that describe how particular ways of measuring a quantity produce good or bad results. Note that these terms apply not to single measurements but to repeated measurements of the same quantity using the same procedure.

Precision describes the consistency of measurements. Measurements that cluster closely together, with little variation from one measurement to the next, have high precision. Measurements that vary a great deal from one measurement to the next have low precision.

Accuracy describes the closeness of measurements to the true value of the quantity being measured or, more technically, the tendency of measurements to cluster around the true value (though we do not usually know the true value; otherwise, we would not bother to measure it).

An analogy from archery may help. An outstanding archer aiming at the center of the target (the true value) produces a tight cluster of arrows around the center (high accuracy, high precision). Under the same circumstances a modestly good archer produces a loose cluster around the center (high accuracy, low precision). If, however, there is a strong crosswind and the archers do nothing to compensate for it, the outstanding archer's shots will cluster tightly around a point away from the center (low accuracy, high precision); and the modestly good archer's shots will cluster loosely around a point away from the center (low accuracy, low precision).

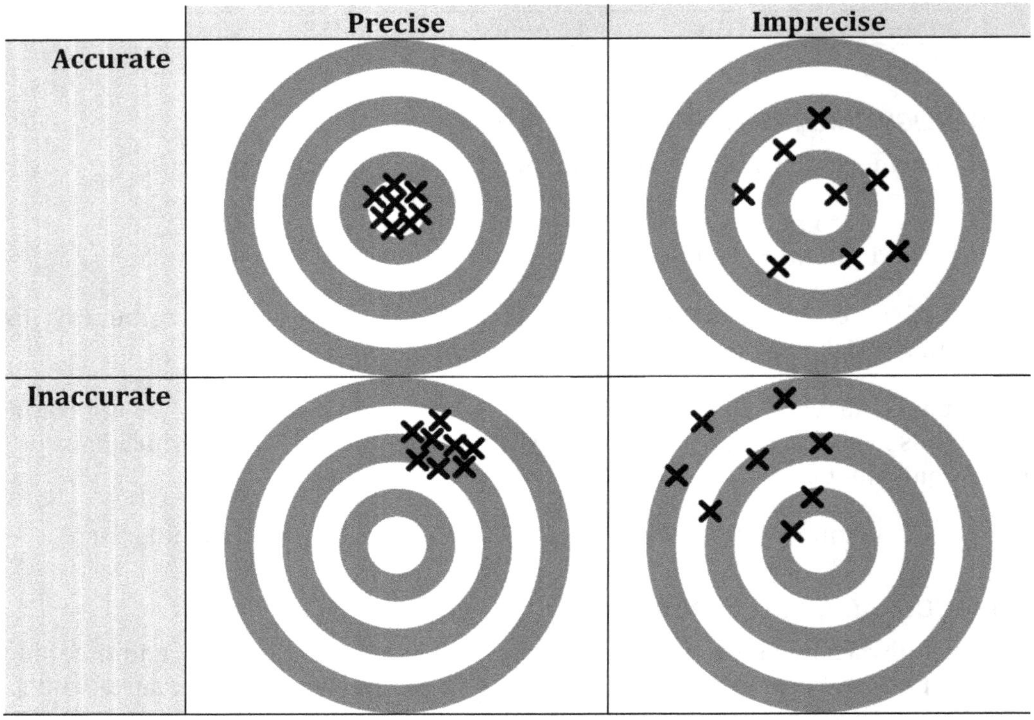

Error in measurement is not a mistake but simply the difference between the true value and the measured value. Since we seldom know the true value, we seldom know the exact error; but sometimes we know a limit on how big the error can be (an error bound), in which case we can use the plus-or-minus sign, $\pm$, to indicate how far the true value might lie from the measured value. For instance, if we measure the length of a metal bar as 25.6 cm and we know the error in our measurement is no more than 0.03 cm (some sources call this **approximate error**), then we might report the length of the bar as 25.6 ± 0.03 cm, which means it lies between 25.57 cm and 25.63 cm.

A particularly important application of this comes from the necessity of rounding whenever we make a measurement. For instance, if we measure the length of a pencil using a ruler marked off in whole inches (no fractions), then we must round our measurement to the nearest inch. Thus, if we report the length as 8 inches, it means that the actual value lies somewhere between 7.5 inches and 8.5 inches, which is to say 8 ± 0.5 inches. This built-in error bound due to rounding is called the **maximum possible error**, and it is always half the magnitude of the smallest unit used in the measurement. Determining the smallest unit can be somewhat subtle, involving the rules for significant figures commonly used in physics, chemistry, and other fields.

Review Video: Precision, Accuracy, and Error
Visit mometrix.com/academy and enter code: 520377

Points, Lines, and Planes

POINTS, LINES, AND PLANES

POINTS AND LINES

A **point** is a fixed location in space, has no size or dimensions, and is commonly represented by a dot. A **line** is a set of points that extends infinitely in two opposite directions. It has length, but no width or depth. A line can be defined by any two distinct points that it contains. A **line segment** is a portion of a line that has definite endpoints. A **ray** is a portion of a line that extends from a single point on that line in one direction along the line. It has a definite beginning, but no ending.

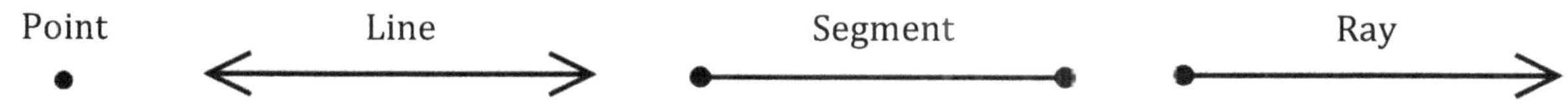

Points are **collinear** if there is a single line that passes through all of them. Otherwise, they are noncollinear. Two points are always collinear since two points define a line. Three points may be noncollinear. For example, the three vertices of a triangle are noncollinear since there is no line that goes through all three of them.

INTERACTIONS BETWEEN LINES

Intersecting lines are lines that have exactly one point in common. **Concurrent lines** are multiple lines that intersect at a single point. **Perpendicular lines** are lines that intersect at right angles. They are represented by the symbol ⊥. The shortest distance from a line to a point not on the line is a perpendicular segment from the point to the line. **Parallel lines** are lines in the same plane that have no points in common and never meet. Two distinct lines in a given plane are always either intersecting or parallel. **Skew lines** are two distinct lines in a three dimensional space that do not intersect and may also not be parallel because there is no single plane that contains them both.

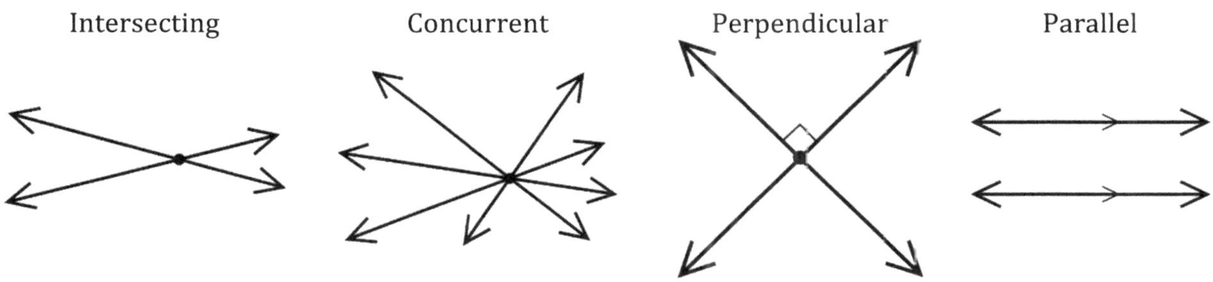

Review Video: Parallel and Perpendicular Lines
Visit mometrix.com/academy and enter code: 815923

A **transversal** is a line that intersects at least two other lines, which may or may not be parallel to one another. A transversal that intersects parallel lines is a common occurrence in geometry. A **bisector** is a line or line segment that divides another line segment into two equal lengths. A

perpendicular bisector of a line segment is composed of points that are equidistant from the endpoints of the segment it is dividing.

Transversal

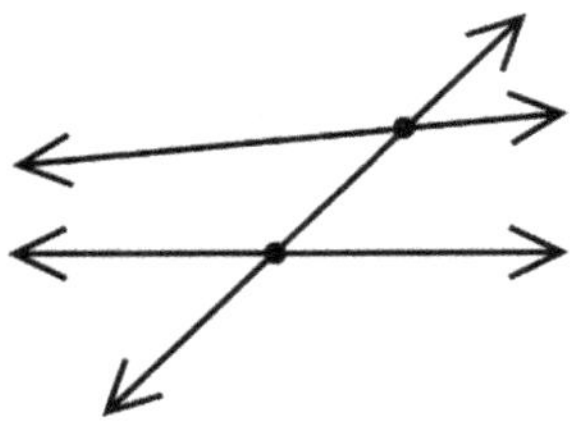

Bisector

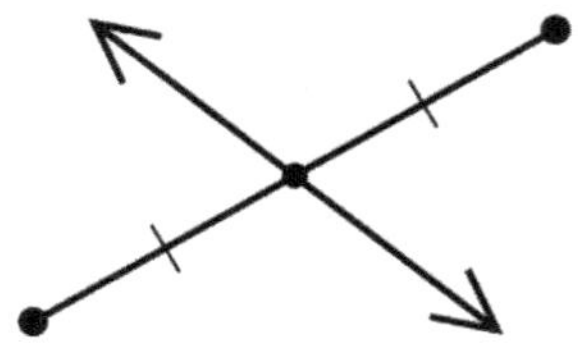

Perpendicular bisector

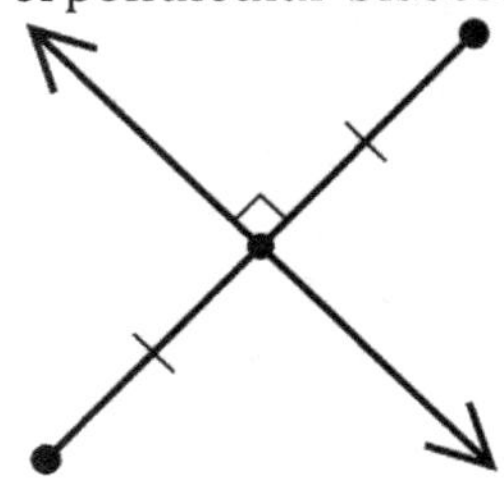

The **projection of a point on a line** is the point at which a perpendicular line drawn from the given point to the given line intersects the line. This is also the shortest distance from the given point to the line. The **projection of a segment on a line** is a segment whose endpoints are the points formed when perpendicular lines are drawn from the endpoints of the given segment to the given line. This is similar to the length a diagonal line appears to be when viewed from above.

Projection of a point on a line

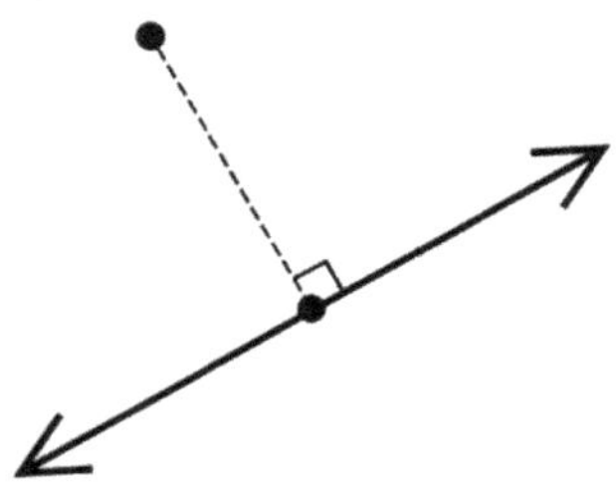

Projection of a segment on a line

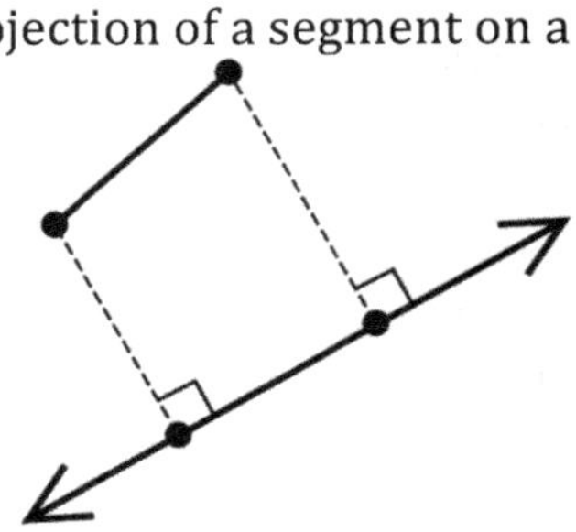

PLANES

A **plane** is a two-dimensional flat surface defined by three non-collinear points. A plane extends an infinite distance in all directions in those two dimensions. It contains an infinite number of points, parallel lines and segments, intersecting lines and segments, as well as parallel or intersecting rays. A plane will never contain a three-dimensional figure or skew lines. Two given planes are either parallel or they intersect at a line. A plane may intersect a circular conic surface to form **conic sections**, such as a parabola, hyperbola, circle or ellipse.

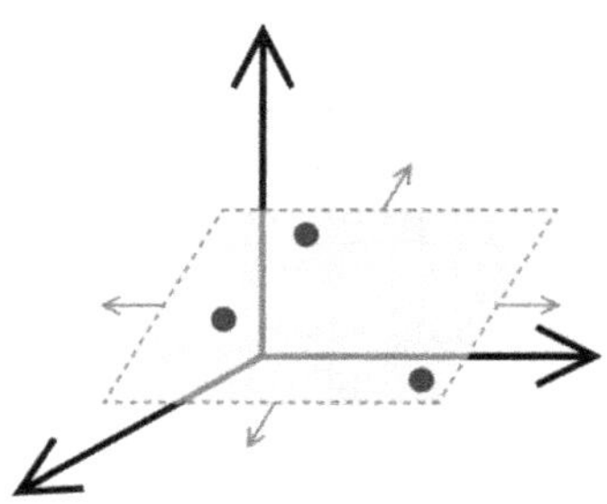

Review Video: Lines and Planes
Visit mometrix.com/academy and enter code: 554267

Angles

ANGLES

ANGLES AND VERTICES

An **angle** is formed when two lines or line segments meet at a common point. It may be a common starting point for a pair of segments or rays, or it may be the intersection of lines. Angles are represented by the symbol $\angle$.

The **vertex** is the point at which two segments or rays meet to form an angle. If the angle is formed by intersecting rays, lines, and/or line segments, the vertex is the point at which four angles are formed. The pairs of angles opposite one another are called vertical angles, and their measures are equal.

- An **acute** angle is an angle with a degree measure less than 90°.
- A **right** angle is an angle with a degree measure of exactly 90°.
- An **obtuse** angle is an angle with a degree measure greater than 90° but less than 180°.
- A **straight angle** is an angle with a degree measure of exactly 180°.
- A **reflex angle** is an angle with a degree measure greater than 180° but less than 360°.
- A **full angle** is an angle with a degree measure of exactly 360°.

Review Video: Angles
Visit mometrix.com/academy and enter code: 264624

RELATIONSHIPS BETWEEN ANGLES

Two angles whose sum is exactly 90° are said to be **complementary**. The two angles may or may not be adjacent. In a right triangle, the two acute angles are complementary.

Two angles whose sum is exactly 180° are said to be **supplementary**. The two angles may or may not be adjacent. Two intersecting lines always form two pairs of supplementary angles. Adjacent supplementary angles will always form a straight line.

Two angles that have the same vertex and share a side are said to be **adjacent**. Vertical angles are not adjacent because they share a vertex but no common side.

Review Video: Adjacent Angles
Visit mometrix.com/academy and enter code: 100375

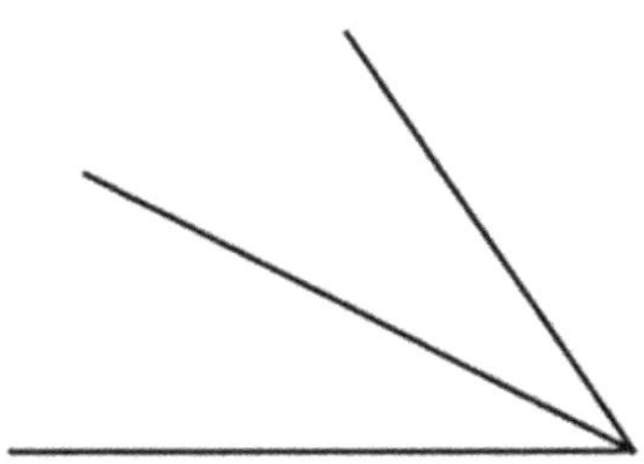

Adjacent
Share vertex and side

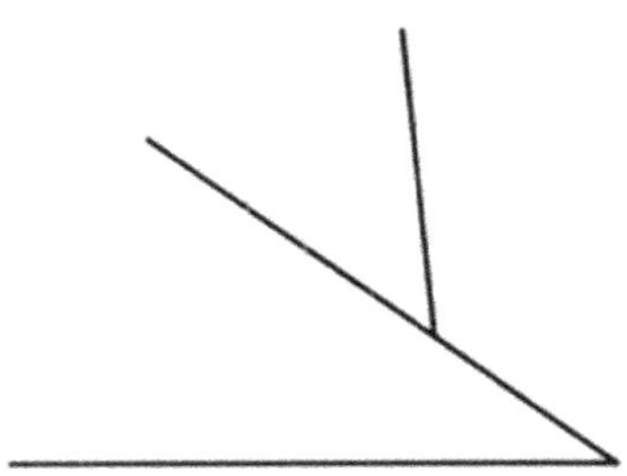

Not adjacent
Share part of a side, but not vertex

Geometry

When two lines are cut by a transversal, the angles that are between the two lines are **interior angles**. In the diagram below, angles 3, 4, 5, and 6 are interior angles.

When two lines are cut by a transversal, the angles that are outside the lines are **exterior angles**. In the diagram below, angles 1, 2, 7, and 8 are exterior angles.

When two lines are cut by a transversal, the angles that are in the same position relative to the transversal and the cut lines are **corresponding angles**. The diagram below has four pairs of corresponding angles: angles 1 and 5, angles 2 and 6, angles 3 and 7, and angles 4 and 8. Corresponding angles formed by parallel lines are congruent.

When two lines are cut by a transversal, the two interior angles that are on opposite sides of the transversal are called **alternate interior angles**. In the diagram below, there are two pairs of alternate interior angles: angles 3 and 6, and angles 4 and 5. Alternate interior angles formed by parallel lines are congruent. Similarly, the two interior angles on the same side of the transversal (angles 3 and 5, and angles 4 and 6) are supplementary when the transversed lines are parallel. Some books call these angles **same side interior angles**.

When two lines are cut by a transversal, the two exterior angles that are on opposite sides of the transversal are called **alternate exterior angles**. In the diagram below, there are two pairs of alternate exterior angles: angles 1 and 8, and angles 2 and 7. Alternate exterior angles formed by parallel lines are congruent. Similarly, the two exterior angles on the same side of the transversal (angles 1 and 7, and angles 2 and 8) are supplementary when the transversed lines are parallel. Some books call these angles **same side exterior angles**.

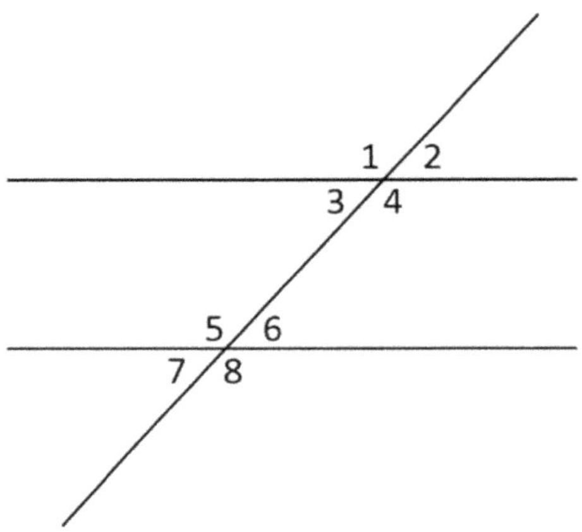

When two lines intersect, four angles are formed. The non-adjacent angles at this vertex are called vertical angles. Vertical angles are congruent. In the diagram, $\angle ABD \cong \angle CBE$ and $\angle ABC \cong \angle DBE$. The other pairs of angles, $(\angle ABC, \angle CBE)$ and $(\angle ABD, \angle DBE)$, are supplementary, meaning the pairs sum to 180°.

Review Video: Congruent Angles
Visit mometrix.com/academy and enter code: 642874

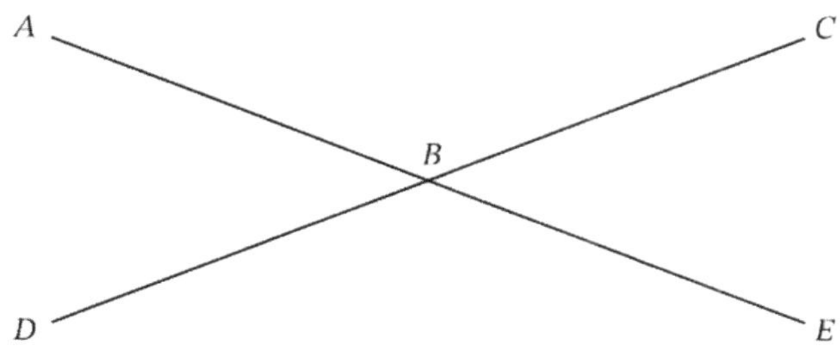

Polygons

Polygons

A **polygon** is a closed, two-dimensional figure with three or more straight line segments called **sides**. The point at which two sides of a polygon intersect is called the **vertex**. In a polygon, the number of sides is always equal to the number of vertices. A polygon with all sides congruent and all angles equal is called a **regular polygon**. Common polygons are:

$$\begin{aligned} \text{Triangle} &= 3 \text{ sides} \\ \text{Quadrilateral} &= 4 \text{ sides} \\ \text{Pentagon} &= 5 \text{ sides} \\ \text{Hexagon} &= 6 \text{ sides} \\ \text{Heptagon} &= 7 \text{ sides} \\ \text{Octagon} &= 8 \text{ sides} \\ \text{Nonagon} &= 9 \text{ sides} \\ \text{Decagon} &= 10 \text{ sides} \\ \text{Dodecagon} &= 12 \text{ sides} \end{aligned}$$

More generally, an n-gon is a polygon that has n angles and n sides.

Review Video: Intro to Polygons
Visit mometrix.com/academy and enter code: 271869

The sum of the interior angles of an n-sided polygon is $(n-2) \times 180°$. For example, in a triangle $n = 3$. So the sum of the interior angles is $(3-2) \times 180° = 180°$. In a quadrilateral, $n = 4$, and the sum of the angles is $(4-2) \times 180° = 360°$.

Review Video: Sum of Interior Angles
Visit mometrix.com/academy and enter code: 984991

Convex and Concave Polygons

A **convex polygon** is a polygon whose diagonals all lie within the interior of the polygon. A **concave polygon** is a polygon with at least one diagonal that is outside the polygon. In the diagram below, quadrilateral $ABCD$ is concave because diagonal $\overline{AC}$ lies outside the polygon and quadrilateral $EFGH$ is convex because both diagonals lie inside the polygon.

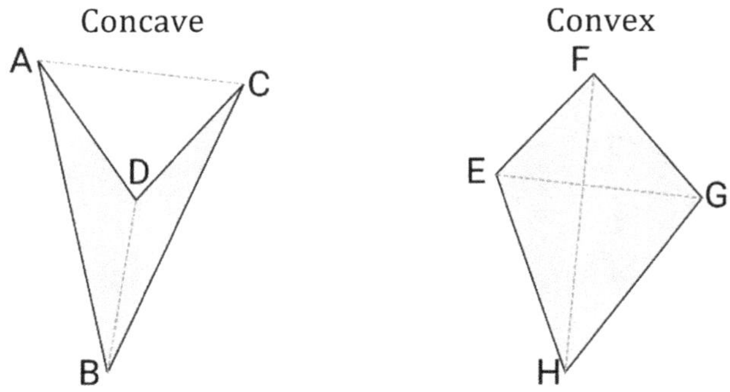

APOTHEM AND RADIUS

A line segment from the center of a regular polygon that is perpendicular to a side of the polygon is called the **apothem**. A line segment from the center of a regular polygon to a vertex of the polygon is called a **radius**. In a regular polygon, the apothem can be used to find the area of the polygon using the formula $A = \frac{1}{2}ap$, where a is the apothem, and p is the perimeter.

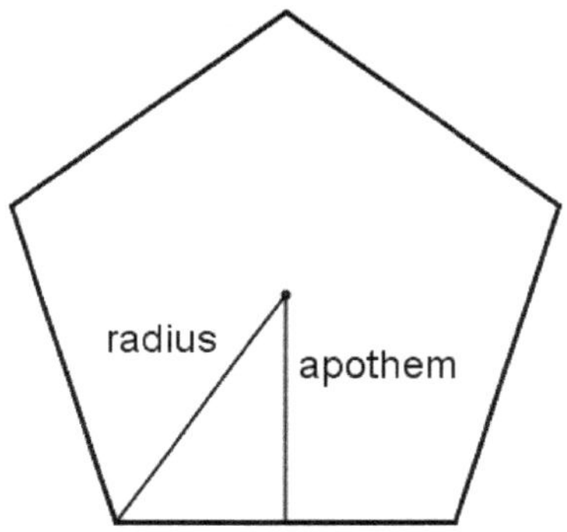

A **diagonal** is a line segment that joins two non-adjacent vertices of a polygon. The number of diagonals a polygon has can be found by using the formula:

$$\text{number of diagonals} = \frac{n(n-3)}{2}$$

Note that n is the number of sides in the polygon. This formula works for all polygons, not just regular polygons.

CONGRUENCE AND SIMILARITY

Congruent figures are geometric figures that have the same size and shape. For congruent polygons all corresponding angle measures are equal, and all corresponding side lengths are equal. Congruence is indicated by the symbol $\cong$. For instance, the expression $ABC \cong DEF$ indicates that the triangles below are congruent. The order of the letters is important, indicating which parts of the polygons correspond to each other. For example, since the letters A and D both come first, $\angle A$ and $\angle D$ have the same measure.

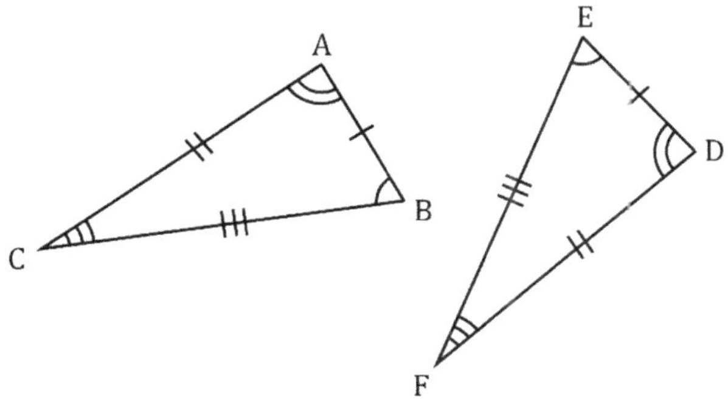

Similar figures are geometric figures that have the same shape, but do not necessarily have the same size. For similar polygons all corresponding angle measures are equal, and all corresponding side lengths are proportional, but they do not have to be equal. It is indicated by the symbol $\sim$. For instance, the expression $ABC \sim DEF$ indicates that the triangles below are similar. Again, the order of the letters indicates which parts of the polygons correspond to each other.

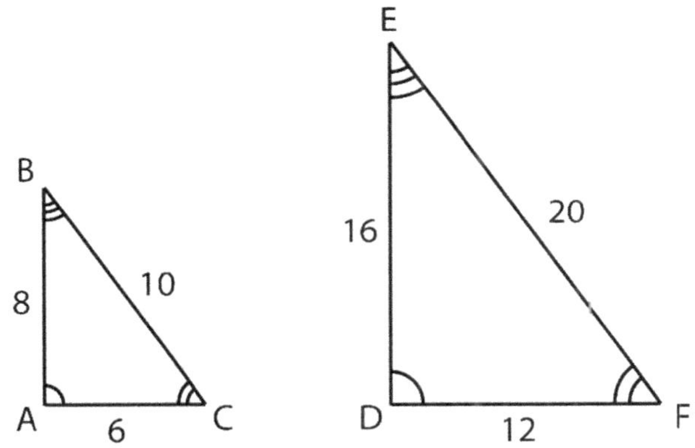

Note that all congruent figures are also similar, but not all similar figures are congruent.

Review Video: Congruent Shapes
Visit mometrix.com/academy and enter code: 492281

LINE OF SYMMETRY

A line that divides a figure or object into congruent parts that are mirror images of each other across the line is called a **line of symmetry**. An object may have no lines of symmetry, one line of symmetry, or multiple (i.e., more than one) lines of symmetry.

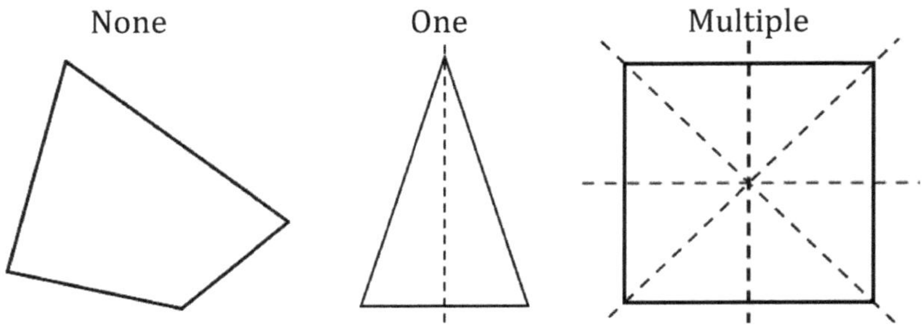

Review Video: Symmetry
Visit mometrix.com/academy and enter code: 528106

Quadrilaterals

QUADRILATERALS

A **quadrilateral** is a closed two-dimensional geometric figure that has four straight sides. The sum of the interior angles of any quadrilateral is 360°.

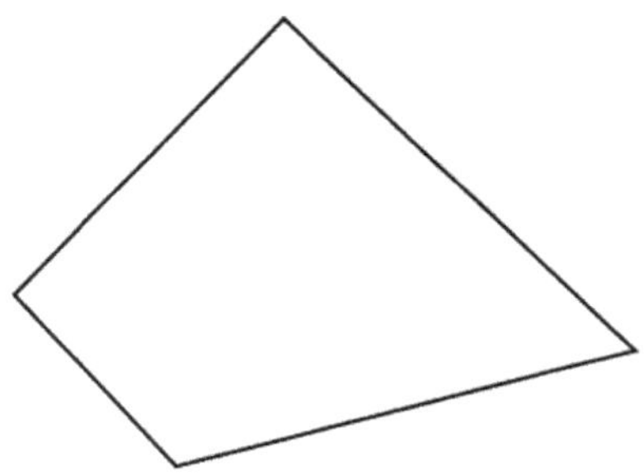

Review Video: Diagonals of Parallelograms, Rectangles, and Rhombi
Visit mometrix.com/academy and enter code: 320040

KITE

A **kite** is a quadrilateral with two pairs of adjacent sides that are congruent. A result of this is perpendicular diagonals. A kite can be concave or convex and has one line of symmetry.

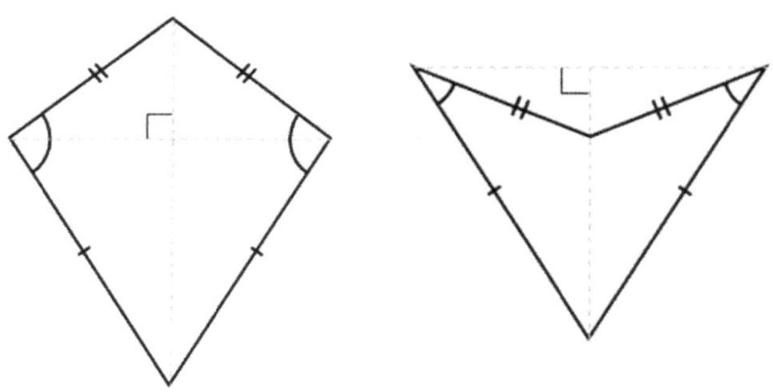

TRAPEZOID

Trapezoid: A trapezoid is defined as a quadrilateral that has at least one pair of parallel sides. There are no rules for the second pair of sides. So, there are no rules for the diagonals and no lines of symmetry for a trapezoid.

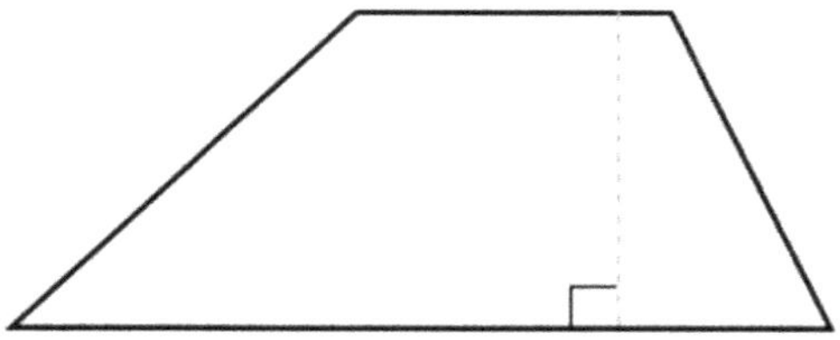

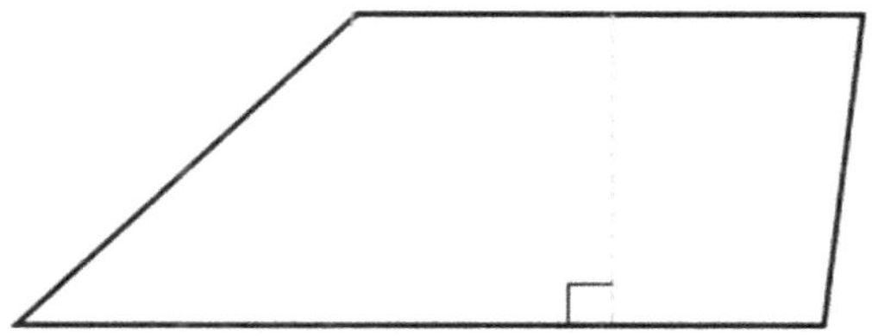

The **area of a trapezoid** is found by the formula $A = \frac{1}{2}h(b_1 + b_2)$, where h is the height (segment joining and perpendicular to the parallel bases), and b_1 and b_2 are the two parallel sides (bases). Do not use one of the other two sides as the height unless that side is also perpendicular to the parallel bases.

The **perimeter of a trapezoid** is found by the formula $P = a + b_1 + c + b_2$, where a, b_1, c, and b_2 are the four sides of the trapezoid.

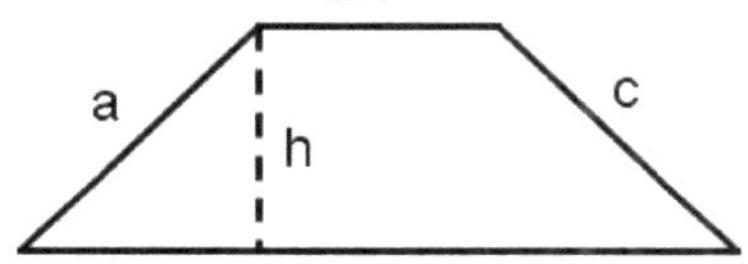

Review Video: Area and Perimeter of a Trapezoid
Visit mometrix.com/academy and enter code: 587523

Isosceles trapezoid: A trapezoid with equal base angles. This gives rise to other properties including: the two nonparallel sides have the same length, the two non-base angles are also equal, and there is one line of symmetry through the midpoints of the parallel sides.

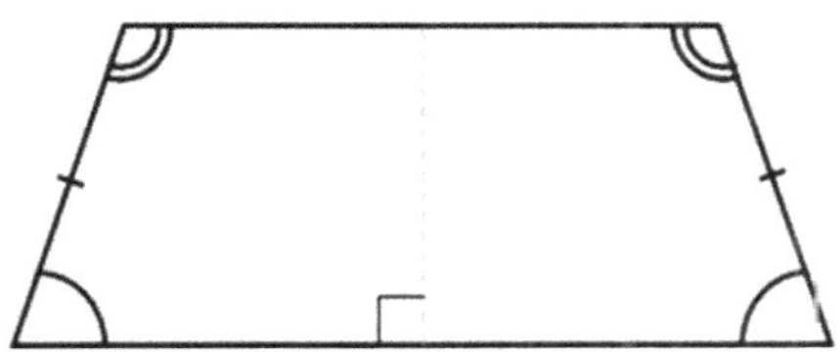

PARALLELOGRAM

A **parallelogram** is a quadrilateral that has two pairs of opposite parallel sides. As such it is a special type of trapezoid. The sides that are parallel are also congruent. The opposite interior angles are always congruent, and the consecutive interior angles are supplementary. The diagonals of a parallelogram divide each other. Each diagonal divides the parallelogram into two congruent triangles. A parallelogram has no line of symmetry, but does have 180-degree rotational symmetry about the midpoint.

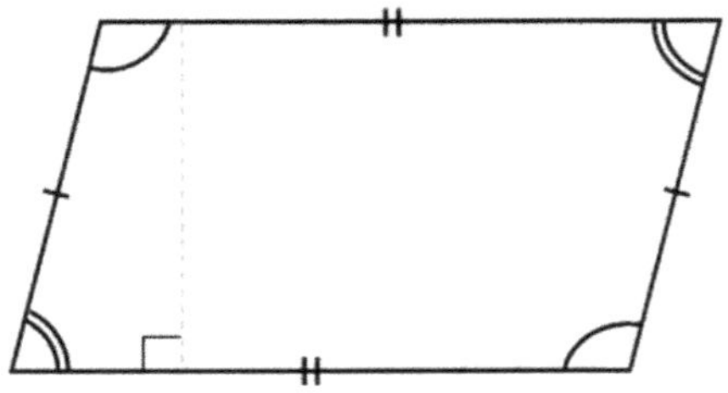

The **area of a parallelogram** is found by the formula $A = bh$, where b is the length of the base, and h is the height. Note that the base and height correspond to the length and width in a rectangle, so this formula would apply to rectangles as well. Do not confuse the height of a parallelogram with the length of the second side. The two are only the same measure in the case of a rectangle.

The **perimeter of a parallelogram** is found by the formula $P = 2a + 2b$ or $P = 2(a + b)$, where a and b are the lengths of the two sides.

Review Video: Area and Perimeter of a Parallelogram
Visit mometrix.com/academy and enter code: 718313

Rectangle

A **rectangle** is a quadrilateral with four right angles. All rectangles are parallelograms and trapezoids, but not all parallelograms or trapezoids are rectangles. The diagonals of a rectangle are congruent. Rectangles have two lines of symmetry (through each pair of opposing midpoints) and 180-degree rotational symmetry about the midpoint.

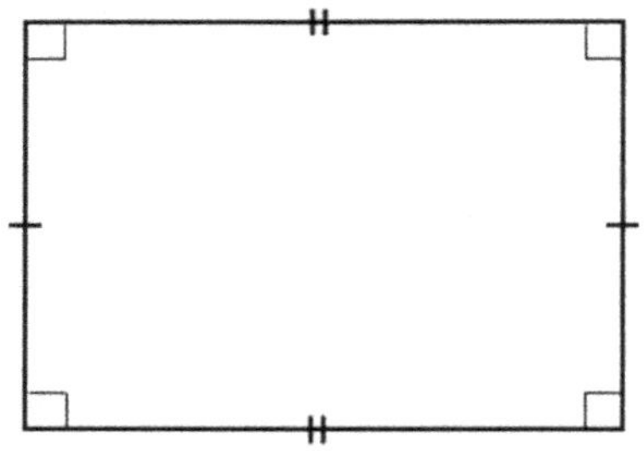

The **area of a rectangle** is found by the formula $A = lw$, where A is the area of the rectangle, l is the length (usually considered to be the longer side) and w is the width (usually considered to be the shorter side). The numbers for l and w are interchangeable.

The **perimeter of a rectangle** is found by the formula $P = 2l + 2w$ or $P = 2(l + w)$, where l is the length, and w is the width. It may be easier to add the length and width first and then double the result, as in the second formula.

Rhombus

A **rhombus** is a quadrilateral with four congruent sides. All rhombuses are parallelograms and kites; thus, they inherit all the properties of both types of quadrilaterals. The diagonals of a rhombus are perpendicular to each other. Rhombi have two lines of symmetry (along each of the diagonals) and 180° rotational symmetry. The **area of a rhombus** is half the product of the diagonals: $A = \frac{d_1 d_2}{2}$ and the perimeter of a rhombus is: $P = 2\sqrt{(d_1)^2 + (d_2)^2}$.

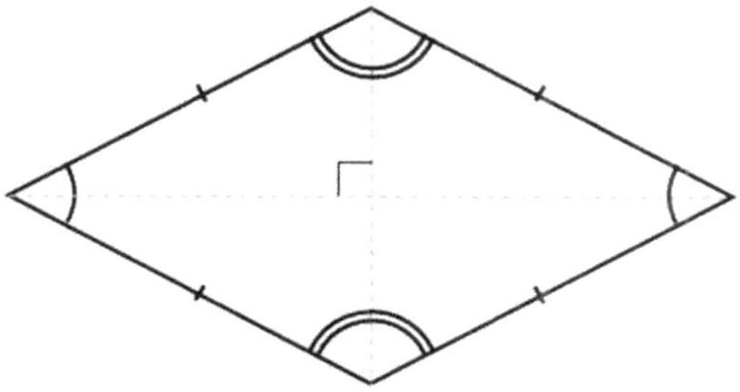

SQUARE

A **square** is a quadrilateral with four right angles and four congruent sides. Squares satisfy the criteria of all other types of quadrilaterals. The diagonals of a square are congruent and perpendicular to each other. Squares have four lines of symmetry (through each pair of opposing midpoints and along each of the diagonals) as well as 90° rotational symmetry about the midpoint.

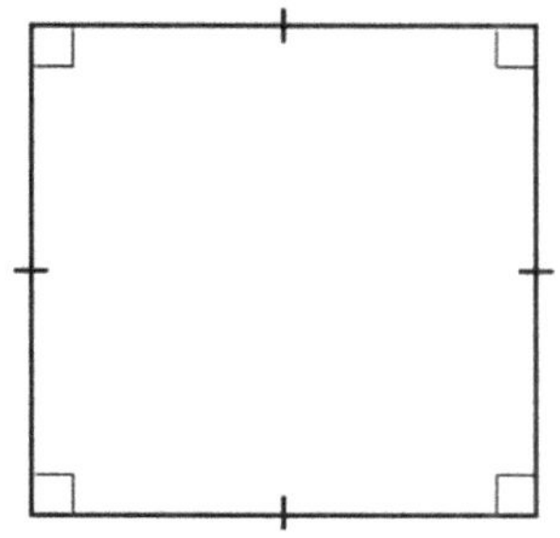

The **area of a square** is found by using the formula $A = s^2$, where s is the length of one side. The **perimeter of a square** is found by using the formula $P = 4s$, where s is the length of one side. Because all four sides are equal in a square, it is faster to multiply the length of one side by 4 than to add the same number four times. You could use the formulas for rectangles and get the same answer.

Review Video: Area and Perimeter of Rectangles and Squares
Visit mometrix.com/academy and enter code: 428109

Geometry

HIERARCHY OF QUADRILATERALS

The hierarchy of quadrilaterals is as follows:

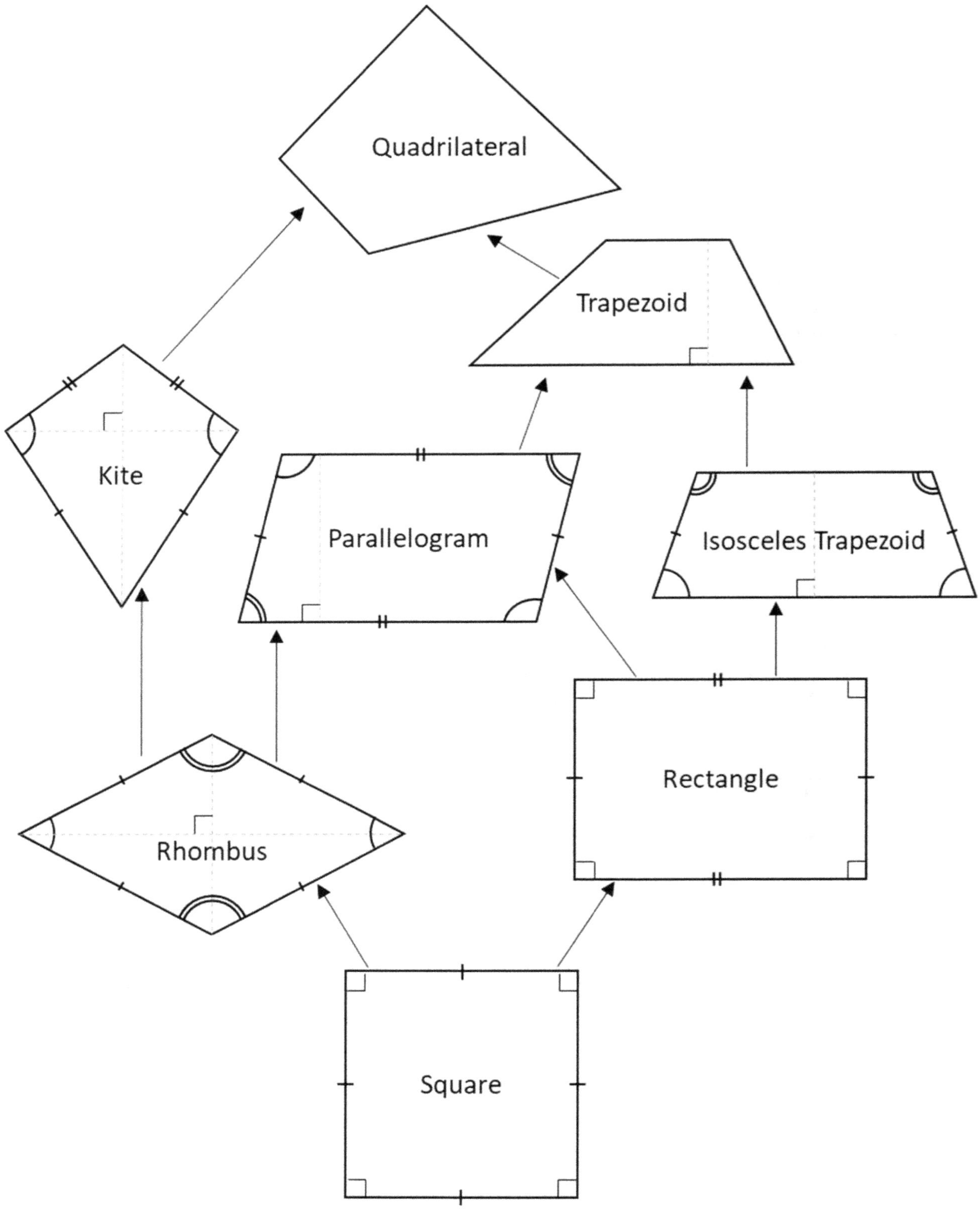

Triangles

TRIANGLES

A triangle is a three-sided figure with the sum of its interior angles being 180°. The **perimeter of any triangle** is found by summing the three side lengths; $P = a + b + c$. For an equilateral triangle, this is the same as $P = 3a$, where a is any side length, since all three sides are the same length.

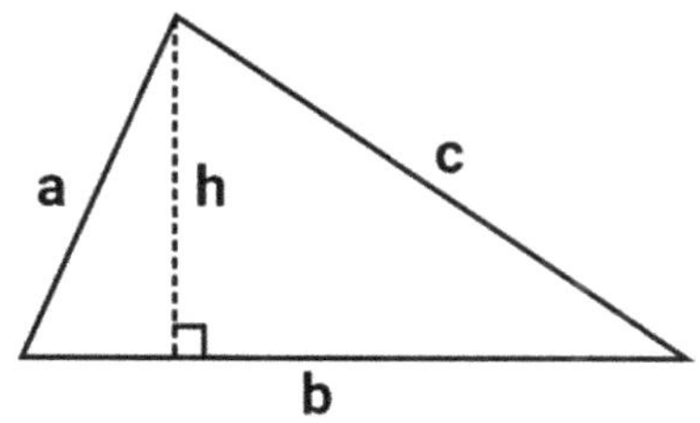

Review Video: Proof that a Triangle is 180 Degrees
Visit mometrix.com/academy and enter code: 687591

Review Video: Area and Perimeter of a Triangle
Visit mometrix.com/academy and enter code: 853779

The **area of any triangle** can be found by taking half the product of one side length referred to as the base, often given the variable b and the perpendicular distance from that side to the opposite vertex called the altitude or height and given the variable h. In equation form that is $A = \frac{1}{2}bh$. Another formula that works for any triangle is $A = \sqrt{s(s-a)(s-b)(s-c)}$, where s is the semiperimeter: $\frac{a+b+c}{2}$, and a, b, and c are the lengths of the three sides. Special cases include isosceles triangles, $A = \frac{1}{2}b\sqrt{a^2 - \frac{b^2}{4}}$, where b is the unique side and a is the length of one of the two congruent sides, and equilateral triangles, $A = \frac{\sqrt{3}}{4}a^2$, where a is the length of a side.

Review Video: Area of Any Triangle
Visit mometrix.com/academy and enter code: 138510

PARTS OF A TRIANGLE

An **altitude** of a triangle is a line segment drawn from one vertex perpendicular to the opposite side. In the diagram that follows, $\overline{BE}$, $\overline{AD}$, and $\overline{CF}$ are altitudes. The length of an altitude is also called the height of the triangle. The three altitudes in a triangle are always concurrent. The point of concurrency of the altitudes of a triangle, O, is called the **orthocenter**. Note that in an obtuse triangle, the orthocenter will be outside the triangle, and in a right triangle, the orthocenter is the vertex of the right angle.

A **median** of a triangle is a line segment drawn from one vertex to the midpoint of the opposite side. In the diagram that follows, $\overline{BH}$, $\overline{AG}$, and $\overline{CI}$ are medians. This is not the same as the altitude, except the altitude to the base of an isosceles triangle and all three altitudes of an equilateral triangle. The point of concurrency of the medians of a triangle, T, is called the **centroid**. This is the same point as the orthocenter only in an equilateral triangle. Unlike the orthocenter, the centroid is always inside the triangle. The centroid can also be considered the exact center of the triangle. Any shape triangle can be perfectly balanced on a tip placed at the centroid. The centroid is also the point that is two-thirds the distance from the vertex to the opposite side.

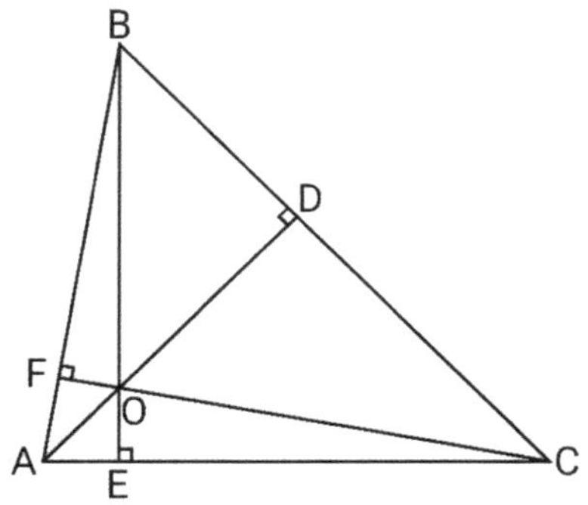

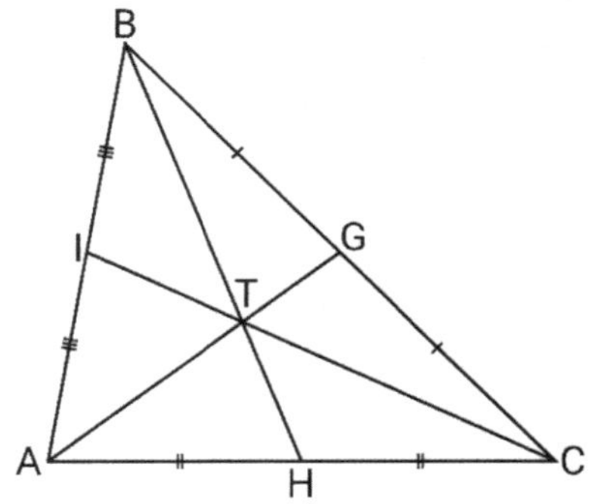

Review Video: Centroid, Incenter, Circumcenter, and Orthocenter
Visit mometrix.com/academy and enter code: 598260

TRIANGLE PROPERTIES

CLASSIFICATIONS OF TRIANGLES

A **scalene triangle** is a triangle with no congruent sides. A scalene triangle will also have three angles of different measures. The angle with the largest measure is opposite the longest side, and the angle with the smallest measure is opposite the shortest side. An **acute triangle** is a triangle whose three angles are all less than 90°. If two of the angles are equal, the acute triangle is also an **isosceles triangle**. An isosceles triangle will also have two congruent angles opposite the two congruent sides. If the three angles are all equal, the acute triangle is also an **equilateral triangle**. An equilateral triangle will also have three congruent angles, each 60°. All equilateral triangles are also acute triangles. An **obtuse triangle** is a triangle with exactly one angle greater than 90°. The other two angles may or may not be equal. If the two remaining angles are equal, the obtuse triangle is also an isosceles triangle. A **right triangle** is a triangle with exactly one angle equal to 90°. All right triangles follow the Pythagorean theorem. A right triangle can never be acute or obtuse.

The table below illustrates how each descriptor places a different restriction on the triangle:

Angles / **Sides**	**Acute:** **All angles < 90°**	**Obtuse:** **One angle > 90°**	**Right:** **One angle = 90°**
Scalene: **No equal side lengths**	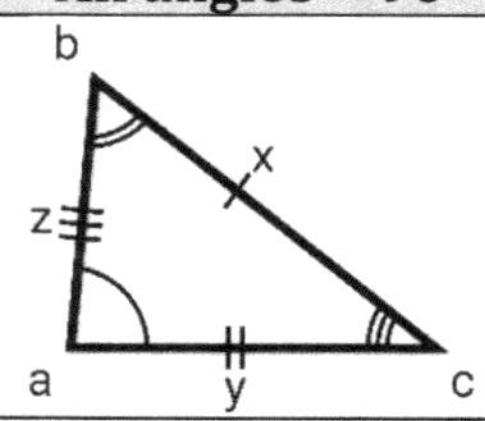	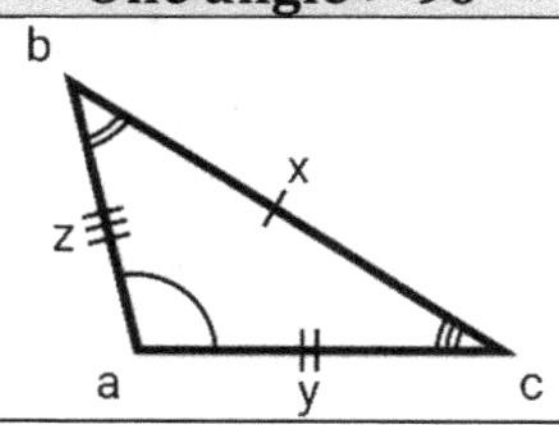	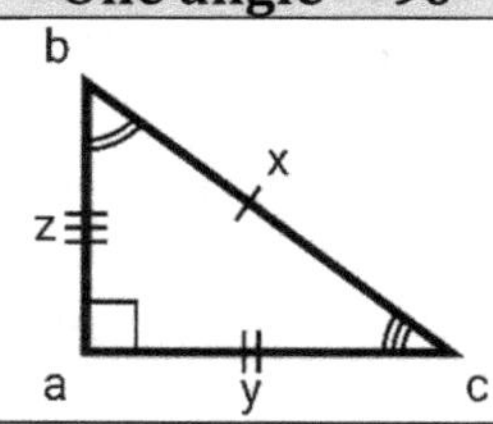
	$90° > \angle a > \angle b > \angle c$ $x > y > z$	$\angle a > 90° > \angle b > \angle c$ $x > y > z$	$90° = \angle a > \angle b > \angle c$ $x > y > z$

Angles / Sides	Acute: All angles < 90°	Obtuse: One angle > 90°	Right: One angle = 90°
Isosceles: Two equal side lengths	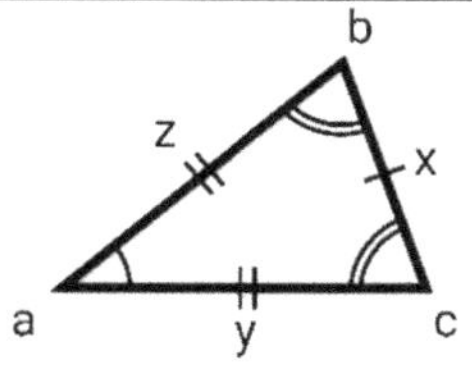	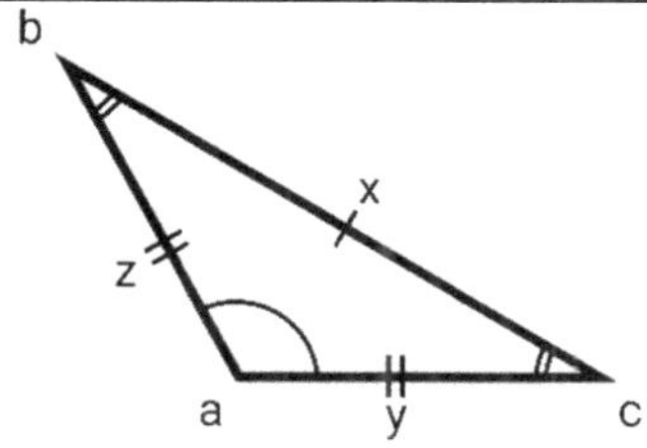	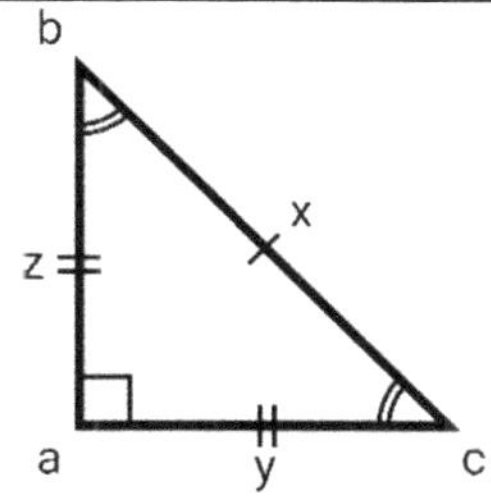
	$90° > \angle a, \angle b, or\ \angle c$ $\angle b = \angle c, \quad y = z$	$\angle a > 90° > \angle b = \angle c$ $x > y = z$	$\angle a = 90°$ $\angle b = \angle c = 45°$ $x > y = z$
Equilateral: Three equal side lengths	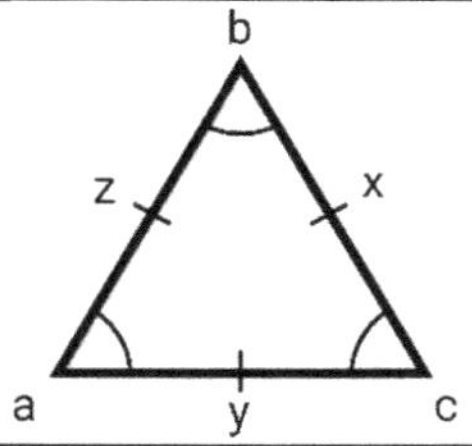		
	$60° = \angle a = \angle b = \angle c$ $x = y = z$		

Review Video: Introduction to Types of Triangles
Visit mometrix.com/academy and enter code: 511711

General Rules for Triangles

The **triangle inequality theorem** states that the sum of the measures of any two sides of a triangle is always greater than the measure of the third side. If the sum of the measures of two sides were equal to the third side, a triangle would be impossible because the two sides would lie flat across the third side and there would be no vertex. If the sum of the measures of two of the sides was less than the third side, a closed figure would be impossible because the two shortest sides would never meet. In other words, for a triangle with sides lengths A, B, and C: $A + B > C$, $B + C > A$, and $A + C > B$.

The sum of the measures of the interior angles of a triangle is always 180°. Therefore, a triangle can never have more than one angle greater than or equal to 90°.

In any triangle, the angles opposite congruent sides are congruent, and the sides opposite congruent angles are congruent. The largest angle is always opposite the longest side, and the smallest angle is always opposite the shortest side.

The line segment that joins the midpoints of any two sides of a triangle is always parallel to the third side and exactly half the length of the third side.

Review Video: General Rules (Triangle Inequality Theorem)
Visit mometrix.com/academy and enter code: 166488

Similarity and Congruence Rules

Similar triangles are triangles whose corresponding angles are equal and whose corresponding sides are proportional. Represented by AAA. Similar triangles whose corresponding sides are congruent are also congruent triangles.

Triangles can be shown to be **congruent** in 5 ways:

- **SSS**: Three sides of one triangle are congruent to the three corresponding sides of the second triangle.
- **SAS**: Two sides and the included angle (the angle formed by those two sides) of one triangle are congruent to the corresponding two sides and included angle of the second triangle.
- **ASA**: Two angles and the included side (the side that joins the two angles) of one triangle are congruent to the corresponding two angles and included side of the second triangle.
- **AAS**: Two angles and a non-included side of one triangle are congruent to the corresponding two angles and non-included side of the second triangle.
- **HL**: The hypotenuse and leg of one right triangle are congruent to the corresponding hypotenuse and leg of the second right triangle.

Review Video: Similar Triangles
Visit mometrix.com/academy and enter code: 398538

Circles

Circles

The **center** of a circle is the single point from which every point on the circle is **equidistant**. The **radius** is a line segment that joins the center of the circle and any one point on the circle. All radii of a circle are equal. Circles that have the same center but not the same length of radii are **concentric**. The **diameter** is a line segment that passes through the center of the circle and has both endpoints on the circle. The length of the diameter is exactly twice the length of the radius. Point O in the diagram below is the center of the circle, segments $\overline{OX}$, $\overline{OY}$, and $\overline{OZ}$ are radii; and segment $\overline{XZ}$ is a diameter.

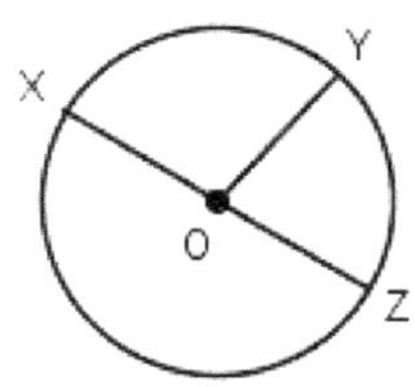

Review Video: Points of a Circle
Visit mometrix.com/academy and enter code: 420746

Review Video: Diameter, Radius, and Circumference
Visit mometrix.com/academy and enter code: 448988

The **area of a circle** is found by the formula $A = \pi r^2$, where r is the length of the radius. If the diameter of the circle is given, remember to divide it in half to get the length of the radius before proceeding.

The **circumference** of a circle is found by the formula $C = 2\pi r$, where r is the radius. Again, remember to convert the diameter if you are given that measure rather than the radius.

Review Video: Area and Circumference of a Circle
Visit mometrix.com/academy and enter code: 243015

Inscribed and Circumscribed Figures

These terms can both be used to describe a given arrangement of figures, depending on perspective. If each of the vertices of figure A lie on figure B, then it can be said that figure A is **inscribed** in figure B, but it can also be said that figure B is **circumscribed** about figure A. The following table and examples help to illustrate the concept. Note that the figures cannot both be circles, as they would be completely overlapping and neither would be inscribed or circumscribed.

Given	Description	Equivalent Description	Figures
Each of the sides of a pentagon is tangent to a circle	The circle is inscribed in the pentagon	The pentagon is circumscribed about the circle	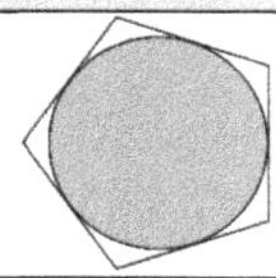
Each of the vertices of a pentagon lie on a circle	The pentagon is inscribed in the circle	The circle is circumscribed about the pentagon	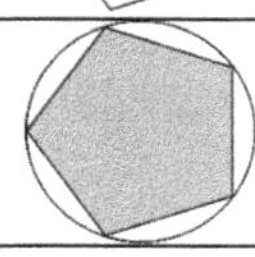

Circle Properties

Arcs

An **arc** is a portion of a circle. Specifically, an arc is the set of points between and including two points on a circle. An arc does not contain any points inside the circle. When a segment is drawn from the endpoints of an arc to the center of the circle, a sector is formed. A **minor arc** is an arc that has a measure less than 180°. A **major arc** is an arc that has a measure of at least 180°. Every minor arc has a corresponding major arc that can be found by subtracting the measure of the minor arc from 360°. A **semicircle** is an arc whose endpoints are the endpoints of the diameter of a circle. A semicircle is exactly half of a circle.

Arc length is the length of that portion of the circumference between two points on the circle. The formula for arc length is $s = \frac{\pi r\theta}{180°}$, where s is the arc length, r is the length of the radius, and θ is the angular measure of the arc in degrees, or $s = r\theta$, where θ is the angular measure of the arc in radians (2π radians $= 360$ degrees).

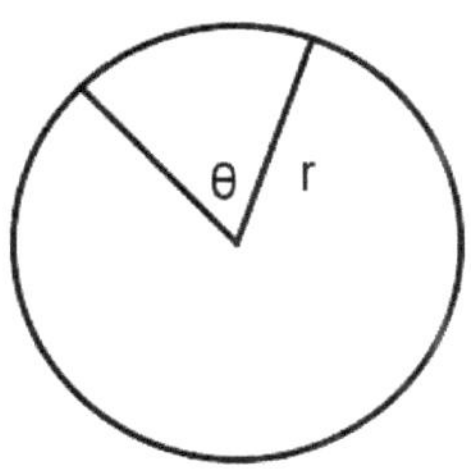

ANGLES OF CIRCLES

A **central angle** is an angle whose vertex is the center of a circle and whose legs intercept an arc of the circle. The measure of a central angle is equal to the measure of the minor arc it intercepts.

An **inscribed angle** is an angle whose vertex lies on a circle and whose legs contain chords of that circle. The portion of the circle intercepted by the legs of the angle is called the intercepted arc. The measure of the intercepted arc is exactly twice the measure of the inscribed angle. In the following diagram, angle ABC is an inscribed angle. $\widehat{AC} = 2(\text{m}\angle ABC)$.

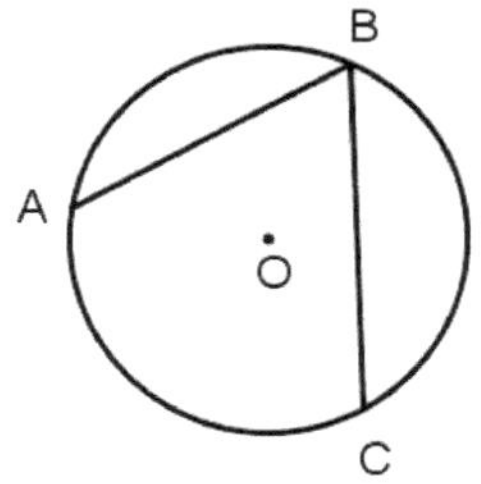

Any angle inscribed in a semicircle is a right angle. The intercepted arc is 180°, making the inscribed angle half that, or 90°. In the diagram below, angle ABC is inscribed in semicircle ABC, making angle ABC equal to 90°.

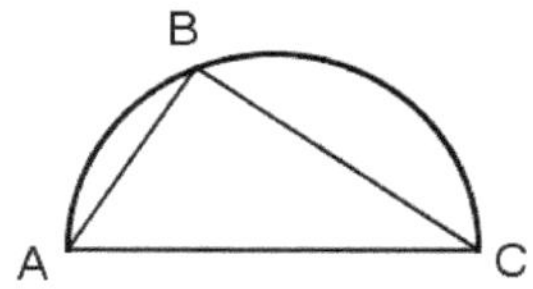

Review Video: Arcs and Angles of Circles
Visit mometrix.com/academy and enter code: 652838

SECANTS, CHORDS, AND TANGENTS

A **secant** is a line that intersects a circle in two points. The segment of a secant line that is contained within the circle is called a **chord**. Two secants may intersect inside the circle, on the circle, or outside the circle. When the two secants intersect on the circle, an inscribed angle is formed. When two secants intersect inside a circle, the measure of each of two vertical angles is equal to half the sum of the two intercepted arcs. Consider the following diagram where $\text{m}\angle AEB = \frac{1}{2}(\widehat{AB} + \widehat{CD})$ and $\text{m}\angle BEC = \frac{1}{2}(\widehat{BC} + \widehat{AD})$.

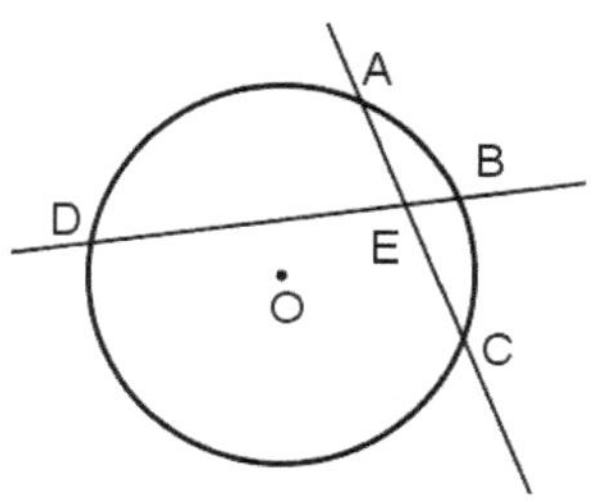

When two secants intersect outside a circle, the measure of the angle formed is equal to half the difference of the two arcs that lie between the two secants. In the diagram below, $m\angle AEB = \frac{1}{2}(\widehat{AB} - \widehat{CD})$.

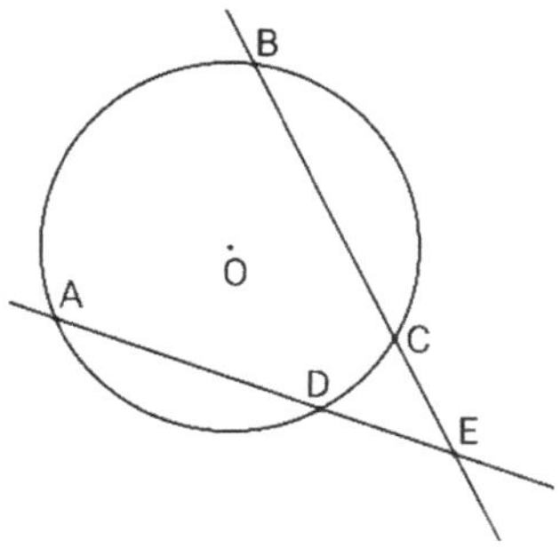

A **tangent** is a line in the same plane as a circle that touches the circle in exactly one point. The point at which a tangent touches a circle is called the **point of tangency**. While a line segment can be tangent to a circle as part of a line that is tangent, it is improper to say a tangent can be simply a line segment that touches the circle in exactly one point.

In the diagram below, $\overleftrightarrow{EB}$ is a secant and contains chord $\overline{EB}$, and $\overleftrightarrow{CD}$ is tangent to circle A. Notice that $\overline{FB}$ is not tangent to the circle. $\overline{FB}$ is a line segment that touches the circle in exactly one point, but if the segment were extended, it would touch the circle in a second point. In the diagram below, point B is the point of tangency.

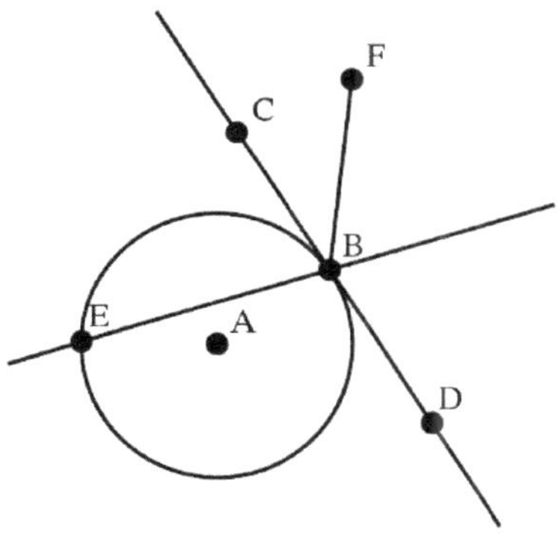

Review Video: Secants, Chords, and Tangents
Visit mometrix.com/academy and enter code: 258360

Review Video: Tangent Lines of a Circle
Visit mometrix.com/academy and enter code: 780167

SECTORS

A **sector** is the portion of a circle formed by two radii and their intercepted arc. While the arc length is exclusively the points that are also on the circumference of the circle, the sector is the entire area bounded by the arc and the two radii.

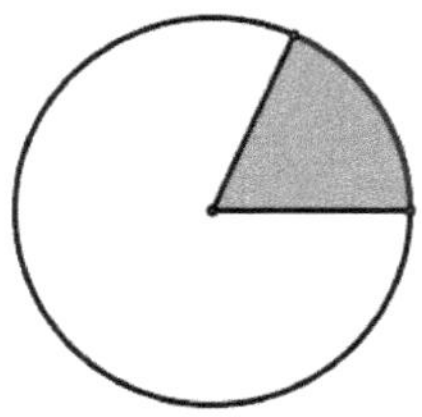

The **area of a sector** of a circle is found by the formula, $A = \frac{\theta r^2}{2}$, where A is the area, θ is the measure of the central angle in radians, and r is the radius. To find the area with the central angle in degrees, use the formula, $A = \frac{\theta \pi r^2}{360}$, where θ is the measure of the central angle and r is the radius.

Transformations

TRANSFORMATIONS

ROTATION

A **rotation** is a transformation that turns a figure around a point called the **center of rotation**, which can lie anywhere in the plane. If a line is drawn from a point on a figure to the center of rotation, and another line is drawn from the center to the rotated image of that point, the angle between the two lines is the **angle of rotation**. The vertex of the angle of rotation is the center of rotation.

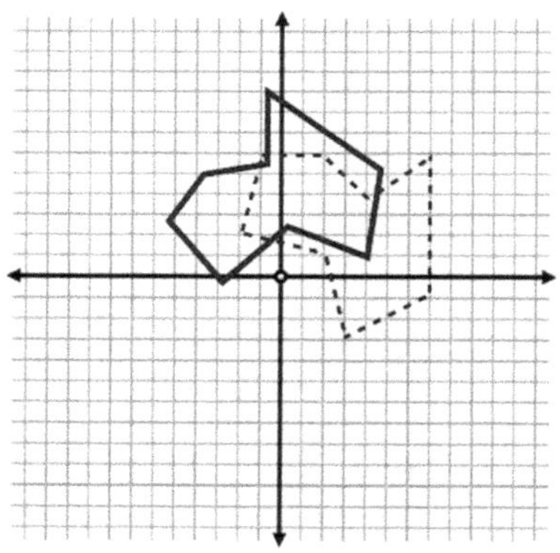

Review Video: Rotation
Visit mometrix.com/academy and enter code: 602600

TRANSLATION AND DILATION

A **translation** is a transformation which slides a figure from one position in the plane to another position in the plane. The original figure and the translated figure have the same size, shape, and orientation. A **dilation** is a transformation which proportionally stretches or shrinks a figure by a

scale factor. The dilated image is the same shape and orientation as the original image but a different size. A polygon and its dilated image are similar.

Translation

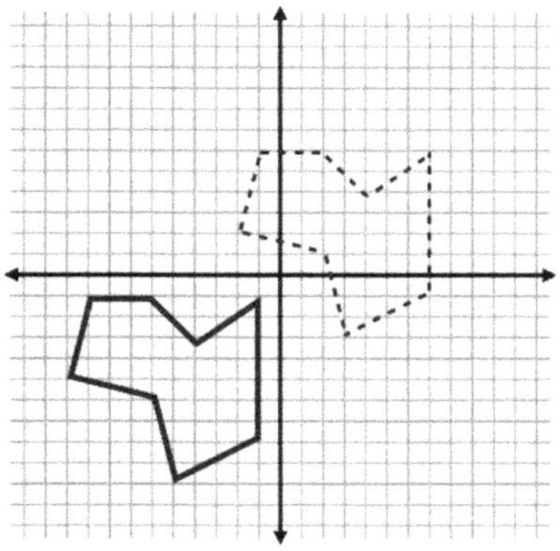

Dilation

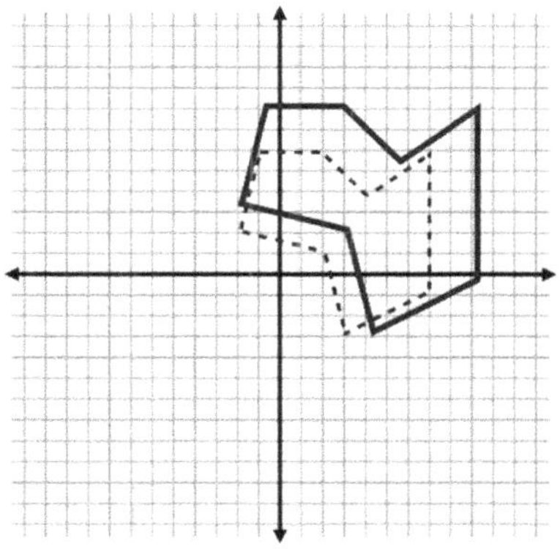

Review Video: Translation
Visit mometrix.com/academy and enter code: 718628

Review Video: Dilation
Visit mometrix.com/academy and enter code: 471630

A **reflection of a figure over a line** (a "flip") creates a congruent image that is the same distance from the line as the original figure but on the opposite side. The **line of reflection** is the perpendicular bisector of any line segment drawn from a point on the original figure to its reflected image (unless the point and its reflected image happen to be the same point, which happens when a figure is reflected over one of its own sides). A **reflection of a figure over a point** (an inversion) in two dimensions is the same as the rotation of the figure 180° about that point. The image of the figure is congruent to the original figure. The **point of reflection** is the midpoint of a line segment which connects a point in the figure to its image (unless the point and its reflected image happen to be the same point, which happens when a figure is reflected in one of its own points).

Reflection of a figure over a line

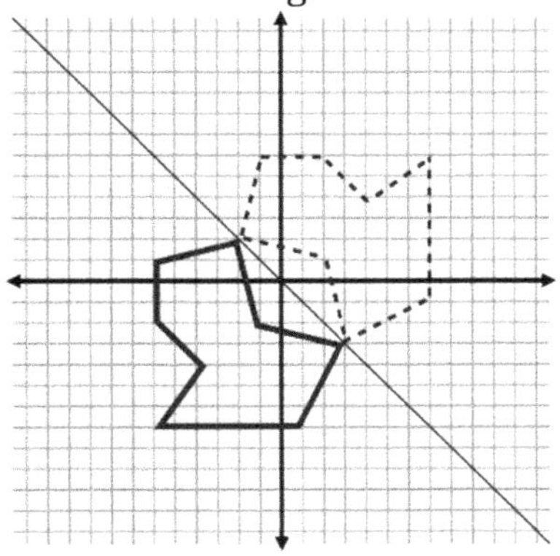

Reflection of a figure over a point

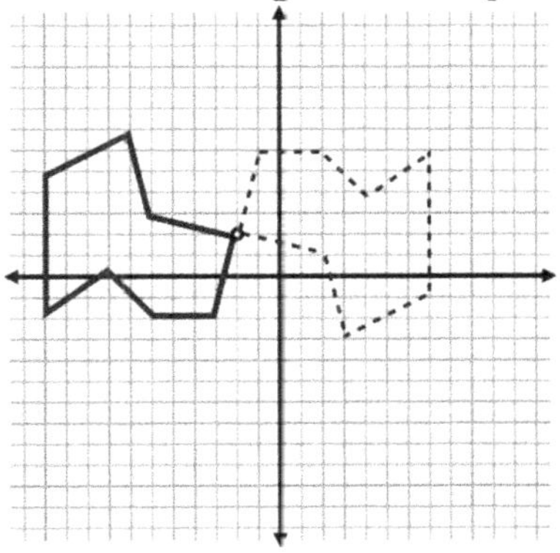

Review Video: Reflection
Visit mometrix.com/academy and enter code: 955068

3D Shapes

3D Shapes

Solids

The **surface area of a solid object** is the area of all sides or exterior surfaces. For objects such as prisms and pyramids, a further distinction is made between base surface area (B) and lateral surface area (LA). For a prism, the total surface area (SA) is $SA = LA + 2B$. For a pyramid or cone, the total surface area is $SA = LA + B$.

The **surface area of a sphere** can be found by the formula $A = 4\pi r^2$, where r is the radius. The volume is given by the formula $V = \frac{4}{3}\pi r^3$, where r is the radius. Both quantities are generally given in terms of π.

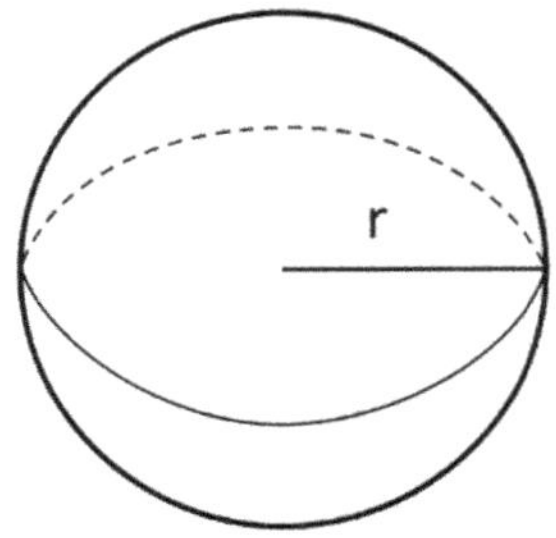

Review Video: Volume and Surface Area of a Sphere
Visit mometrix.com/academy and enter code: 786928

Review Video: How to Calculate the Volume of 3D Objects
Visit mometrix.com/academy and enter code: 163343

The **volume of any prism** is found by the formula $V = Bh$, where B is the area of the base, and h is the height (perpendicular distance between the bases). The surface area of any prism is the sum of the areas of both bases and all sides. It can be calculated as $SA = 2B + Ph$, where P is the perimeter of the base.

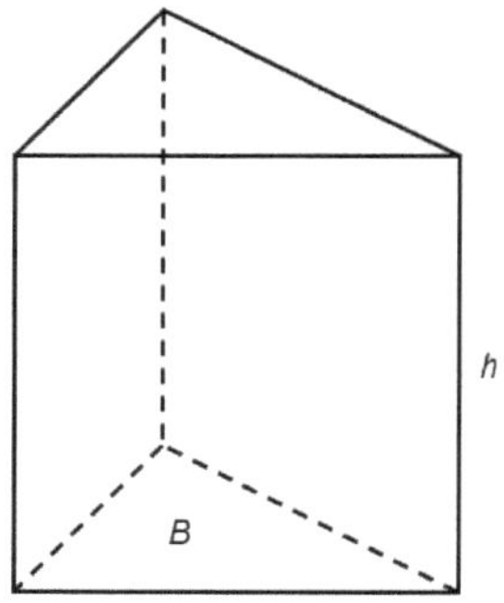

Review Video: Volume and Surface Area of a Prism
Visit mometrix.com/academy and enter code: 420158

For a **rectangular prism**, the volume can be found by the formula $V = lwh$, where V is the volume, l is the length, w is the width, and h is the height. The surface area can be calculated as $SA = 2lw + 2hl + 2wh$ or $SA = 2(lw + hl + wh)$.

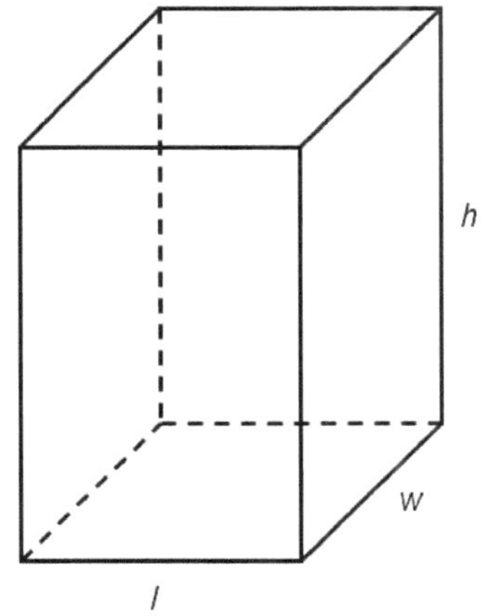

Review Video: Volume and Surface Area of a Rectangular Prism
Visit mometrix.com/academy and enter code: 282814

The **volume of a cube** can be found by the formula $V = s^3$, where s is the length of a side. The surface area of a cube is calculated as $SA = 6s^2$, where SA is the total surface area and s is the length of a side. These formulas are the same as the ones used for the volume and surface area of a rectangular prism, but simplified since all three quantities (length, width, and height) are the same.

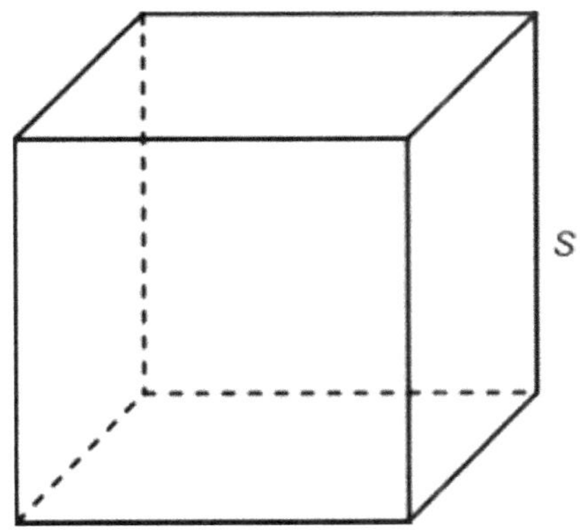

Review Video: Volume and Surface Area of a Cube
Visit mometrix.com/academy and enter code: 664455

The **volume of a cylinder** can be calculated by the formula $V = \pi r^2 h$, where r is the radius, and h is the height. The surface area of a cylinder can be found by the formula $SA = 2\pi r^2 + 2\pi rh$. The first term is the base area multiplied by two, and the second term is the perimeter of the base multiplied by the height.

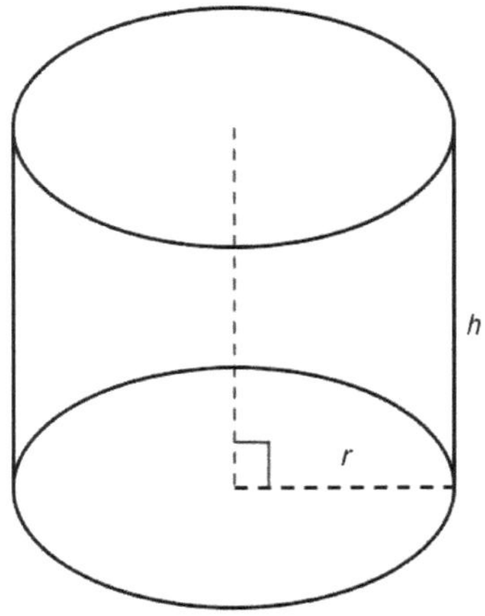

Review Video: Volume and Surface Area of a Right Circular Cylinder
Visit mometrix.com/academy and enter code: 226463

The **volume of a pyramid** is found by the formula $V = \frac{1}{3}Bh$, where B is the area of the base, and h is the height (perpendicular distance from the vertex to the base). Notice this formula is the same as $\frac{1}{3}$ times the volume of a prism. Like a prism, the base of a pyramid can be any shape.

Finding the **surface area of a pyramid** is not as simple as the other shapes we've looked at thus far. If the pyramid is a right pyramid, meaning the base is a regular polygon and the vertex is directly over the center of that polygon, the surface area can be calculated as $SA = B + \frac{1}{2}Ph_s$, where P is the perimeter of the base, and h_s is the slant height (distance from the vertex to the midpoint of one side of the base). If the pyramid is irregular, the area of each triangle side must be calculated individually and then summed, along with the base.

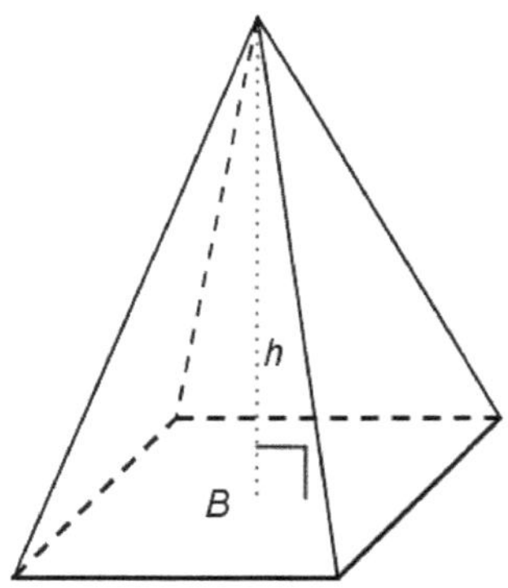

Review Video: Volume and Surface Area of a Pyramid
Visit mometrix.com/academy and enter code: 621932

The **volume of a cone** is found by the formula $V = \frac{1}{3}\pi r^2 h$, where r is the radius, and h is the height. Notice this is the same as $\frac{1}{3}$ times the volume of a cylinder. The surface area can be calculated as

$SA = \pi r^2 + \pi rs$, where s is the slant height. The slant height can be calculated using the Pythagorean theorem to be $\sqrt{r^2 + h^2}$, so the surface area formula can also be written as $SA = \pi r^2 + \pi r\sqrt{r^2 + h^2}$.

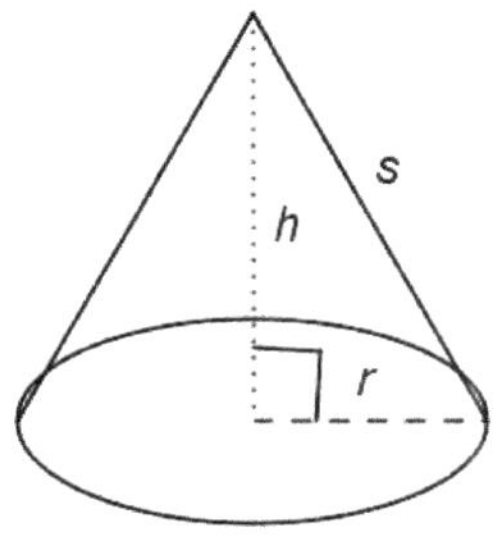

Review Video: Volume and Surface Area of a Right Circular Cone
Visit mometrix.com/academy and enter code: 573574

Conic Sections

Conic sections are a family of shapes that can be thought of as cross sections of a pair of infinite right cones stacked vertex to vertex. This is easiest to see with a visual representation:

A three-dimensional look at representative conic sections. (Note that a hyperbola intersects both cones.)

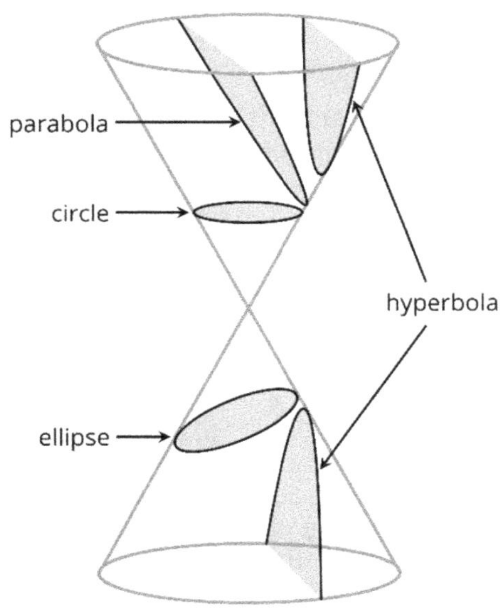

A side-on look at representative conic sections. (Note that the parabola is parallel to the slant of the cones.)

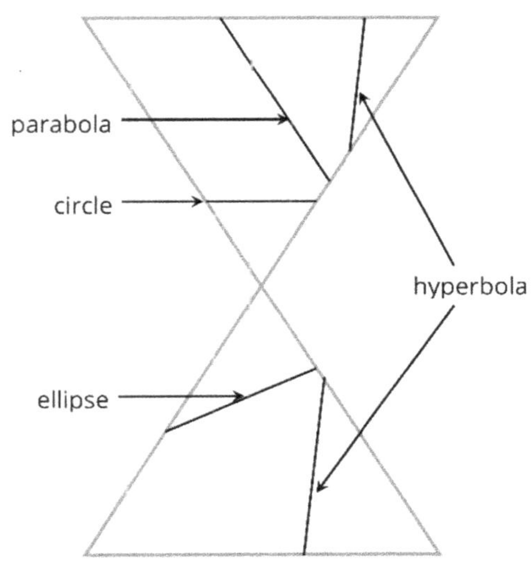

In short, a circle is a horizontal cross section, a parabola is a cross section parallel to the slant of the cone, an ellipse is a cross section at an angle *less than* the slant of the cone, and a hyperbola is a cross section at an angle *greater than* the slant of the cone.

Basic Trigonometry

PARABOLAS

A **parabola** is the set of all points in a plane that are equidistant from a fixed line, called the **directrix**, and a fixed point not on the line, called the **focus**. The **axis** is the line perpendicular to the directrix that passes through the focus.

For parabolas that open up or down, the standard equation is $(x - h)^2 = 4c(y - k)$, where h, c, and k are coefficients. If c is positive, the parabola opens up. If c is negative, the parabola opens down. The vertex is the point (h, k). The directrix is the line having the equation $y = -c + k$, and the focus is the point $(h, c + k)$.

For parabolas that open left or right, the standard equation is $(y - k)^2 = 4c(x - h)$, where k, c, and h are coefficients. If c is positive, the parabola opens to the right. If c is negative, the parabola opens to the left. The vertex is the point (h, k). The directrix is the line having the equation $x = -c + h$, and the focus is the point $(c + h, k)$.

Review Video: Parabolas
Visit mometrix.com/academy and enter code: 129187

Review Video: Vertex of a Parabola
Visit mometrix.com/academy and enter code: 272300

Review Video: How to Convert Between Standard Form and Vertex Form
Visit mometrix.com/academy and enter code: 916601

PYTHAGOREAN THEOREM

The side of a triangle opposite the right angle is called the **hypotenuse**. The other two sides are called the legs. The Pythagorean theorem states a relationship among the legs and hypotenuse of a right triangle: $(a^2 + b^2 = c^2)$, where a and b are the lengths of the legs of a right triangle, and c is the length of the hypotenuse. Note that this formula will only work with right triangles.

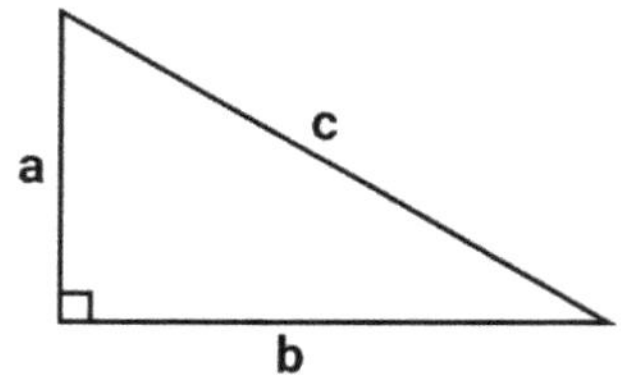

Review Video: Pythagorean Theorem
Visit mometrix.com/academy and enter code: 906576

TRIGONOMETRIC FORMULAS

In the diagram below, angle C is the right angle, and side c is the hypotenuse. Side a is the side opposite to angle A and side b is the side opposite to angle B. Using ratios of side lengths as a means to calculate the sine, cosine, and tangent of an acute angle only works for right triangles.

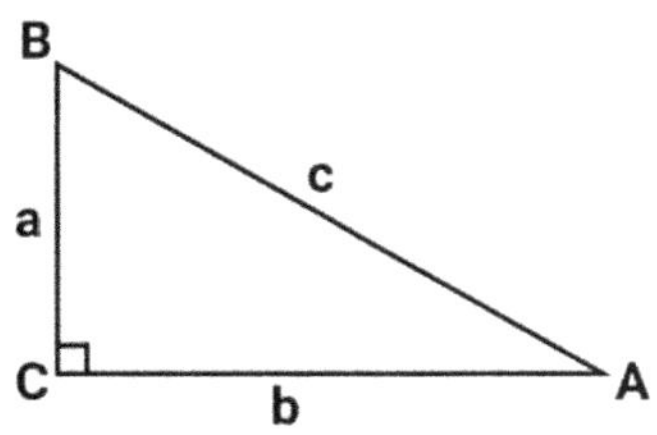

$$\sin A = \frac{\text{opposite side}}{\text{hypotenuse}} = \frac{a}{c} \qquad \csc A = \frac{1}{\sin A} = \frac{\text{hypotenuse}}{\text{opposite side}} = \frac{c}{a}$$

$$\cos A = \frac{\text{adjacent side}}{\text{hypotenuse}} = \frac{b}{c} \qquad \sec A = \frac{1}{\cos A} = \frac{\text{hypotenuse}}{\text{adjacent side}} = \frac{c}{b}$$

$$\tan A = \frac{\text{opposite side}}{\text{adjacent side}} = \frac{a}{b} \qquad \cot A = \frac{1}{\tan A} = \frac{\text{adjacent side}}{\text{opposite side}} = \frac{b}{a}$$

LAWS OF SINES AND COSINES

The **law of sines** states that $\frac{\sin A}{a} = \frac{\sin B}{b} = \frac{\sin C}{c}$, where A, B, and C are the angles of a triangle, and a, b, and c are the sides opposite their respective angles. This formula will work with all triangles, not just right triangles.

The **law of cosines** is given by the formula $c^2 = a^2 + b^2 - 2ab(\cos C)$, where a, b, and c are the sides of a triangle, and C is the angle opposite side c. This is a generalized form of the Pythagorean theorem that can be used on any triangle.

> **Review Video: Law of Sines**
> Visit mometrix.com/academy and enter code: 206844
>
> **Review Video: Law of Cosines**
> Visit mometrix.com/academy and enter code: 158911

ELLIPSES AND HYPERBOLAS

ELLIPSE

An **ellipse** is the set of all points in a plane whose total distance from two fixed points called the **foci** (singular: focus) is constant, and whose center is the midpoint between the foci.

The standard equation of an ellipse that is wider than it is tall is $\frac{(x-h)^2}{a^2} + \frac{(y-k)^2}{b^2} = 1$, where a and b are coefficients. The center is the point (h, k) and the foci are the points $(h + c, k)$ and $(h - c, k)$, where $c^2 = a^2 - b^2$ and $a^2 > b^2$.

The major axis has length $2a$, and the minor axis has length $2b$.

Eccentricity (e) is a measure of how elongated an ellipse is, and is the ratio of the distance between the foci to the length of the major axis. Eccentricity will have a value between 0 and 1. The closer to 0 the eccentricity is, the closer the ellipse is to being a circle. The formula for eccentricity is $e = \frac{c}{a}$.

> **Review Video: Ellipse**
> Visit mometrix.com/academy and enter code: 703899

HYPERBOLA

A **hyperbola** is the set of all points in a plane, whose distance from two fixed points, called foci, has a constant difference.

The standard equation of a horizontal hyperbola is $\frac{(x-h)^2}{a^2} - \frac{(y-k)^2}{b^2} = 1$, where a, b, h, and k are real numbers. The center is the point (h, k), the vertices are the points $(h + a, k)$ and $(h - a, k)$, and the foci are the points that every point on one of the parabolic curves is equidistant from. The foci are found using the formulas $(h + c, k)$ and $(h - c, k)$, where $c^2 = a^2 + b^2$. The asymptotes are two lines the graph of the hyperbola approaches but never reaches, and are given by the equations $y = \left(\frac{b}{a}\right)(x - h) + k$ and $y = -\left(\frac{b}{a}\right)(x - h) + k$.

The standard equation of a vertical hyperbola is $\frac{(y-k)^2}{a^2} - \frac{(x-h)^2}{b^2} = 1$, where a, b, k, and h are real numbers. The center is the point (h, k), the vertices are the points $(h, k + a)$ and $(h, k - a)$, and the foci are the points that every point on one of the hyperbolic curves is equidistant from and are found using the formulas $(h, k + c)$ and $(h, k - c)$, where $c^2 = a^2 + b^2$. The asymptotes are two lines the graph of the hyperbola approaches but never reaches, and are given by the equations $y = \left(\frac{a}{b}\right)(x - h) + k$ and $y = -\left(\frac{a}{b}\right)(x - h) + k$.

Review Video: Hyperbolas
Visit mometrix.com/academy and enter code: 509429

Chapter Quiz

Ready to see how well you retained what you just read? Scan the QR code to go directly to the chapter quiz interface for this study guide. If you're using a computer, simply visit the online resources page at **mometrix.com/resources719/priimath5161** and click the Chapter Quizzes link.

Probability and Statistics

Transform passive reading into active learning! After immersing yourself in this chapter, put your comprehension to the test by taking a quiz. The insights you gained will stay with you longer this way. Scan the QR code to go directly to the chapter quiz interface for this study guide. If you're using a computer, simply visit the online resources page at **mometrix.com/resources719/priimath5161** and click the Chapter Quizzes link.

Displaying Data

Displaying Information

Frequency Tables

Frequency tables show how frequently each unique value appears in a set. A **relative frequency table** is one that shows the proportions of each unique value compared to the entire set. Relative frequencies are given as percentages; however, the total percent for a relative frequency table will not necessarily equal 100 percent due to rounding. An example of a frequency table with relative frequencies is below.

Favorite Color	Frequency	Relative Frequency
Blue	4	13%
Red	7	22%
Green	3	9%
Purple	6	19%
Cyan	12	38%

Review Video: Data Interpretation of Graphs
Visit mometrix.com/academy and enter code: 200439

Circle Graphs

Circle graphs, also known as *pie charts*, provide a visual depiction of the relationship of each type of data compared to the whole set of data. The circle graph is divided into sections by drawing radii to create central angles whose percentage of the circle is equal to the individual data's percentage of the whole set. Each 1% of data is equal to 3.6° in the circle graph. Therefore, data represented by a 90° section of the circle graph makes up 25% of the whole. When complete, a circle graph often looks like a pie cut into uneven wedges. The pie chart below shows the data from the frequency table referenced earlier where people were asked their favorite color.

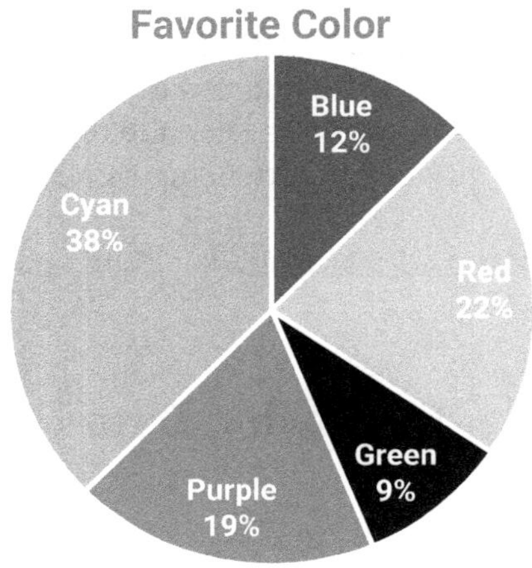

PICTOGRAPHS

A **pictograph** is a graph, generally in the horizontal orientation, that uses pictures or symbols to represent the data. Each pictograph must have a key that defines the picture or symbol and gives the quantity each picture or symbol represents. Pictures or symbols on a pictograph are not always shown as whole elements. In this case, the fraction of the picture or symbol shown represents the same fraction of the quantity a whole picture or symbol stands for. For example, a row with $3\frac{1}{2}$ ears of corn, where each ear of corn represents 100 stalks of corn in a field, would equal $3\frac{1}{2} \times 100 = 350$ stalks of corn in the field.

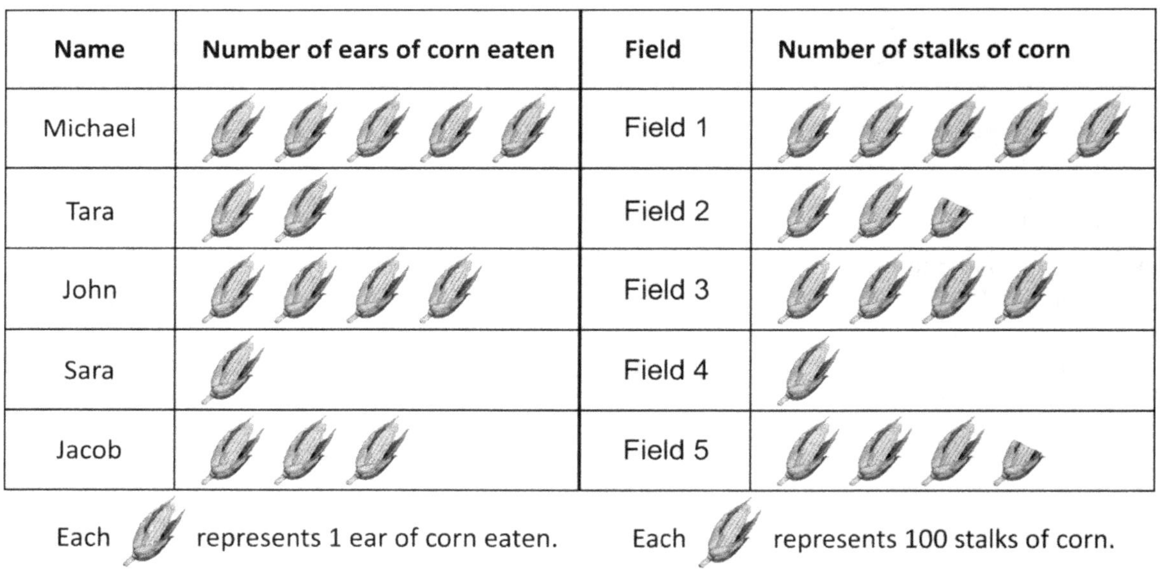

Review Video: Pictographs
Visit mometrix.com/academy and enter code: 147860

LINE GRAPHS

Line graphs have one or more lines of varying styles (solid or broken) to show the different values for a set of data. The individual data are represented as ordered pairs, much like on a Cartesian plane. In this case, the x- and y-axes are defined in terms of their units, such as dollars or time. The individual plotted points are joined by line segments to show whether the value of the data is increasing (line sloping upward), decreasing (line sloping downward), or staying the same (horizontal line). Multiple sets of data can be graphed on the same line graph to give an easy visual comparison. An example of this would be graphing achievement test scores for different groups of students over the same time period to see which group had the greatest increase or decrease in performance from year to year (as shown below).

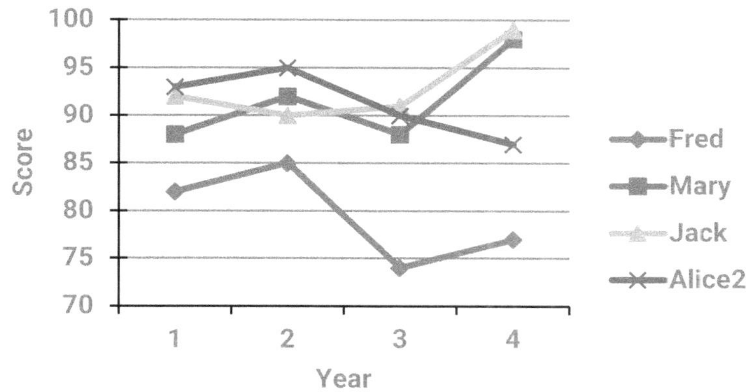

Review Video: How to Create a Line Graph
Visit mometrix.com/academy and enter code: 480147

LINE PLOTS

A **line plot**, also known as a *dot plot*, has plotted points that are not connected by line segments. In this graph, the horizontal axis lists the different possible values for the data, and the vertical axis lists the number of times the individual value occurs. A single dot is graphed for each value to show the number of times it occurs. This graph is more closely related to a bar graph than a line graph. Do not connect the dots in a line plot or it will misrepresent the data.

Review Video: Line Plot
Visit mometrix.com/academy and enter code: 754610

STEM AND LEAF PLOTS

A **stem and leaf plot** is useful for depicting groups of data that fall into a range of values. Each piece of data is separated into two parts: the first, or left, part is called the stem; the second, or right, part is called the leaf. Each stem is listed in a column from smallest to largest. Each leaf that has the common stem is listed in that stem's row from smallest to largest. For example, in a set of two-digit numbers, the digit in the tens place is the stem, and the digit in the ones place is the leaf. With a stem and leaf plot, you can easily see which subset of numbers (10s, 20s, 30s, etc.) is the largest. This information is also readily available by looking at a histogram, but a stem and leaf plot also allows you to look closer and see exactly which values fall in that range. Using a sample set of test scores (82, 88, 92, 93, 85, 90, 92, 95, 74, 88, 90, 91, 78, 87, 98, 99), we can assemble a stem and leaf plot like the one below.

Test Scores									
7	4	8							
8	2	5	7	8	8				
9	0	0	1	2	2	3	5	8	9

Review Video: Stem and Leaf Plots
Visit mometrix.com/academy and enter code: 302339

Bar Graphs

A **bar graph** is one of the few graphs that can be drawn correctly in two different configurations – both horizontally and vertically. A bar graph is similar to a line plot in the way the data is organized on the graph. Both axes must have their categories defined for the graph to be useful. Rather than placing a single dot to mark the point of the data's value, a bar, or thick line, is drawn from zero to the exact value of the data, whether it is a number, percentage, or other numerical value. Longer bar lengths correspond to greater data values. To read a bar graph, read the labels for the axes to find the units being reported. Then, look where the bars end in relation to the scale given on the corresponding axis and determine the associated value.

The bar chart below represents the responses from our favorite-color survey.

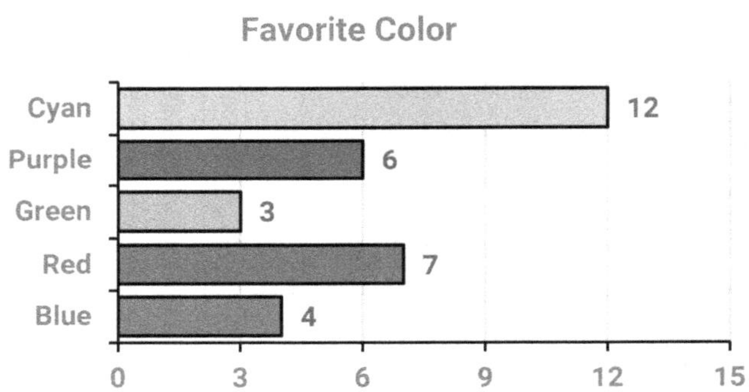

Histograms

At first glance, a **histogram** looks like a vertical bar graph. The difference is that a bar graph has a separate bar for each piece of data and a histogram has one continuous bar for each *range* of data. For example, a histogram may have one bar for the range 0–9, one bar for 10–19, etc. While a bar graph has numerical values on one axis, a histogram has numerical values on both axes. Each range is of equal size, and they are ordered left to right from lowest to highest. The height of each column on a histogram represents the number of data values within that range. Like a stem and leaf plot, a histogram makes it easy to glance at the graph and quickly determine which range has the greatest quantity of values. A simple example of a histogram is below.

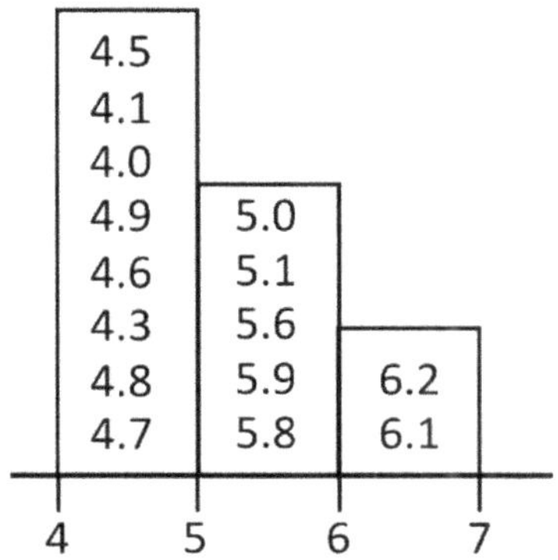

5-Number Summary

The **5-number summary** of a set of data gives a very informative picture of the set. The five numbers in the summary include the minimum value, maximum value, and the three quartiles. This information gives the reader the range and median of the set, as well as an indication of how the data is spread about the median.

Box and Whisker Plots

A **box-and-whiskers plot** is a graphical representation of the 5-number summary. To draw a box-and-whiskers plot, plot the points of the 5-number summary on a number line. Draw a box whose ends are through the points for the first and third quartiles. Draw a vertical line in the box through the median to divide the box in half. Draw a line segment from the first quartile point to the minimum value, and from the third quartile point to the maximum value.

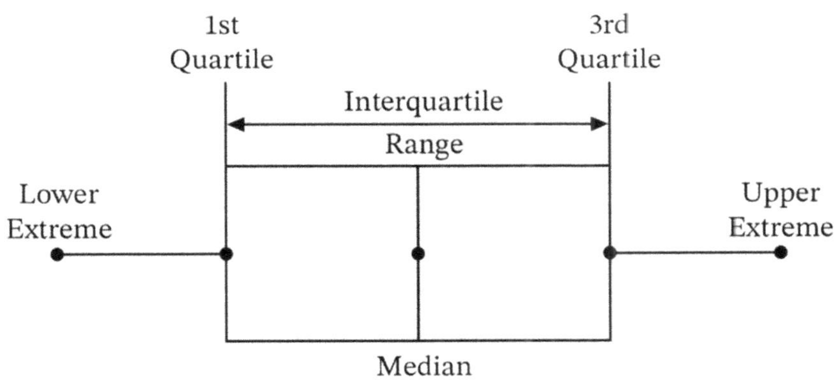

Review Video: Box and Whisker Plots
Visit mometrix.com/academy and enter code: 810817

Example

Given the following data (32, 28, 29, 26, 35, 27, 30, 31, 27, 32), we first sort it into numerical order: 26, 27, 27, 28, 29, 30, 31, 32, 32, 35. We can then find the median. Since there are ten values, we take the average of the 5th and 6th values to get 29.5. We find the lower quartile by taking the median of the data smaller than the median. Since there are five values, we take the 3rd value, which is 27. We find the upper quartile by taking the median of the data larger than the overall median, which is 32. Finally, we note our minimum and maximum, which are simply the smallest and largest values in the set: 26 and 35, respectively. Now we can create our box plot:

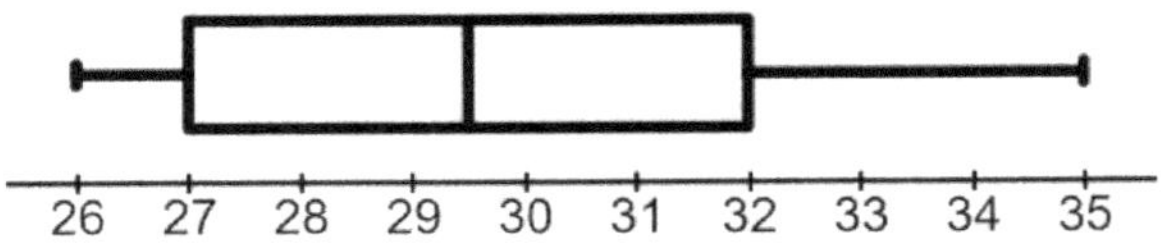

This plot is fairly "long" on the right whisker, showing one or more unusually high values (but not quite outliers). The other quartiles are similar in length, showing a fairly even distribution of data.

INTERQUARTILE RANGE

The **interquartile range, or IQR**, is the difference between the upper and lower quartiles. It measures how the data is dispersed: a high IQR means that the data is more spread out, while a low IQR means that the data is clustered more tightly around the median. To find the IQR, subtract the lower quartile value (Q_1) from the upper quartile value (Q_3).

EXAMPLE

To find the upper and lower quartiles, we first find the median and then take the median of all values above it and all values below it. In the following data set (16, 18, 13, 24, 16, 51, 32, 21, 27, 39), we first rearrange the values in numerical order: 13, 16, 16, 18, 21, 24, 27, 32, 39, 51. There are 10 values, so the median is the average of the 5th and 6th: $\frac{21+24}{2} = \frac{45}{2} = 22.5$. We do not actually need this value to find the upper and lower quartiles. We look at the set of numbers below the median: 13, 16, 16, 18, 21. There are five values, so the 3rd is the median (16), or the value of the lower quartile (Q_1). Then we look at the numbers above the median: 24, 27, 32, 39, 51. Again there are five values, so the 3rd is the median (32), or the value of the upper quartile (Q_3). We find the IQR by subtracting Q_1 from Q_3: $32 - 16 = 16$.

68-95-99.7 RULE

The **68–95–99.7 rule** describes how a normal distribution of data should appear when compared to the mean. This is also a description of a normal bell curve. According to this rule, 68 percent of the data values in a normally distributed set should fall within one standard deviation of the mean (34 percent above and 34 percent below the mean), 95 percent of the data values should fall within two standard deviations of the mean (47.5 percent above and 47.5 percent below the mean), and 99.7 percent of the data values should fall within three standard deviations of the mean, again, equally distributed on either side of the mean. This means that only 0.3 percent of all data values should fall more than three standard deviations from the mean. On the graph below, the normal curve is centered on the y-axis. The x-axis labels are how many standard deviations away from the center you are. Therefore, it is easy to see how the 68-95-99.7 rule can apply.

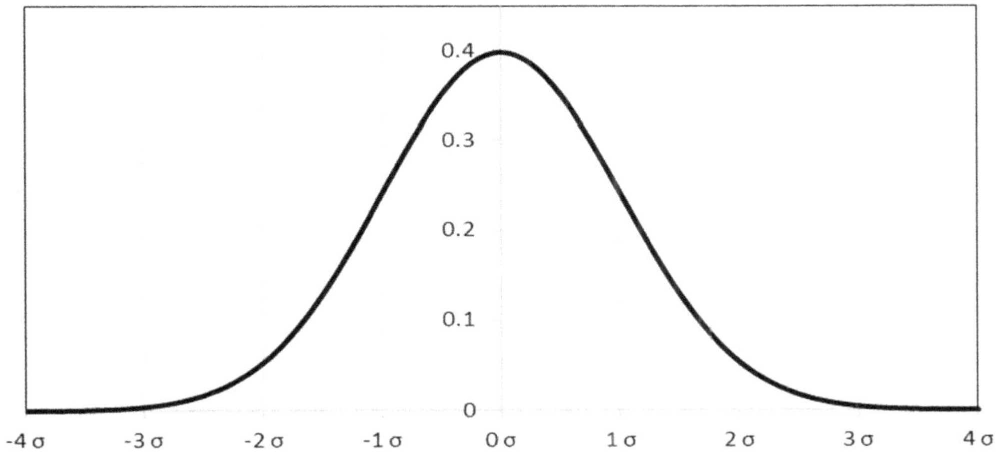

Scatter Plots

Bivariate Data

Bivariate data is simply data from two different variables. (The prefix *bi-* means *two.*) In a *scatter plot*, each value in the set of data is plotted on a grid similar to a Cartesian plane, where each axis represents one of the two variables. By looking at the pattern formed by the points on the grid, you can often determine whether or not there is a relationship between the two variables, and what that relationship is, if it exists. The variables may be directly proportionate, inversely proportionate, or show no proportion at all. It may also be possible to determine if the data is linear, and if so, to find an equation to relate the two variables. The following scatter plot shows the relationship between preference for brand "A" and the age of the consumers surveyed.

Scatter Plots

Scatter plots are also useful in determining the type of function represented by the data and finding the simple regression. Linear scatter plots may be positive or negative. Nonlinear scatter plots are generally exponential or quadratic. Below are some common types of scatter plots:

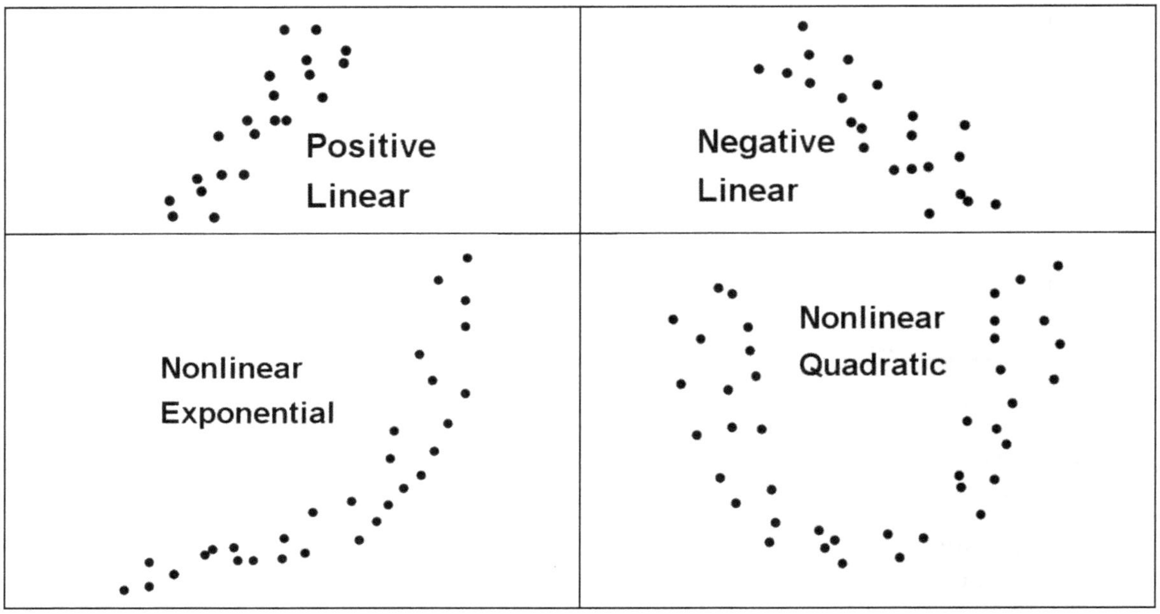

Review Video: Scatter Plot
Visit mometrix.com/academy and enter code: 596526

Statistics

Introduction to Statistics

Statistics is the branch of mathematics that deals with collecting, recording, interpreting, illustrating, and analyzing large amounts of **data**. The following terms are often used in the discussion of data and **statistics**:

- **Data** – the collective name for pieces of information (singular is datum)
- **Quantitative data** – measurements (such as length, mass, and speed) that provide information about quantities in numbers
- **Qualitative data** – information (such as colors, scents, tastes, and shapes) that cannot be measured using numbers
- **Discrete data** – information that can be expressed only by a specific value, such as whole or half numbers. (e.g., since people can be counted only in whole numbers, a population count would be discrete data.)
- **Continuous data** – information (such as time and temperature) that can be expressed by any value within a given range
- **Primary data** – information that has been collected directly from a survey, investigation, or experiment, such as a questionnaire or the recording of daily temperatures. (Primary data that has not yet been organized or analyzed is called **raw data**.)
- **Secondary data** – information that has been collected, sorted, and processed by the researcher

- **Ordinal data** – information that can be placed in numerical order, such as age or weight
- **Nominal data** – information that *cannot* be placed in numerical order, such as names or places

Data Collection

Population

In statistics, the **population** is the entire collection of people, plants, etc., that data can be collected from. For example, a study to determine how well students in local schools perform on a standardized test would have a population of all the students enrolled in those schools, although a study may include just a small sample of students from each school. A **parameter** is a numerical value that gives information about the population, such as the mean, median, mode, or standard deviation. Remember that the symbol for the mean of a population is μ and the symbol for the standard deviation of a population is σ.

Sample

A **sample** is a portion of the entire population. Whereas a parameter helped describe the population, a **statistic** is a numerical value that gives information about the sample, such as mean, median, mode, or standard deviation. Keep in mind that the symbols for mean and standard deviation are different when they are referring to a sample rather than the entire population. For a sample, the symbol for mean is $\bar{x}$ and the symbol for standard deviation is s. The mean and standard deviation of a sample may or may not be identical to that of the entire population due to a sample only being a subset of the population. However, if the sample is random and large enough, statistically significant values can be attained. Samples are generally used when the population is too large to justify including every element or when acquiring data for the entire population is impossible.

Inferential Statistics

Inferential statistics is the branch of statistics that uses samples to make predictions about an entire population. This type of statistic is often seen in political polls, where a sample of the population is questioned about a particular topic or politician to gain an understanding of the attitudes of the entire population of the country. Often, exit polls are conducted on election days using this method. Inferential statistics can have a large margin of error if you do not have a valid sample.

Sampling Distribution

Statistical values calculated from various samples of the same size make up the **sampling distribution**. For example, if several samples of identical size are randomly selected from a large population and then the mean of each sample is calculated, the distribution of values of the means would be a sampling distribution.

The **sampling distribution of the mean** is the distribution of the sample mean, $\bar{x}$, derived from random samples of a given size. It has three important characteristics. First, the mean of the sampling distribution of the mean is equal to the mean of the population that was sampled. Second, assuming the standard deviation is non-zero, the standard deviation of the sampling distribution of the mean equals the standard deviation of the sampled population divided by the square root of the sample size. This is sometimes called the standard error. Finally, as the sample size gets larger, the sampling distribution of the mean gets closer to a normal distribution via the central limit theorem.

Survey Study

A **survey study** is a method of gathering information from a small group in an attempt to gain enough information to make accurate general assumptions about the population. Once a survey study is completed, the results are then put into a summary report.

Survey studies are generally in the format of surveys, interviews, or questionnaires as part of an effort to find opinions of a particular group or to find facts about a group.

It is important to note that the findings from a survey study are only as accurate as the sample chosen from the population.

Correlational Studies

Correlational studies seek to determine how much one variable is affected by changes in a second variable. For example, correlational studies may look for a relationship between the amount of time a student spends studying for a test and the grade that student earned on the test or between student scores on college admissions tests and student grades in college.

It is important to note that correlational studies cannot show a cause and effect, but rather can show only that two variables are or are not potentially correlated.

Experimental Studies

Experimental studies take correlational studies one step farther, in that they attempt to prove or disprove a cause-and-effect relationship. These studies are performed by conducting a series of experiments to test the hypothesis. For a study to be scientifically accurate, it must have both an experimental group that receives the specified treatment and a control group that does not get the treatment. This is the type of study pharmaceutical companies do as part of drug trials for new medications. Experimental studies are only valid when the proper scientific method has been followed. In other words, the experiment must be well-planned and executed without bias in the testing process, all subjects must be selected at random, and the process of determining which subject is in which of the two groups must also be completely random.

Observational Studies

Observational studies are the opposite of experimental studies. In observational studies, the tester cannot change or in any way control all of the variables in the test. For example, a study to determine which gender does better in math classes in school is strictly observational. You cannot change a person's gender, and you cannot change the subject being studied. The big downfall of observational studies is that you have no way of proving a cause-and-effect relationship because you cannot control outside influences. Events outside of school can influence a student's performance in school, and observational studies cannot take that into consideration.

Random Samples

For most studies, a **random sample** is necessary to produce valid results. Random samples should not have any particular influence to cause sampled subjects to behave one way or another. The goal is for the random sample to be a **representative sample**, or a sample whose characteristics give an accurate picture of the characteristics of the entire population. To accomplish this, you must make sure you have a proper **sample size**, or an appropriate number of elements in the sample.

Biases

In statistical studies, biases must be avoided. **Bias** is an error that causes the study to favor one set of results over another. For example, if a survey to determine how the country views the president's job performance only speaks to registered voters in the president's party, the results will be skewed because a disproportionately large number of responders would tend to show approval, while a disproportionately large number of people in the opposite party would tend to express disapproval. **Extraneous variables** are, as the name implies, outside influences that can affect the outcome of a study. They are not always avoidable but could trigger bias in the result.

DATA ANALYSIS

DISPERSION

A **measure of dispersion** is a single value that helps to "interpret" the measure of central tendency by providing more information about how the data values in the set are distributed about the measure of central tendency. The measure of dispersion helps to eliminate or reduce the disadvantages of using the mean, median, or mode as a single measure of central tendency, and give a more accurate picture of the dataset as a whole. To have a measure of dispersion, you must know or calculate the range, standard deviation, or variance of the data set.

RANGE

The **range** of a set of data is the difference between the greatest and lowest values of the data in the set. To calculate the range, you must first make sure the units for all data values are the same, and then identify the greatest and lowest values. If there are multiple data values that are equal for the highest or lowest, just use one of the values in the formula. Write the answer with the same units as the data values you used to do the calculations.

Review Video: Statistical Range
Visit mometrix.com/academy and enter code: 778541

SAMPLE STANDARD DEVIATION

Standard deviation is a measure of dispersion that compares all the data values in the set to the mean of the set to give a more accurate picture. To find the **standard deviation of a sample**, use the formula

$$s = \sqrt{\frac{\sum_{i=1}^{n}(x_i - \bar{x})^2}{n-1}}$$

Note that s is the standard deviation of a sample, x_i represents the individual values in the data set, $\bar{x}$ is the mean of the data values in the set, and n is the number of data values in the set. The higher the value of the standard deviation is, the greater the variance of the data values from the mean. The units associated with the standard deviation are the same as the units of the data values.

Review Video: Standard Deviation
Visit mometrix.com/academy and enter code: 419469

SAMPLE VARIANCE

The **variance of a sample** is the square of the sample standard deviation (denoted s^2). While the mean of a set of data gives the average of the set and gives information about where a specific data value lies in relation to the average, the variance of the sample gives information about the degree to which the data values are spread out and tells you how close an individual value is to the average compared to the other values. The units associated with variance are the same as the units of the data values squared.

PERCENTILE

Percentiles and quartiles are other methods of describing data within a set. **Percentiles** tell what percentage of the data in the set fall below a specific point. For example, achievement test scores are often given in percentiles. A score at the 80th percentile is one which is equal to or higher than 80 percent of the scores in the set. In other words, 80 percent of the scores were lower than that score.

Quartiles are percentile groups that make up quarter sections of the data set. The first quartile is the 25th percentile. The second quartile is the 50th percentile; this is also the median of the dataset. The third quartile is the 75th percentile.

SKEWNESS

Skewness is a way to describe the symmetry or asymmetry of the distribution of values in a dataset. If the distribution of values is symmetrical, there is no skew. In general the closer the mean of a data set is to the median of the data set, the less skew there is. Generally, if the mean is to the right of the median, the data set is *positively skewed*, or right-skewed, and if the mean is to the left of the median, the data set is *negatively skewed*, or left-skewed. However, this rule of thumb is not infallible. When the data values are graphed on a curve, a set with no skew will be a perfect bell curve.

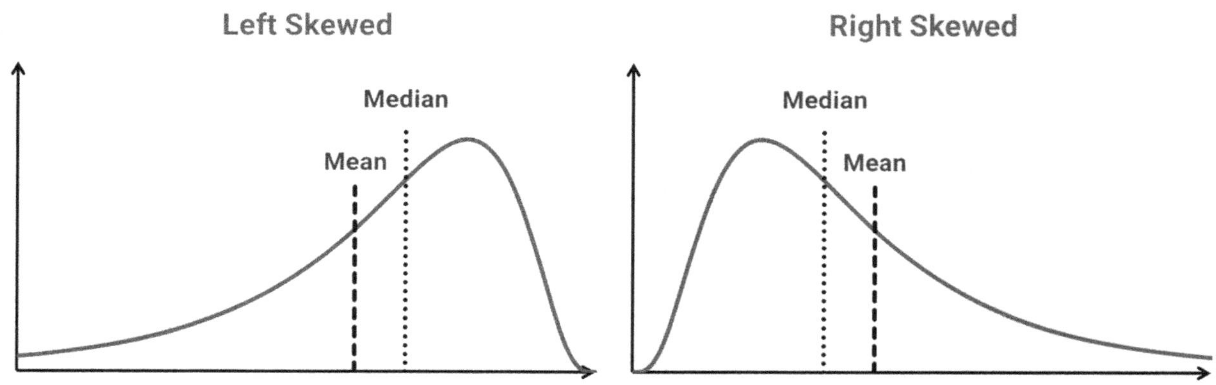

To estimate skew, use the formula:

$$\text{skew} = \frac{\sqrt{n(n-1)}}{n-2}\left(\frac{\frac{1}{n}\sum_{i=1}^{n}(x_i-\bar{x})^3}{\left(\frac{1}{n}\sum_{i=1}^{n}(x_i-\bar{x})^2\right)^{\frac{3}{2}}}\right)$$

Note that n is the datapoints in the set, x_i is the ith value in the set, and $\bar{x}$ is the mean of the set.

Review Video: Skew
Visit mometrix.com/academy and enter code: 661486

UNIMODAL VS. BIMODAL

If a distribution has a single peak, it would be considered **unimodal**. If it has two discernible peaks it would be considered **bimodal**. Bimodal distributions may be an indication that the set of data being considered is actually the combination of two sets of data with significant differences. A

uniform distribution is a distribution in which there is *no distinct peak or variation* in the data. No values or ranges are particularly more common than any other values or ranges.

OUTLIER

An outlier is an extremely high or extremely low value in the data set. It may be the result of measurement error, in which case, the outlier is not a valid member of the data set. However, it may also be a valid member of the distribution. Unless a measurement error is identified, the experimenter cannot know for certain if an outlier is or is not a member of the distribution. There are arbitrary methods that can be employed to designate an extreme value as an outlier. One method designates an outlier (or possible outlier) to be any value less than $Q_1 - 1.5(IQR)$ or any value greater than $Q_3 + 1.5(IQR)$.

DATA ANALYSIS

SIMPLE REGRESSION

In statistics, **simple regression** is using an equation to represent a relation between independent and dependent variables. The independent variable is also referred to as the explanatory variable or the predictor and is generally represented by the variable x in the equation. The dependent variable, usually represented by the variable y, is also referred to as the response variable. The equation may be any type of function – linear, quadratic, exponential, etc. The best way to handle this task is to use the regression feature of your graphing calculator. This will easily give you the curve of best fit and provide you with the coefficients and other information you need to derive an equation.

LINE OF BEST FIT

In a scatter plot, the **line of best fit** is the line that best shows the trends of the data. The line of best fit is given by the equation $\hat{y} = ax + b$, where a and b are the regression coefficients. The regression coefficient a is also the slope of the line of best fit, and b is also the y-coordinate of the point at which the line of best fit crosses the y-axis. Not every point on the scatter plot will be on the line of best fit. The differences between the y-values of the points in the scatter plot and the corresponding y-values according to the equation of the line of best fit are the residuals. The line of best fit is also called the least-squares regression line because it is also the line that has the lowest sum of the squares of the residuals.

CORRELATION COEFFICIENT

The **correlation coefficient** is the numerical value that indicates how strong the relationship is between the two variables of a linear regression equation. A correlation coefficient of –1 is a perfect negative correlation. A correlation coefficient of +1 is a perfect positive correlation. Correlation coefficients close to –1 or +1 are very strong correlations. A correlation coefficient equal to zero indicates there is no correlation between the two variables. This test is a good indicator of whether or not the equation for the line of best fit is accurate. The formula for the correlation coefficient is

$$r = \frac{\sum_{i=1}^{n}(x_i - \bar{x})(y_i - \bar{y})}{\sqrt{\sum_{i=1}^{n}(x_i - \bar{x})^2}\sqrt{\sum_{i=1}^{n}(y_i - \bar{y})^2}}$$

where r is the correlation coefficient, n is the number of data values in the set, (x_i, y_i) is a point in the set, and $\bar{x}$ and $\bar{y}$ are the means.

Z-SCORE

A **z-score** is an indication of how many standard deviations a given value falls from the sample mean. To calculate a z-score, use the formula:

$$\frac{x - \bar{x}}{\sigma}$$

In this formula x is the data value, $\bar{x}$ is the mean of the sample data, and σ is the standard deviation of the population. If the z-score is positive, the data value lies above the mean. If the z-score is negative, the data value falls below the mean. These scores are useful in interpreting data such as standardized test scores, where every piece of data in the set has been counted, rather than just a small random sample. In cases where standard deviations are calculated from a random sample of the set, the z-scores will not be as accurate.

CENTRAL LIMIT THEOREM

According to the **central limit theorem**, regardless of what the original distribution of a sample is, the distribution of the means tends to get closer and closer to a normal distribution as the sample size gets larger and larger (this is necessary because the sample is becoming more all-encompassing of the elements of the population). As the sample size gets larger, the distribution of the sample mean will approach a normal distribution with a mean of the population mean and a variance of the population variance divided by the sample size.

MEASURES OF CENTRAL TENDENCY

A **measure of central tendency** is a statistical value that gives a reasonable estimate for the center of a group of data. There are several different ways of describing the measure of central tendency. Each one has a unique way it is calculated, and each one gives a slightly different perspective on the data set. Whenever you give a measure of central tendency, always make sure the units are the same. If the data has different units, such as hours, minutes, and seconds, convert all the data to the same unit, and use the same unit in the measure of central tendency. If no units are given in the data, do not give units for the measure of central tendency.

MEAN

The **statistical mean** of a group of data is the same as the arithmetic average of that group. To find the mean of a set of data, first convert each value to the same units, if necessary. Then find the sum of all the values, and count the total number of data values, making sure you take into consideration each individual value. If a value appears more than once, count it more than once. Divide the sum of the values by the total number of values and apply the units, if any. Note that the mean does not have to be one of the data values in the set, and may not divide evenly.

$$\text{mean} = \frac{\text{sum of the data values}}{\text{quantity of data values}}$$

For instance, the mean of the data set $\{88, 72, 61, 90, 97, 68, 88, 79, 86, 93, 97, 71, 80, 84, 89\}$ would be the sum of the fifteen numbers divided by 15:

$$\frac{88 + 72 + 61 + 90 + 97 + 68 + 88 + 79 + 86 + 93 + 97 + 71 + 80 + 84 + 89}{15} = \frac{1242}{15} = 82.8$$

While the mean is relatively easy to calculate and averages are understood by most people, the mean can be very misleading if it is used as the sole measure of central tendency. If the data set has outliers (data values that are unusually high or unusually low compared to the rest of the data values), the mean can be very distorted, especially if the data set has a small number of values. If unusually high values are countered with unusually low values, the mean is not affected as much. For example, if five of twenty students in a class get a 100 on a test, but the other 15 students have an average of 60 on the same test, the class average would appear as 70. Whenever the mean is skewed by outliers, it is always a good idea to include the median as an alternate measure of central tendency.

A **weighted mean**, or weighted average, is a mean that uses "weighted" values. The formula is weighted mean $= \frac{w_1x_1+w_2x_2+w_3x_3...+w_nx_n}{w_1+w_2+w_3+\cdots+w_n}$. Weighted values, such as $w_1, w_2, w_3, ... w_n$ are assigned to each member of the set $x_1, x_2, x_3, ... x_n$. When calculating the weighted mean, make sure a weight value for each member of the set is used.

Review Video: All About Averages
Visit mometrix.com/academy and enter code: 176521

MEDIAN

The **statistical median** is the value in the middle of the set of data. To find the median, list all data values in order from smallest to largest or from largest to smallest. Any value that is repeated in the set must be listed the number of times it appears. If there are an odd number of data values, the median is the value in the middle of the list. If there is an even number of data values, the median is the arithmetic mean of the two middle values.

For example, the median of the data set {88, 72, 61, 90, 97, 68, 88, 79, 86, 93, 97, 71, 80, 84, 88} is 86 since the ordered set is {61, 68, 71, 72, 79, 80, 84, **86**, 88, 88, 88, 90, 93, 97, 97}.

The big disadvantage of using the median as a measure of central tendency is that is relies solely on a value's relative size as compared to the other values in the set. When the individual values in a set of data are evenly dispersed, the median can be an accurate tool. However, if there is a group of rather large values or a group of rather small values that are not offset by a different group of values, the information that can be inferred from the median may not be accurate because the distribution of values is skewed.

MODE

The **statistical mode** is the data value that occurs the greatest number of times in the data set. It is possible to have exactly one mode, more than one mode, or no mode. To find the mode of a set of data, arrange the data like you do to find the median (all values in order, listing all multiples of data values). Count the number of times each value appears in the data set. If all values appear an equal number of times, there is no mode. If one value appears more than any other value, that value is the mode. If two or more values appear the same number of times, but there are other values that appear fewer times and no values that appear more times, all of those values are the modes.

For example, the mode of the data set {**88**, 72, 61, 90, 97, 68, **88**, 79, 86, 93, 97, 71, 80, 84, **88**} is 88.

The main disadvantage of the mode is that the values of the other data in the set have no bearing on the mode. The mode may be the largest value, the smallest value, or a value anywhere in between in the set. The mode only tells which value or values, if any, occurred the greatest number of times. It does not give any suggestions about the remaining values in the set.

Review Video: Mean, Median, and Mode
Visit mometrix.com/academy and enter code: 286207

Probability

PROBABILITY

Probability is the likelihood of a certain outcome occurring for a given event. An **event** is any situation that produces a result. It could be something as simple as flipping a coin or as complex as launching a rocket. Determining the probability of an outcome for an event can be equally simple or complex. As such, there are specific terms used in the study of probability that need to be understood:

- **Compound event**—an event that involves two or more independent events (rolling a pair of dice and taking the sum)
- **Desired outcome** (or success)—an outcome that meets a particular set of criteria (a roll of 1 or 2 if we are looking for numbers less than 3)
- **Independent events**—two or more events whose outcomes do not affect one another (two coins tossed at the same time)
- **Dependent events**—two or more events whose outcomes affect one another (two cards drawn consecutively from the same deck)
- **Certain outcome**—probability of outcome is 100% or 1
- **Impossible outcome**—probability of outcome is 0% or 0
- **Mutually exclusive outcomes**—two or more outcomes whose criteria cannot all be satisfied in a single event (a coin coming up heads and tails on the same toss)
- **Random variable**—refers to all possible outcomes of a single event which may be discrete or continuous.

Review Video: Intro to Probability
Visit mometrix.com/academy and enter code: 212374

SAMPLE SPACE

The total set of all possible results of a test or experiment is called a **sample space**, or sometimes a universal sample space. The sample space, represented by one of the variables S, Ω, or U (for universal sample space) has individual elements called outcomes. Other terms for outcome that may be used interchangeably include elementary outcome, simple event, or sample point. The number of outcomes in a given sample space could be infinite or finite, and some tests may yield multiple unique sample sets. For example, tests conducted by drawing playing cards from a standard deck would have one sample space of the card values, another sample space of the card suits, and a third sample space of suit-denomination combinations. For most tests, the sample spaces considered will be finite.

An **event**, represented by the variable E, is a portion of a sample space. It may be one outcome or a group of outcomes from the same sample space. If an event occurs, then the test or experiment will generate an outcome that satisfies the requirement of that event. For example, given a standard deck of 52 playing cards as the sample space, and defining the event as the collection of face cards, then the event will occur if the card drawn is a J, Q, or K. If any other card is drawn, the event is said to have not occurred.

For every sample space, each possible outcome has a specific likelihood, or probability, that it will occur. The probability measure, also called the **distribution**, is a function that assigns a real number probability, from zero to one, to each outcome. For a probability measure to be accurate, every outcome must have a real number probability measure that is greater than or equal to zero and less than or equal to one. Also, the probability measure of the sample space must equal one, and the probability measure of the union of multiple outcomes must equal the sum of the individual probability measures.

Probabilities of events are expressed as real numbers from zero to one. They give a numerical value to the chance that a particular event will occur. The probability of an event occurring is the sum of the probabilities of the individual elements of that event. For example, in a standard deck of 52 playing cards as the sample space and the collection of face cards as the event, the probability of drawing a specific face card is $\frac{1}{52} = 0.019$, but the probability of drawing any one of the twelve face cards is $12(0.019) = 0.228$. Note that rounding of numbers can generate different results. If you multiplied 12 by the fraction $\frac{1}{52}$ before converting to a decimal, you would get the answer $\frac{12}{52} = 0.231$.

Theoretical and Experimental Probability

Theoretical probability can usually be determined without actually performing the event. The likelihood of an outcome occurring, or the probability of an outcome occurring, is given by the formula:

$$P(A) = \frac{\text{Number of acceptable outcomes}}{\text{Number of possible outcomes}}$$

Note that $P(A)$ is the probability of an outcome A occurring, and each outcome is just as likely to occur as any other outcome. If each outcome has the same probability of occurring as every other possible outcome, the outcomes are said to be equally likely to occur. The total number of acceptable outcomes must be less than or equal to the total number of possible outcomes. If the two are equal, then the outcome is certain to occur and the probability is 1. If the number of acceptable outcomes is zero, then the outcome is impossible and the probability is 0. For example, if there are 20 marbles in a bag and 5 are red, then the theoretical probability of randomly selecting a red marble is 5 out of 20, $\left(\frac{5}{20} = \frac{1}{4}, 0.25, \text{or } 25\%\right)$.

If the theoretical probability is unknown or too complicated to calculate, it can be estimated by an experimental probability. **Experimental probability**, also called empirical probability, is an estimate of the likelihood of a certain outcome based on repeated experiments or collected data. In other words, while theoretical probability is based on what *should* happen, experimental probability is based on what *has* happened. Experimental probability is calculated in the same way as theoretical probability, except that actual outcomes are used instead of possible outcomes. The more experiments performed or datapoints gathered, the better the estimate should be.

Theoretical and experimental probability do not always line up with one another. Theoretical probability says that out of 20 coin-tosses, 10 should be heads. However, if we were actually to toss 20 coins, we might record just 5 heads. This doesn't mean that our theoretical probability is incorrect; it just means that this particular experiment had results that were different from what was predicted. A practical application of empirical probability is the insurance industry. There are no set functions that define lifespan, health, or safety. Insurance companies look at factors from hundreds of thousands of individuals to find patterns that they then use to set the formulas for insurance premiums.

Review Video: Empirical Probability
Visit mometrix.com/academy and enter code: 513468

Objective and Subjective Probability

Objective probability is based on mathematical formulas and documented evidence. Examples of objective probability include raffles or lottery drawings where there is a pre-determined number of possible outcomes and a predetermined number of outcomes that correspond to an event. Other cases of objective probability include probabilities of rolling dice, flipping coins, or drawing cards. Most gambling games are based on objective probability.

In contrast, **subjective probability** is based on personal or professional feelings and judgments. Often, there is a lot of guesswork following extensive research. Areas where subjective probability is applicable include sales trends and business expenses. Attractions set admission prices based on subjective probabilities of attendance based on varying admission rates in an effort to maximize their profit.

Complement of an Event

Sometimes it may be easier to calculate the possibility of something not happening, or the **complement of an event**. Represented by the symbol $\bar{A}$, the complement of A is the probability that event A does not happen. When you know the probability of event A occurring, you can use the formula $P(\bar{A}) = 1 - P(A)$, where $P(\bar{A})$ is the probability of event A not occurring, and $P(A)$ is the probability of event A occurring.

Addition Rule

The **addition rule** for probability is used for finding the probability of a compound event. Use the formula $P(A \cup B) = P(A) + P(B) - P(A \cap B)$, where $P(A \cap B)$ is the probability of both events occurring to find the probability of a compound event. The probability of both events occurring at the same time must be subtracted to eliminate any overlap in the first two probabilities.

Conditional Probability

Given two events A and B, the **conditional probability** $P(A|B)$ is the probability that event A will occur, given that event B has occurred. The conditional probability cannot be calculated simply from $P(A)$ and $P(B)$; these probabilities alone do not give sufficient information to determine the conditional probability. It can, however, be determined if you are also given the probability of the intersection of events A and B, $P(A \cap B)$, the probability that events A and B both occur. Specifically, $P(A|B) = \frac{P(A \cap B)}{P(B)}$. For instance, suppose you have a jar containing two red marbles and two blue marbles, and you draw two marbles at random. Consider event A being the event that the first marble drawn is red, and event B being the event that the second marble drawn is blue. If we want to find the probability that B occurs given that A occurred, $P(B|A)$, then we can compute it using the fact that $P(A)$ is $\frac{1}{2}$, and $P(A \cap B)$ is $\frac{1}{3}$. (The latter may not be obvious, but may be determined by finding the product of $\frac{1}{2}$ and $\frac{2}{3}$). Therefore $P(B|A) = \frac{P(A \cap B)}{P(A)} = \frac{1/3}{1/2} = \frac{2}{3}$.

Conditional Probability in Everyday Situations

Conditional probability often arises in everyday situations in, for example, estimating the risk or benefit of certain activities. The conditional probability of having a heart attack given that you exercise daily may be smaller than the overall probability of having a heart attack. The conditional probability of having lung cancer given that you are a smoker is larger than the overall probability of having lung cancer. Note that changing the order of the conditional probability changes the meaning: the conditional probability of having lung cancer given that you are a smoker is a very different thing from the probability of being a smoker given that you have lung cancer. In an extreme case, suppose that a certain rare disease is caused only by eating a certain food, but even then, it is unlikely. Then the conditional probability of having that disease given that you eat the dangerous food is nonzero but low, but the conditional probability of having eaten that food given that you have the disease is 100%!

Review Video: Conditional Probability
Visit mometrix.com/academy and enter code: 397924

Independence

The conditional probability $P(A|B)$ is the probability that event A will occur given that event B occurs. If the two events are independent, we do not expect that whether or not event B occurs should have any effect on whether or not event A occurs. In other words, we expect $P(A|B) = P(A)$.

This can be proven using the usual equations for conditional probability and the joint probability of independent events. The conditional probability $P(A|B) = \frac{P(A \cap B)}{P(B)}$. If A and B are independent, then $P(A \cap B) = P(A)P(B)$. So $P(A|B) = \frac{P(A)P(B)}{P(B)} = P(A)$. By similar reasoning, if A and B are independent then $P(B|A) = P(B)$.

Multiplication Rule

The **multiplication rule** can be used to find the probability of two independent events occurring using the formula $P(A \cap B) = P(A) \times P(B)$, where $P(A \cap B)$ is the probability of two independent events occurring, $P(A)$ is the probability of the first event occurring, and $P(B)$ is the probability of the second event occurring.

The multiplication rule can also be used to find the probability of two dependent events occurring using the formula $P(A \cap B) = P(A) \times P(B|A)$, where $P(A \cap B)$ is the probability of two dependent events occurring and $P(B|A)$ is the probability of the second event occurring after the first event has already occurred.

Use a **combination of the multiplication** rule and the rule of complements to find the probability that at least one outcome of the element will occur. This is given by the general formula $P(\text{at least one event occurring}) = 1 - P(\text{no outcomes occurring})$. For example, to find the probability that at least one even number will show when a pair of dice is rolled, find the probability that two odd numbers will be rolled (no even numbers) and subtract from one. You can always use a tree diagram or make a chart to list the possible outcomes when the sample space is small, such as in the dice-rolling example, but in most cases it will be much faster to use the multiplication and complement formulas.

Review Video: Multiplication Rule
Visit mometrix.com/academy and enter code: 782598

Union and Intersection of Two Sets of Outcomes

If A and B are each a set of elements or outcomes from an experiment, then the **union** (symbol $\cup$) of the two sets is the set of elements found in set A or set B. For example, if $A = \{2, 3, 4\}$ and $B = \{3, 4, 5\}$, $A \cup B = \{2, 3, 4, 5\}$. Note that the outcomes 3 and 4 appear only once in the union. For statistical events, the union is equivalent to "or"; $P(A \cup B)$ is the same thing as $P(A \text{ or } B)$. The **intersection** (symbol $\cap$) of two sets is the set of outcomes common to both sets. For the above sets

A and B, $A \cap B = \{3, 4\}$. For statistical events, the intersection is equivalent to "and"; $P(A \cap B)$ is the same thing as $P(A \text{ and } B)$. It is important to note that union and intersection operations commute. That is:

$$A \cup B = B \cup A \text{ and } A \cap B = B \cap A$$

Permutations and Combinations in Probability

When trying to calculate the probability of an event using the $\frac{\text{desired outcomes}}{\text{total outcomes}}$ formula, you may frequently find that there are too many outcomes to individually count them. **Permutation** and **combination formulas** offer a shortcut to counting outcomes. A permutation is an arrangement of a specific number of a set of objects in a specific order. The number of **permutations** of r items given a set of n items can be calculated as ${}_nP_r = \frac{n!}{(n-r)!}$. Combinations are similar to permutations, except there are no restrictions regarding the order of the elements. While ABC is considered a different permutation than BCA, ABC and BCA are considered the same combination. The number of **combinations** of r items given a set of n items can be calculated as ${}_nC_r = \frac{n!}{r!(n-r)!}$ or ${}_nC_r = \frac{{}_nP_r}{r!}$.

Suppose you want to calculate how many different 5-card hands can be drawn from a deck of 52 cards. This is a combination since the order of the cards in a hand does not matter. There are 52 cards available, and 5 to be selected. Thus, the number of different hands is ${}_{52}C_5 = \frac{52!}{5! \times 47!} = 2{,}598{,}960$.

Review Video: Probability - Permutation and Combination
Visit mometrix.com/academy and enter code: 907664

Tree Diagrams

For a simple sample space, possible outcomes may be determined by using a **tree diagram** or an organized chart. In either case, you can easily draw or list out the possible outcomes. For example, to determine all the possible ways three objects can be ordered, you can draw a tree diagram:

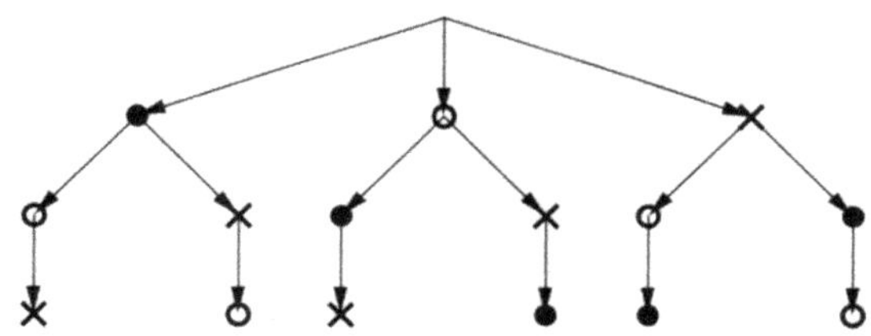

Review Video: Tree Diagrams
Visit mometrix.com/academy and enter code: 829158

You can also make a chart to list all the possibilities:

First object	Second object	Third object
●	x	o
●	o	x
o	●	x
o	x	●
x	●	o
x	o	●

Either way, you can easily see there are six possible ways the three objects can be ordered.

If two events have no outcomes in common, they are said to be **mutually exclusive**. For example, in a standard deck of 52 playing cards, the event of all card suits is mutually exclusive to the event of all card values. If two events have no bearing on each other so that one event occurring has no influence on the probability of another event occurring, the two events are said to be independent. For example, rolling a standard six-sided die multiple times does not change that probability that a particular number will be rolled from one roll to the next. If the outcome of one event does affect the probability of the second event, the two events are said to be dependent. For example, if cards are drawn from a deck, the probability of drawing an ace after an ace has been drawn is different than the probability of drawing an ace if no ace (or no other card, for that matter) has been drawn.

In probability, the **odds in favor of an event** are the number of times the event will occur compared to the number of times the event will not occur. To calculate the odds in favor of an event, use the formula $\frac{P(A)}{1-P(A)}$, where $P(A)$ is the probability that the event will occur. Many times, odds in favor is given as a ratio in the form $\frac{a}{b}$ or $a:b$, where a is the probability of the event occurring and b is the complement of the event, the probability of the event not occurring. If the odds in favor are given as 2:5, that means that you can expect the event to occur two times for every 5 times that it does not occur. In other words, the probability that the event will occur is $\frac{2}{2+5}=\frac{2}{7}$.

In probability, the **odds against an event** are the number of times the event will not occur compared to the number of times the event will occur. To calculate the odds against an event, use the formula $\frac{1-P(A)}{P(A)}$, where $P(A)$ is the probability that the event will occur. Many times, odds against is given as a ratio in the form $\frac{b}{a}$ or $b:a$, where b is the probability the event will not occur (the complement of the event) and a is the probability the event will occur. If the odds against an event are given as 3:1, that means that you can expect the event to not occur 3 times for every one time it does occur. In other words, 3 out of every 4 trials will fail.

Two-Way Frequency Tables

If we have a two-way frequency table, it is generally a straightforward matter to read off the probabilities of any two events A and B, as well as the joint probability of both events occurring, $P(A \cap B)$. We can then find the conditional probability $P(A|B)$ by calculating $P(A|B)=\frac{P(A\cap B)}{P(B)}$. We could also check whether or not events are independent by verifying whether $P(A)P(B)=P(A\cap B)$.

For example, a certain store's recent T-shirt sales:

	Small	Medium	Large	Total
Blue	25	40	35	100
White	27	25	22	74
Black	8	23	15	46
Total	60	88	72	220

Suppose we want to find the conditional probability that a customer buys a black shirt (event A), given that the shirt he buys is size small (event B). From the table, the probability $P(B)$ that a customer buys a small shirt is $\frac{60}{220} = \frac{3}{11}$. The probability $P(A \cap B)$ that he buys a small, black shirt is $\frac{8}{220} = \frac{2}{55}$. The conditional probability $P(A|B)$ that he buys a black shirt, given that he buys a small shirt, is therefore $P(A|B) = \frac{2/55}{3/11} = \frac{2}{15}$.

Similarly, if we want to check whether the event a customer buys a blue shirt, A, is independent of the event that a customer buys a medium shirt, B. From the table, $P(A) = \frac{100}{220} = \frac{5}{11}$ and $P(B) = \frac{88}{220} = \frac{4}{10}$. Also, $P(A \cap B) = \frac{40}{220} = \frac{2}{11}$. Since $\left(\frac{5}{11}\right)\left(\frac{4}{10}\right) = \frac{20}{110} = \frac{2}{11}$, $P(A)P(B) = P(A \cap B)$ and these two events are indeed independent.

Expected Value

Expected value is a method of determining the expected outcome in a random situation. It is a sum of the weighted probabilities of the possible outcomes. Multiply the probability of an event occurring by the weight assigned to that probability (such as the amount of money won or lost). A practical application of the expected value is to determine whether a game of chance is really fair. If the sum of the weighted probabilities is equal to zero, the game is generally considered fair because the player has a fair chance to at least break even. If the expected value is less than zero, then players are expected to lose more than they win. For example, a lottery drawing might allow the player to choose any three-digit number, 000–999. The probability of choosing the winning number is 1:1000. If it costs \$1 to play, and a winning number receives \$500, the expected value is $\left(-\$1 \times \frac{999}{1{,}000}\right) + \left(\$499 \times \frac{1}{1{,}000}\right) = -\$0.50$. You can expect to lose on average 50 cents for every dollar you spend.

Review Video: Expected Value
Visit mometrix.com/academy and enter code: 643554

Expected Value and Simulators

A die roll simulator will show the results of n rolls of a die. The result of each die roll may be recorded. For example, suppose a die is rolled 100 times. All results may be recorded. The numbers of 1s, 2s, 3s, 4s, 5s, and 6s, may be counted. The experimental probability of rolling each number will equal the ratio of the frequency of the rolled number to the total number of rolls. As the number of rolls increases, or approaches infinity, the experimental probability will approach the theoretical probability of $\frac{1}{6}$. Thus, the expected value for the roll of a die is shown to be $\left(1 \times \frac{1}{6}\right) + \left(2 \times \frac{1}{6}\right) + \left(3 \times \frac{1}{6}\right) + \left(4 \times \frac{1}{6}\right) + \left(5 \times \frac{1}{6}\right) + \left(6 \times \frac{1}{6}\right)$, or 3.5.

Chapter Quiz

Ready to see how well you retained what you just read? Scan the QR code to go directly to the chapter quiz interface for this study guide. If you're using a computer, simply visit the online resources page at **mometrix.com/resources719/priimath5161** and click the Chapter Quizzes link.

Discrete Mathematics

Set Theory

Discrete Mathematics

Among mathematicians, there is not an agreed-upon definition of discrete math. What is agreed upon is the fact that discrete math deals with processes that use a finite, or countable, number of elements. In discrete math, the elements will be discontinuous, as this branch of mathematics does not involve the continuity that processes of calculus do. Generally, discrete math uses countable sets of rational numbers, although they do not use the set of all real numbers, as that would then make the math continuous and put it in the category of algebra or calculus. Discrete math has numerous applications in the fields of computer science and business.

Review Video: Introduction to Set Theory
Visit mometrix.com/academy and enter code: 397496

Element of a Set

A set is a mathematical collection of items, and an element is an item that is included in the set. These items are typically sets of numbers, or sets of geometrical points, or even sets of sets.

Whether an item is an **element of a set** is a binary property. In other words, a particular item either is or is not an element of a given set. There are not different degrees of belonging to a set, and one element of a set cannot be more an element than another. Additionally, the elements do not have any particular order, and there is no count of how many times an element appears in a set. For example, in the set $\{0, 2, 1, 2, 4\}$, the numbers 0, 1, 2, and 4 are elements of the set. Two is not more of an element than the others because it is listed twice. Any number other than these, such as 3, is not an element of the set.

The mathematical symbol $\in$ means "is an element of." for instance, "$1 \in A$" means that "the number 1 is an element of set A". The symbol can be negated with a diagonal slash, so "$1 \notin A$" means "the number 1 is *not* an element of set A."

Empty Set

The empty set contains no elements. The symbol for the **empty set** is a circle with a line through it, $\emptyset$, or the empty set can be written in roster form as $\{\quad\}$. Although it may seem trivial, the empty set is important for the same reason that zero is an important number—without the empty set, many set operations and definitions would be incomplete. For instance, the intersection of two non-overlapping sets is the empty set. The set of all prime numbers that are perfect squares is the empty set, because there are no such numbers. Other sets can be constructed using the empty set. For instance, a set *containing* the empty set, $\{\{\quad\}\}$ or $\{\emptyset\}$, is different from the empty set itself, since it's not actually empty—it contains one element, the empty set.

The empty set is, by definition, a subset of every set, including itself. This is because every element that is an element of the empty set is also an element of any other set, since the empty set contains no elements that could make this statement false.

Subset, Proper Subset, and Superset

Set A is a **subset** of set B if set A is contained entirely within set B. More formally, set A is a subset of set B if every element of set A is also in set B. We can write this as $\forall x(x \in A \Rightarrow x \in B)$. The symbol for subset is $\subseteq$, so we can write "$A \subseteq B$" to mean "set A is a subset of set B," or "$A \nsubseteq B$" to mean "set A is NOT a subset of set B".

Any set is a subset of itself, since naturally any element in a set is in that set. A **proper subset** means a subset that is not equal to the other set. In other words, set A is a proper subset of set B if set A is a subset of set B but they are not the same, so $A \subseteq B$ but $B \nsubseteq A$. For example, $\{0, 1, 2\}$ is a proper subset of $\{0, 1, 2, 3\}$ because every element in the first set is included in the second set but the two sets are not identical. The symbol for a proper subset is $\subset$, so we write "$A \subset B$" to mean "set A is a proper subset of set B," and "$A \not\subset B$" to mean "set A is not a proper subset of set B."

The converse of a subset is a **superset**. If A is a subset of B, then B is a superset of A. In the example above, $\{0, 1, 2, 3\}$ is a superset of $\{0, 1, 2\}$ because the first set contains every element of the second plus more.

Union of Two or More Sets

The **union** of two or more sets includes all the elements that are included in all of the sets. An element is in the union of sets A and B if it is in set A or in set B, or in both. The union is written with the symbol $\cup$, so "$A \cup B$" (read "A union B") means the union of sets A and B.

For example, given the sets $A = \{1, 2, 3, 4\}$ and $B = \{2, 4, 6, 8\}$, then $A \cup B = \{1, 2, 3, 4, 6, 8\}$. 1 and 3 are in $A \cup B$ because they are in set A, and 6 and 8 are in $A \cup B$ because they are in set B. 2 and 4 appear in both A and B, so they are in $A \cup B$, but they only appear once.

The union operation on sets is both commutative and associative. That is, $A \cup B = B \cup A$, and $(A \cup B) \cup C = A \cup (B \cup C)$.

Intersection of Two or More Sets

The intersection of two or more sets is a set including every element that appears in *all* of the sets. An element is in the intersection of sets A and B if and only if, the element appears in both set A and set B. For instance, if set $A = \{0, 1, 2, 3\}$ and set $B = \{2, 3, 4, 5\}$, the intersection of sets A and B includes the elements 2 and 3, because these are the only elements that appear in both sets. The intersection is written with the symbol $\cap$, so "$A \cap B$" (read "A intersection B") means the intersection of sets A and B.

The intersection operation on sets is both commutative and associative. That is, $A \cap B = B \cap A$, and $(A \cap B) \cap C = A \cap (B \cap C)$. The union and intersection operations also jointly have distributive properties: $A \cap (B \cup C) = (A \cap B) \cup (A \cap C)$ and $A \cup (B \cap C) = (A \cup B) \cap (A \cup C)$.

Complement of a Set

The complement of a set is a set containing all the elements that are *not* in the original set. To determine this, we must also define the **universe of discourse**—the set of all possible elements we're considering, abbreviated U. For mathematical applications, the universe of discourse is commonly the set of all integers (abbreviated $\mathbb{Z}$), or the set of all real numbers (abbreviated $\mathbb{R}$). If U is the set of all integers, for example, then the complement of the set $A = \{1, 2, 3\}$ would include elements such as -3, 0, or 6, but it would not include elements such as $\sqrt{7}$ or $\frac{1}{4}$, because these are not in the universe of discourse.

Discrete Mathematics

The **complement of a set** is often written either by drawing a line over the name of the original set, or by adding an apostrophe or a superscripted C. The complement of set A, for instance, could be written as $\bar{A}$, as A', or as A^C.

For example, given set $A = \{2, 4, 6, 8\}$ with a universe of discourse of $U = \{1, 2, 3, 4, 5, 6, 7, 8, 9, 10\}$, the complement of set A would be $\{1, 3, 5, 7, 9, 10\}$—all the elements of U that are not in A.

Difference of Two Sets

The difference of two sets is a set containing all the elements that are in the first set but not in the second. The difference of sets A and B is also referred to as the **relative complement** of set A with respect to set B, and is written $A \setminus B$ or $A - B$. The difference is always a subset of the first set: $A \setminus B \subseteq A$. However, it is *not* necessarily the case that $A \setminus B \subseteq B$. In fact, because any element of $A \setminus B$ is by definition not an element of B, the only time that $A \setminus B$ can be a subset of B is if $A \setminus B$ has no elements, i.e., if $A \setminus B = \emptyset$, which is true if and only if, $A \subseteq B$.

Unlike the union and intersection, the operation of the difference between sets is neither commutative nor associative. In general, $A \setminus B \neq B \setminus A$, and $A \setminus (B \setminus C) \neq (A \setminus B) \setminus C$.

For example, consider the sets $A = \{1, 2, 3\}$, $B = \{2, 4\}$, and $C = \{1, 2\}$. Then $A \setminus B = \{1, 3\}$ since 1 and 3 are in set A but not in set B. Similarly, $B \setminus A = \{4\}$, $B \setminus C = \{4\}$, $C \setminus B = \{1\}$, and $A \setminus C = \{3\}$. However, $C \setminus A = \emptyset$, i.e., set C is a subset of set A, so there are no elements that appear in C but not in A, which means the difference between C and A is the empty set.

Symbols for Sets Commonly Used in Mathematics

Certain sets are used frequently enough in mathematics to have their own symbols. The standard symbols for these sets resemble upper-case letters of the Latin alphabet, but in a typeface with double lines. Among the most commonly used **sets** are:

- $\mathbb{R}$ – the set of all real numbers
- $\mathbb{Z}$ – the set of all integers
- $\mathbb{N}$ – the set of all natural numbers
- $\mathbb{Q}$ – the set of all rational numbers (numbers that can be written as the ratio of two integers, $\frac{p}{q}$, where $q \neq 0$)
- $\mathbb{C}$ – the set of all complex numbers (numbers of the form $a + bi$, where a and b are real numbers and $i = \sqrt{-1}$)

A superscripted plus or minus sign is used to restrict the set to positive or negative numbers. for instance, $\mathbb{R}^+$ would be the set of all positive real numbers, and $\mathbb{Z}^-$ would be the set of all negative integers.

Some of these sets are, of course, subsets of others. Specifically, $\mathbb{N} \subset \mathbb{Z} \subset \mathbb{Q} \subset \mathbb{R} \subset \mathbb{C}$.

Recursive and Explicit Formulas

Sets of numbers can have a consistent relationship between the elements of the set. Some of these relationships can be represented with simple recursive or explicit formulas. For example, consider the set of all positive, even numbers (a) and the set of all positive, odd numbers (b):

Positive, Even Numbers:

Recursive: $a_n = a_{n-1} + 2, n \geq 2, a_1 = 0$

Explicit: $a_n = 2n, n \geq 0$

POSITIVE, ODD NUMBERS:

Recursive: $b_n = b_{n-1} + 2, n \geq 2, b_1 = 1$
Explicit: $b_n = 2n + 1, n \geq 0$

Each set of numbers represents a linear function, with a constant rate of change of 2. The positive, even numbers represent a linear function that is proportional, whereas the positive, odd numbers represent a linear function that is not proportional. The set of even, positive numbers is represented by a function with a y-intercept of 0. The set of odd, positive numbers is represented by a function with a y-intercept of 1.

Venn Diagrams

SET OPERATIONS WITH VENN DIAGRAMS

A Venn diagram is a useful visual tool for representing two or three sets and their common elements. Each set is drawn as a **circle**, with the different circles overlapping. Elements are placed in the circle corresponding to the appropriate set or sets—or placed outside all the circles if they belong to the universe of discourse but not to any of the sets.

For example, suppose our universe of discourse is the integers from 1 to 9, and we have the three sets $A = \{1, 2, 3, 4, 5, 6\}$, $B = \{4, 5, 6, 7\}$, and $C = \{3, 6, 9\}$. This could be illustrated with the following **Venn diagram**:

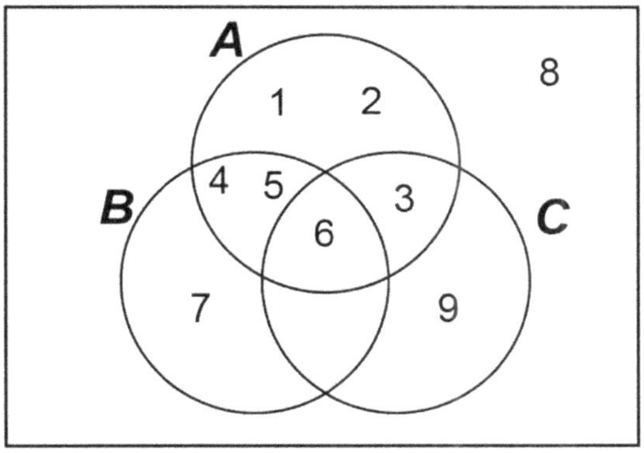

Note, for example, that 6 appears in the center of the diagram where all three circles overlap, because it is an element of all three sets. On the other hand, 8 is placed outside the three circles because it is not an element of any of the sets. Sometimes, instead of writing the elements themselves in the diagram, the total *number* of elements in each part of the Venn diagram is noted. This is especially useful for solving problems involving these numbers of elements.

USING VENN DIAGRAMS TO SOLVE PROBLEMS

Venn diagrams are useful for solving problems involving the numbers of elements in sets and in their intersections. By putting those numbers into a Venn diagram, it's simple to see how many must be in the "leftover" parts.

For example, suppose we're told that 200 voters were polled about two propositions, Proposition 1 and Proposition 2. Further,120 support Proposition 1, 85 support Proposition 2, and 50 support both propositions. To find how many of the voters support neither proposition, we can draw a Venn diagram with a circle representing the supporters of each proposition. We know 50 voters support both propositions, so we can write a 50 in the center of the diagram.

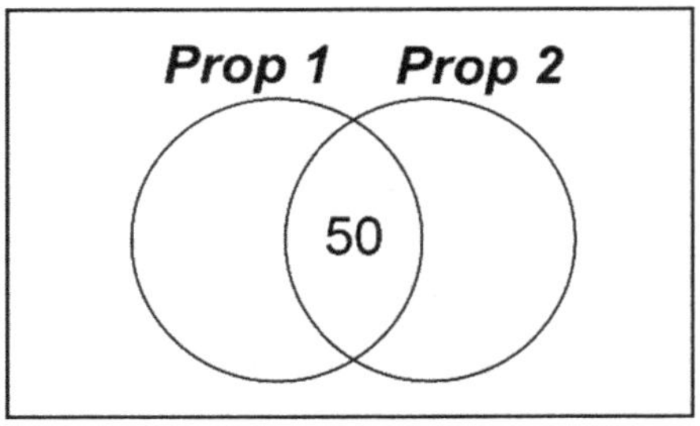

The Proposition 1 circle should contain 120 voters total, so subtracting the 50 voters in the overlap, the other section of the circle must contain $120 - 50 = 70$. Similarly, the nonoverlapping part of the Proposition 2 circle should contain $85 - 50 = 35$.

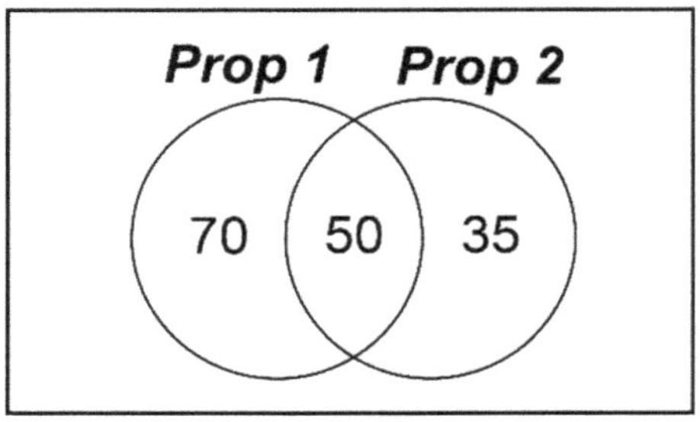

Adding all three sections, the total number of voters supporting either proposition is $70 + 50 + 35 = 155$, so there must be $200 - 155 = 45$ voters who support neither.

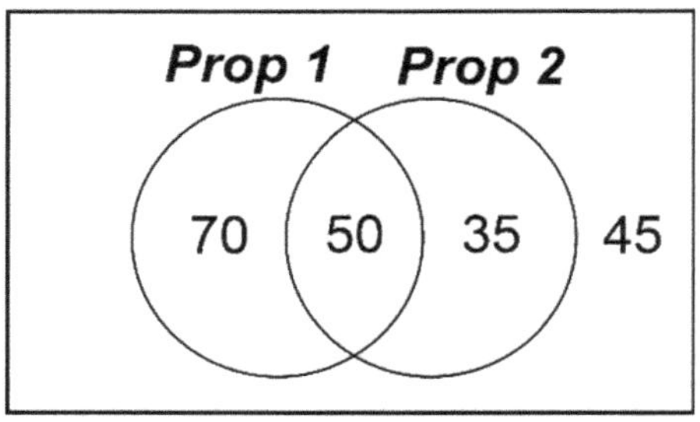

Advanced Set Theory

Properties of Infinite Sets

An infinite set has an infinite number of elements. a more technical definition is that a set is **infinite** if its elements can be put in one-to-one correspondence with the elements of one of its proper subsets. For instance, consider the set of all positive integers, $\mathbb{Z}^+$. If we remove the number 1, we can still match up every positive integer to an element of the remaining subset (by simply matching the number n to the number $n + 1$). $\mathbb{Z}^+$ is therefore an infinite set.

The union of an infinite set and any other set is always infinite. The intersection of two infinite sets may or may not be infinite. For instance, consider the set P of prime numbers, the set O of positive odd integers, and the set E of positive even integers—all infinite sets. $P \cap O$ is another infinite set, because there are infinitely many odd prime numbers. $P \cap E$ is finite, because there is only one even prime number. $E \cap O$ is the empty set. Likewise, the complement of an infinite set may be infinite, finite, or the empty set (if the set equals the universe of discourse).

Cardinality

Although all infinite sets have infinitely many members, they may still have different sizes. Two sets are defined to have the same size—or more technically the same **cardinality**—if each element of one set can be matched up to a unique element of the other, with none left over. For example, the set of all integers, $\mathbb{Z}$, and the set of all *even* integers have the same cardinality, even though the latter is a proper subset of the former: every number n in $\mathbb{Z}$ can be matched to a unique number $2n$ in the set of all *even* integers. Although it's more difficult to prove, the set of all rational numbers, $\mathbb{Q}$, also has the same cardinality as $\mathbb{Z}$. However, the set of all real numbers, $\mathbb{R}$, does *not* have the same cardinality—it is impossible to match each integer to a unique real number without any real numbers left over. The most famous proof of this fact was developed by Georg Cantor, and is called the *diagonal argument*.

Sets with the same cardinality as $\mathbb{Z}$ are said to be "countably infinite," or simply **countable**. Sets with a larger cardinality are said to be "uncountably infinite," or **uncountable**.

Cartesian Products and Binary Relations

A **Cartesian product** is the product of two sets of data, X and Y, such that all elements x are a member of set X, and all elements y are a member of set Y. The product of the two sets, $X \times Y$ is the set of all ordered pairs (x, y). For example, given a standard deck of 52 playing cards, there are four possible suits (hearts, diamonds, clubs, and spades) and thirteen possible card values (the numbers 2 through 10, ace, jack, queen, and king). If the card suits are set X and the card values are set Y, then there are $4 \times 13 = 52$ possible different (x, y) combinations, as seen in the 52 cards of a standard deck.

A **binary relation**, also referred to as a relation, dyadic relation, or 2-place relation, is a subset of a Cartesian product. It shows the relation between one set of objects and a second set of objects, or between one set of objects and itself. The prefix *bi-* means *two*, so there are always two sets involved—either two different sets, or the same set used twice. The ordered pairs of the Cartesian product are used to indicate a binary relation. Relations are possible for situations involving more than two sets, but those are not called binary relations.

The five types of relations are reflexive, symmetric, transitive, antisymmetric, and equivalence. A **reflexive relation** has $x\mathfrak{R}x$ (x related to x) for all values of x in the set. A **symmetric relation** has $x\mathfrak{R}y \Rightarrow y\mathfrak{R}x$ for all values of x and y in the set. A **transitive relation** has ($x\mathfrak{R}y$ and $y\mathfrak{R}z$) $\Rightarrow x\mathfrak{R}z$ for all values of x, y, and z in the set. An **antisymmetric relation** has ($x\mathfrak{R}y$ and $y\mathfrak{R}x$) $\Rightarrow x = y$ for all values of x and y in the set. A relation that is reflexive, symmetric, and transitive is called an **equivalence relation**.

Vertex-Edge Graphs

A **vertex-edge graph** is a set of items or objects connected by pathways or links. As an example, consider the following set of nodes and a few graphs representing ways of connecting them with edges.

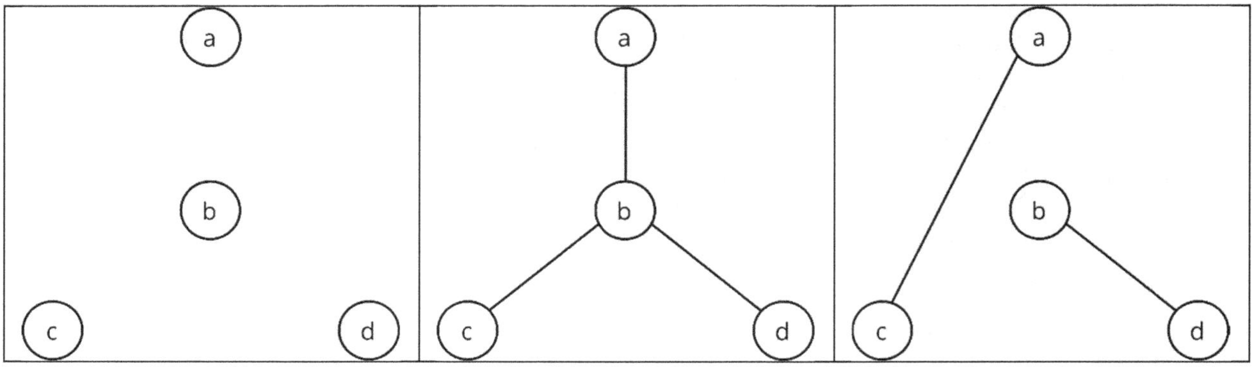

Vertex-edge graphs are useful for solving problems involving schedules, relationships, networks, or paths among a set number of objects. The number of objects may be large, but it will never be infinite. The **vertices** or points on the graph represent the objects and may also be referred to as **nodes**. The nodes are joined by line segments called **edges** or links that show the specific paths that connect the various elements represented by the nodes. The number of nodes does not have to equal the number of edges. There may be more or less, depending on the number of allowable paths.

An **endpoint** on a vertex-edge graph is a vertex on exactly one edge. In the case of a vertex that is an endpoint, the edge that the vertex is on is incident with the vertex. Two edges are considered to be adjacent if they share a vertex. Two vertices are considered to be adjacent if they share an edge.

In a vertex-edge graph, a **loop** is an edge that has the same vertex as both endpoints. To calculate the **degree of a vertex** in a vertex-edge graph, count the number of edges that are incident with the vertex, counting loops twice since they meet the vertex at both ends. The **degree sum formula** states that the sum of the degrees of all vertices on a vertex-edge graph is always equal to twice the number of edges on the graph. Thus, the sum of the degrees will never be odd, even if there are an odd number of vertices. Consider the following graph. Node d is an endpoint, there is a loop on node b, and the degree sum of the graph is $2 \times 5 = 10$.

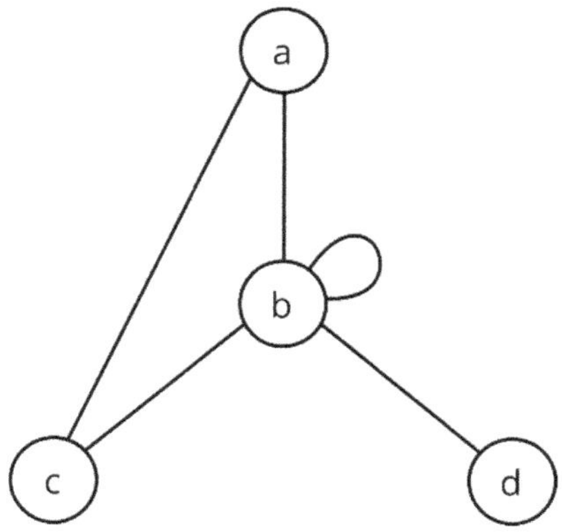

In a vertex-edge graph, a **path** is a given sequence of vertices that follows one or more edges to get from vertex to vertex. There is no jumping over spaces to get from one vertex to the next, although doubling back over an edge already traveled is allowed. A **simple path** is a path that does not repeat an edge in traveling from beginning to end. Think of the vertex-edge graph as a map, with the vertices as cities on the map, and the edges as roads between the cities. To get from one city to another, you must drive on the roads. A simple path allows you to complete your trip without driving on the same road twice.

In a vertex-edge graph, a **circuit** is a path that has the same starting and stopping point. Picturing the vertex-edge graph as a map with cities and roads, a circuit is like leaving home on vacation and then returning home after you have visited your intended destinations. You may go in one direction and then turn around, or you may go in a circle. A **simple circuit** on the graph completes the circuit without repeating an edge. This is like going on vacation and returning home without driving on the same road twice.

Graph

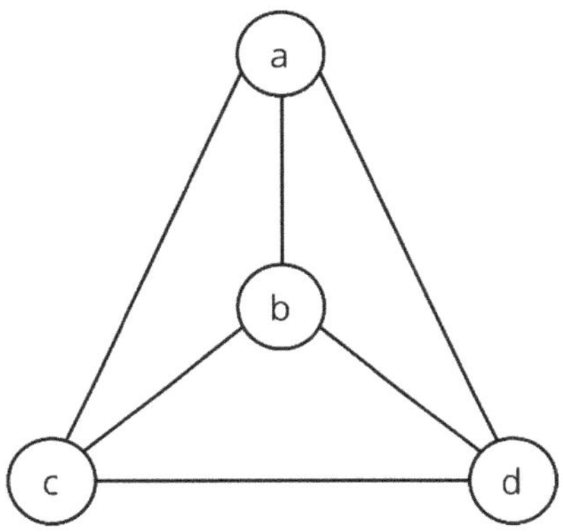

Example Paths

Path:
a-b-c-b-d

Simple path:
a-b-d

Circuit:
b-a-c-a-b

Simple circuit:
a-b-c-a

On a vertex-edge graph, any path that uses each edge exactly one time is called an **Euler path**. One simple way to rule out the possibility of an Euler path is to calculate the degree of each vertex. If more than two vertices have an odd degree, an Euler path is impossible. A path that uses each vertex exactly one time is called a **Hamiltonian path**.

Graph

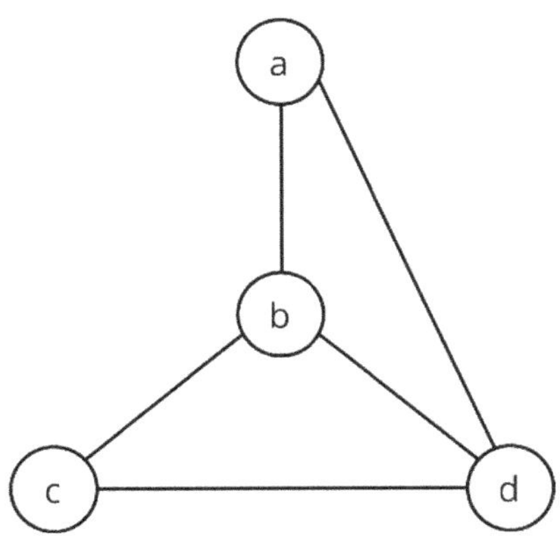

Example Paths

Euler path:
b-c-d-a-b-d

Hamiltonian path:
a-b-c-d

If every pair of vertices is joined by an edge, the vertex-edge graph is said to be **complete**. If the vertex-edge graph has no simple circuits in it, then the graph is said to be a **tree**. If every vertex is

Discrete Mathematics

connected to every other vertex by some *path*, then the graph is said to be **connected**, otherwise it is **disconnected**.

Complete and Connected

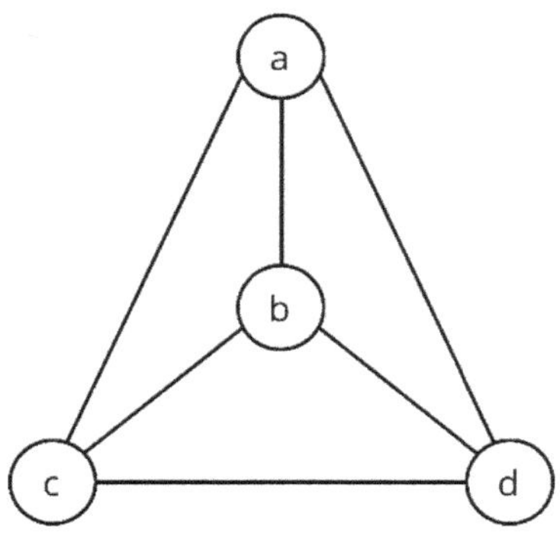

Tree and Connected

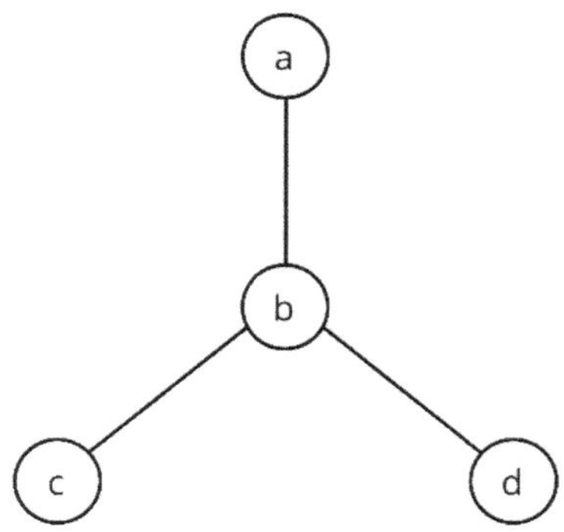

Disconnected

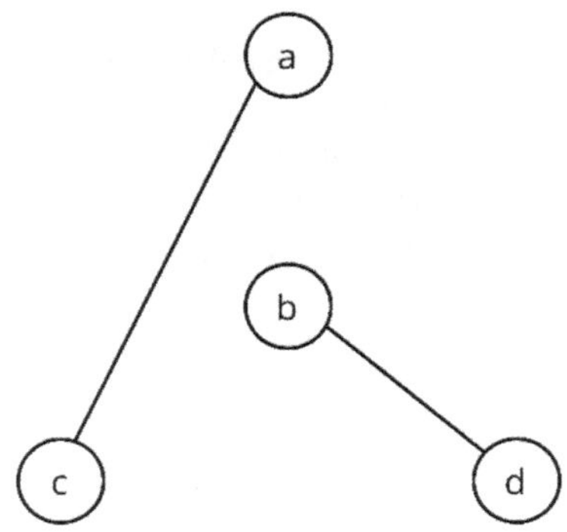

Praxis Practice Test #1

Want to take this practice test in an online interactive format?
Check out the online resources page, which includes interactive practice questions and much more: **mometrix.com/resources719/priimath5161**

1. Ava and Jack packed a picnic lunch and went for a hike. The graph below shows their distance from the start of the trail at times throughout the hike. If the x-axis represents the number of hours they spent and the y-axis represents their distance from the beginning of the trail in miles, which of the following statements is (are) true?

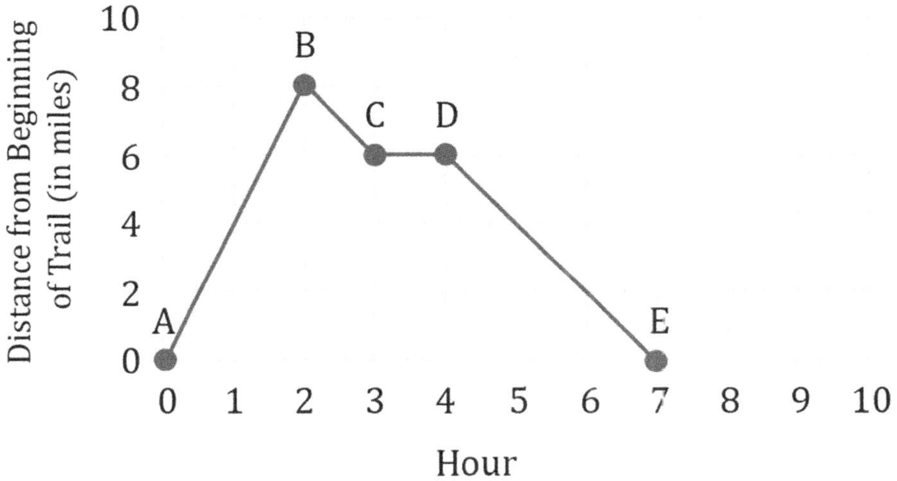

I. Their average speed was 4 mph.
II. Their speed from point B to point C was the same as from point D to E.
III. They walked a total distance of sixteen miles.

a. I only
b. II only
c. II and III
d. I, II, and III

2. Consider the function f, whose graph appears here, and the function g, defined by $g(x) = -x^2 + 8x - 15$. Which of the following statements correctly describes properties these two functions have in common?

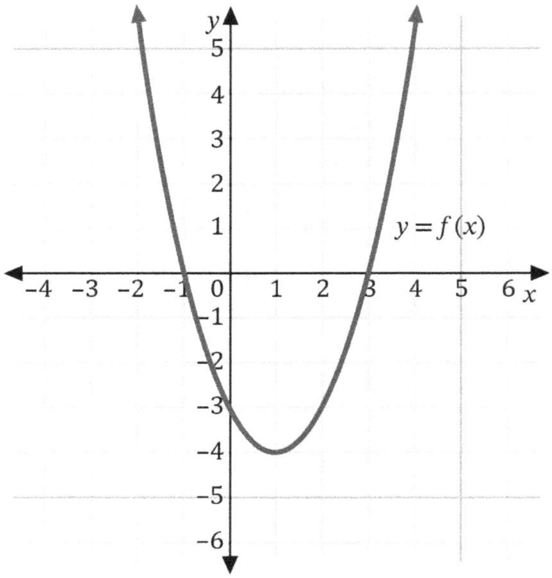

a. They have the same vertex.
b. They have the same two roots.
c. They have the same y-intercept.
d. They have exactly one common root.

3. Which of these does NOT simulate randomly selecting a person from a group of 11?

a. Assigning each student a unique card value of A, 2, 3, 4, 5, 6, 7, 8, 9, 10, or J and drawing a single card from a deck of cards, redrawing if an invalid card is selected.
b. Assigning each student a unique number 2–12 and rolling a pair of dice to determine which student is selected.
c. Assigning each student a unique letter A–K, labeling 11 index cards A–K, and blindly drawing one.
d. Assigning each student a unique numeric value of 1–11 and spinning a spinner mounted on a 11-section wheel.

4. Graph the following piecewise function and use your graph to determine which of the following statements about discontinuities and extrema is correct.

$$f(x) = \begin{cases} 4 - x^2, & \text{if } x < 0 \\ 4 - 2x, & \text{if } 0 \le x < 2 \\ 7 - x, & \text{if } x \ge 2 \end{cases}$$

a. The function has 2 discontinuities and a maximum value of 7.
b. The function has 1 discontinuity and a maximum of 5.
c. The function has 2 discontinuities and no extrema.
d. The function has no discontinuities and a maximum of 4.

5. Find the area between the curves $f(x) = \frac{1}{2}x^2$ and $g(x) = \sqrt{x} - \frac{1}{x^2}$ over the interval $[1, 4]$. Hint: The graphs of these two functions never cross.

a. $6\frac{7}{12}$
b. $5\frac{1}{12}$
c. $14\frac{5}{12}$
d. $15\frac{9}{14}$

6. A cube inscribed in a sphere has a volume of 125 cubic units. What is the volume of the sphere in cubic units?

a. $25\pi\sqrt{3}$
b. $\frac{125\pi\sqrt{3}}{2}$
c. $\frac{375\pi\sqrt{3}}{8}$
d. $\frac{5\pi\sqrt{3}}{12}$

7. A target is made of concentric circles, as shown below. The width of each band is $2m$ and the center is a circle with radius of $2m$.

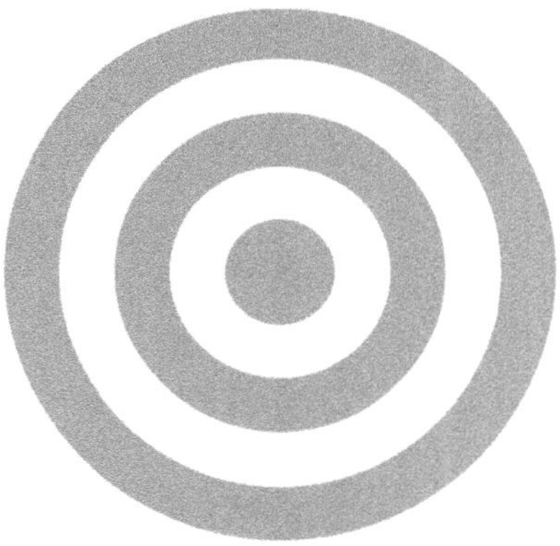

Which of the following gives the area of the two white rings as a function of m?

a. $A = 16\pi m^2$
b. $A = 24\pi m^2$
c. $A = 40\pi m^2$
d. $A = 64\pi m^2$

8. A bird looks down from a tree at a 30° angle of depression and sees a beetle in the grass. When he turns his gaze downward to a 60° angle of depression, he sees a worm. If the beetle is 30 feet from the worm, how far from the bird in the tree is the worm?

a. 26 feet
b. 30 feet
c. 45 feet
d. 52 feet

Praxis Practice Test #1

Refer to the following for questions 9–10:

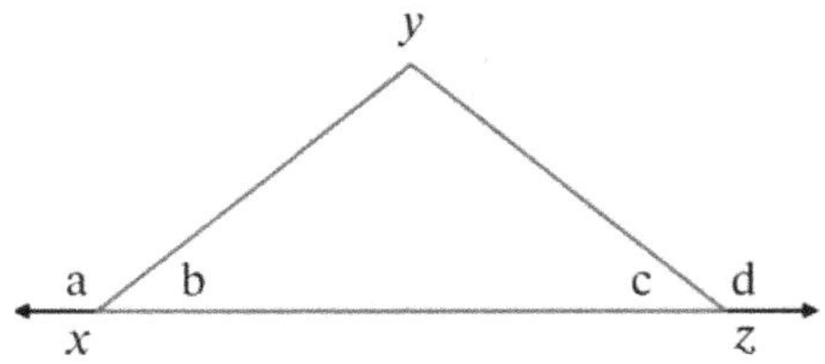

Given: $\angle a \cong \angle d$
Prove: ΔXYZ is isosceles

Statement	Reason
1. $\angle a \cong \angle d$	Given
2. $m\angle a + m\angle b = 180°; m\angle c + m\angle d = 180°$	Linear pair angles are supplementary
3. $m\angle a + m\angle b = m\angle c + m\angle d$	Substitution
4. $m\angle a = m\angle d$	Congruent angles have equal measures
5. $m\angle b = m\angle c$	__________
6. $\angle b \cong \angle c$	Congruent angles have equal measures
7. __________	If two angles of a triangle are congruent, the sides opposite them are congruent
8. ΔXYZ is isosceles	An isosceles triangle has two congruent sides

■

9. Which of the following justifies step 5 in the proof?

a. Linear pair angles are supplementary
b. Subtraction property of equality
c. Definition of supplementary angles
d. An isosceles triangle has two congruent angles

10. Step 7 in the proof should contain which of the following statements?

a. $\overline{XY} \cong \overline{ZY}$
b. $\overline{XZ} \cong \overline{XZ}$
c. $\overline{XY} \cong \overline{XZ}$
d. $\angle YXZ \cong \angle YZX$

11. Zeke drove from his house to a furniture store in Atlanta and then back home along the same route. It took Zeke three hours to drive to the store. By driving an average of 20 mph faster on his return trip, Zeke was able to save an hour of driving time. What was Zeke's average driving speed on his round trip?

12. Which of these graphs is NOT representative of the data set shown below?

Stem	Leaf
3	6799
4	23889
5	011158
6	00123

Key: $2|123 = 21, 22, 23$

a.
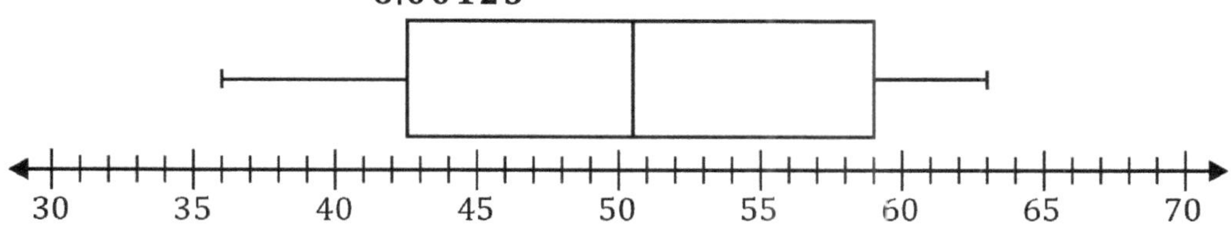

b.
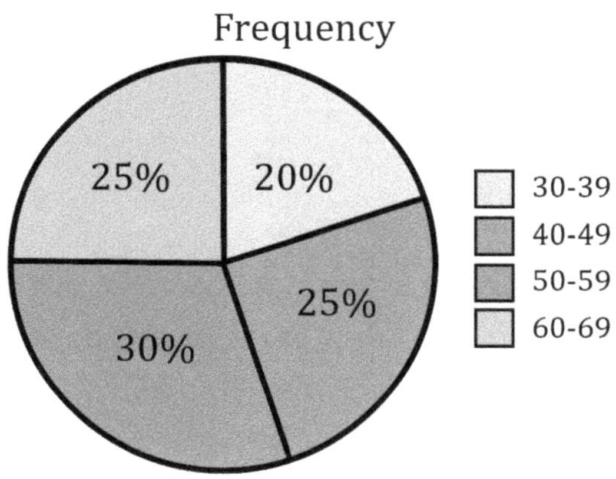

c.
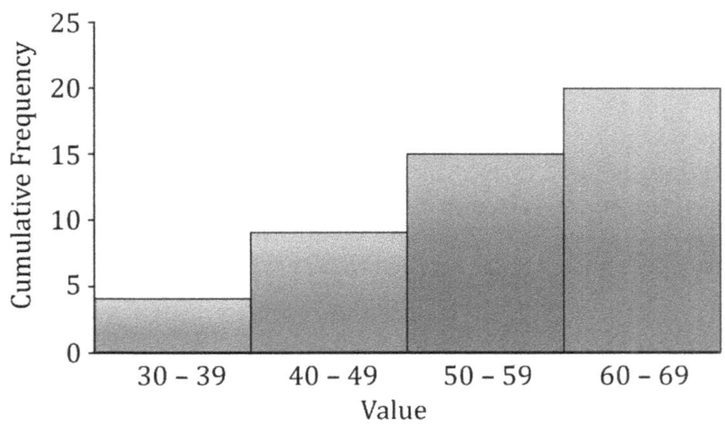

d. All of these graphs represent the data set.

13. Last year, Jenny tutored students in math, in chemistry, and for the ACT. She tutored 10 students in math, 8 students in chemistry, and 7 students for the ACT. She tutored 5 students in math and chemistry, and she tutored 4 students in chemistry and for the ACT, and 5 students in math and for the ACT. She tutored 3 students in all three subjects. How many students did Jenny tutor last year?

14. This question may have more than one correct answer. Select all correct answers.

You ask your students to reduce the fraction $\frac{72}{120}$ to lowest terms in a single step by canceling the greatest common factor of the numerator and denominator. Three students take different approaches to finding the greatest common factor. Which of their approaches are mathematically correct? Select all correct approaches.

a.

Student A concludes that the greatest common factor is 24 based on the following calculation.

$$120 - 72 = 48$$
$$72 - 48 = 24$$
$$48 = 2 \cdot 24$$

b.

Student B finds the prime factorizations of 72 and 120.

$$72 = 8 \cdot 9 = 2^3 \cdot 3^2$$
$$120 = 10 \cdot 12 = 2 \cdot 5 \cdot 3 \cdot 4 = 2^3 \cdot 3 \cdot 5$$

Student B then forms a product of the common primes raised to the lower power to which they appear.

$$2^3 \cdot 3 = 24$$

c.

Student C works out that the factors of 72 are 1, 2, 3, 4, 6, 8, 9, 12, 18, 24, 36, and 72. Then finds the factors of 120 are 1, 2, 3, 4, 5, 6, 8, 10, 12, 15, 20, 24, 30, 40, 60, and 120. After some examination, the student finds that the greatest number common to the two lists is 24.

15. The heights of 200 high school males are normally distributed with a mean height of 67.5 inches and a standard deviation of 3.5 inches. In a normal distribution, approximately 68% of values are within one standard deviation of the mean. About how many individuals from the selected group are shorter than 5'11"?

a. 125
b. 142
c. 168
d. 184

16. Solve: $7x^2 + 6x = -2$.

a. $x = \frac{-3\pm\sqrt{23}}{7}$
b. $x = \pm i\sqrt{5}$
c. $x = \pm\frac{2i\sqrt{2}}{7}$
d. $x = \frac{-3\pm i\sqrt{5}}{7}$

17. Calculate $\int\left(6x^2 + \frac{1}{2}x - 3\right)dx$.

a. $x^3 + x^2 - 3x + C$
b. $12x^3 + x^2 - 3x + C$
c. $2x^3 + \frac{1}{4}x^2 - 3x + C$
d. $12x^2 + x + C$

18. Which of these is the least biased sampling technique?

a. A politician interviews every 8th grade student in the district to get the opinion of youth on a new policy.
b. A business directs users to leave anonymous reviews online.
c. A principal goes through the school directory and calls every 5th family to request opinions on a policy change.
d. A political commercial asks voters to call in with ideas for reform so that he can get a better understanding of what his constituents want.

Refer to the following for questions 19–20:

The box-and-whisker plot displays student test scores assessed throughout a semester to see if students were improving.

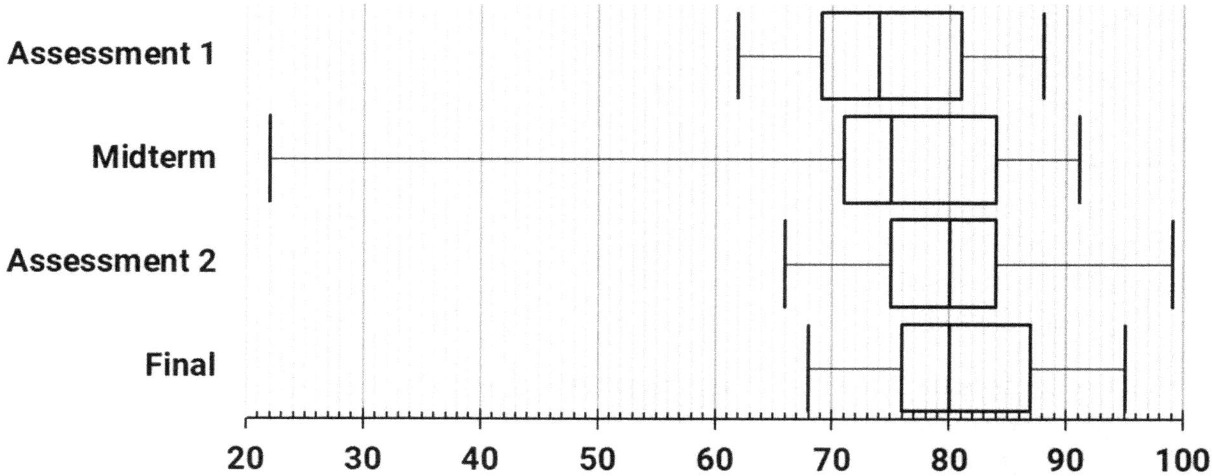

19. Which assessment has the greatest range of test scores?

a. Assessment 1
b. Midterm
c. Assessment 2
d. Final

20. What is the probability that a test score chosen at random from the Final is less than 87?

a. 0.25
b. 0.5
c. 0.6
d. 0.75

21. This question may have more than one correct answer. Select all correct answers. Three students attempt to solve the equation $5^{3x-1} = 96$. Which of their answers are correct? Select all correct answers.

a.

Student A

$$\begin{aligned} 5^{3x-1} &= 96 \\ \log_2 5^{3x-1} &= \log_2 96 \\ (3x-1)\log_2 5 &= \log_2(32 \cdot 3) \\ 3x - 1 &= \frac{\log_2 32 + \log_2 3}{\log_2 5} \\ 3x &= \frac{5 + \log_2 3}{\log_2 5} + 1 \\ x &= \frac{5 + \log_2 3}{3\log_2 5} + \frac{1}{3} \end{aligned}$$

b.

Student B

$$\begin{aligned} 5^{3x-1} &= 96 \\ \log 5^{3x-1} &= \log 96 \\ (3x-1)\log 5 &= \log 96 \\ 3x - 1 &= \frac{\log 96}{\log 5} \\ 3x &= \log 96 - \log 5 + 1 \\ x &= \frac{\log 96 - \log 5 + 1}{3} \end{aligned}$$

c.

Student C

$$\begin{aligned} 5^{3x-1} &= 96 \\ \ln 5^{3x-1} &= \ln 96 \\ (3x-1)\ln 5 &= \ln 96 \\ 3x - 1 &= \frac{\ln 96}{\ln 5} \\ 3x &= \frac{96}{5} + 1 \\ x &= \frac{96}{15} + \frac{1}{3} = \frac{101}{15} \end{aligned}$$

22. This question may have more than one correct answer. Select all correct answers.

You give your students a test on which they are supposed to find the range of the function $f(x) = 3x^2 - 18x + 30$. Three of your students get the correct answer but show different work explaining their answers. Which of your students arrived at their answers in mathematically correct ways? Select all correct approaches.

a. Student A writes, "The leading coefficient is positive, so the parabola opens upward. The GCF of the coefficients of the function is $\text{gcf}(3,18,30) = 3$. The range consists of all values greater than or equal to this GCF. In other words, the range is $[3, \infty)$.

b. Student B writes, "I complete the square to put the function in vertex form. This gives me $f(x) = 3x^2 - 18x + 30 = 3(x^2 - 6x) + 30 = 3(x^2 - 6x + 9) + 30 - 27 = 3(x - 3)^2 + 3$. The vertex of the parabola is at the point (3,3). Since the leading coefficient is positive, the range consists of all real numbers greater than or equal to the y-value of the vertex. Thus, the range of the function is $[3, \infty)$."

c. Student C writes, "The graph of $f(x) = ax^2 + bx + c$ is a parabola with axis of symmetry $-\frac{b}{2a}$. For the given parabola, $a = 3$ and $b = -18$, so the axis of symmetry is $-\frac{-18}{2 \cdot 3} = 3$. The range consists of all values greater than or equal to the axis of symmetry since the leading coefficient is positive. That is, the range is $[3, \infty)$."

23. If 1 inch on a map represents 10 feet, how many yards apart are two points if the distance between the points on the map is 24 inches?

a. 240 yards
b. 80 yards
c. 24 yards
d. 12 yards

24. Which of these equations is represented by the graph below?

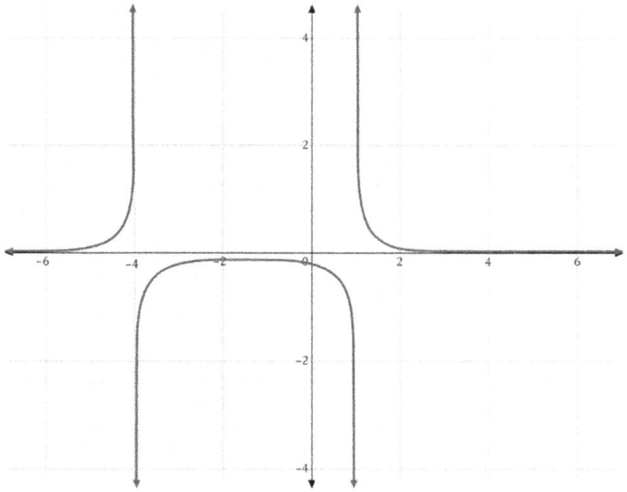

a. $y = \frac{2x-2}{x^2+3x-4}$
b. $y = \frac{1}{x^2+3x-4}$
c. $y = \frac{1}{x+4} + \frac{1}{x-1}$
d. $y = \frac{2}{3x^2+9x-12}$

25. Identify the cross-section polygon formed by a plane containing the given points on the cube.

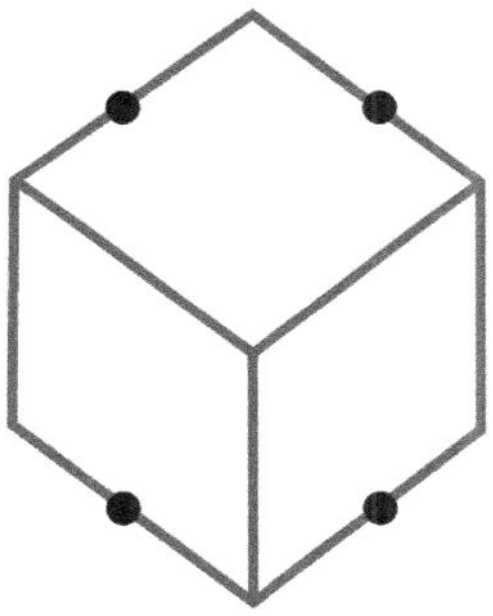

a. Rectangle
b. Trapezoid
c. Pentagon
d. Hexagon

26. Find $(g \circ f)(x)$ when $f(x) = 2x + 4$ and $g(x) = x^2 - 3x + 2$.

a. $4x^2 + 10x + 6$
b. $2x^2 - 6x + 8$
c. $4x^2 + 13x + 18$
d. $2x^2 - 3x + 6$

27. Which of these defines the recursive sequence $a_1 = 2, a_{n+1} = a_n - 3$ explicitly?

a. $a_n = 2n - 3$
b. $a_n = -3n + 5$
c. $a_n = n - 1$
d. $a_n = -2n + 2$

28. Find $f''(x)$ if $f(x) = -3x^3 - x^2 + 2x - 4$.

a. $-9x^2 - 2x + 2$
b. $-18x - 2$
c. $-27x^2 - 2x$
d. $-9x^3 - 2x^2 - 2x$

29. A student begins solving the linear inequality $3x - 7 < 20$ as follows.

$$3x - 7 < 20$$
$$x - 7 < \frac{20}{3}$$

When asked to justify this first step, the student replies, "I divided both sides by 3." How would you evaluate this step and the student's justification?

a. Dividing both sides of the inequality by 3 is an uncommon, yet correct, first step, but the student carried it out incorrectly by failing to apply the distributive property on the left side of the inequality.
b. Dividing both sides of the inequality by 3 is a correct first step, and the student carried it out correctly.
c. Dividing both sides of the inequality by 3 is an incorrect first step. The correct first step is to subtract 7 to both sides of the inequality.
d. Dividing both sides of the inequality by 3 is an incorrect first step. The correct first step is to subtract 7 to both sides of the inequality and reverse the direction of the inequality.

30. A student devises a method for constructing the center of a circle and asks you to evaluate it. The method involves drawing two chords of the circle and constructing their perpendicular bisectors. The point at which the perpendicular bisectors intersect is the center of the circle. How do you evaluate this method?

a. You congratulate the student on finding a clever construction that always works.
b. You congratulate the student on finding a clever construction that works in general, but you explain that it fails in a special case.
c. You explain that the method fails in general, though it often produces a point deceptively close to the center, but works in a special case.
d. You explain that the method always fails though it often produces a point deceptively close to the center.

31. Two teachers from a county are chosen at random to attend a conference for high school science educators. What is the approximate probability that two women from the same department will be chosen?

	Biology	Chemistry	Physics
Women	26	31	20
Men	16	11	25

a. 8.6%
b. 10.7%
c. 11.9%
d. 13.8%

32. Solve $\sec^2 \theta = 2\tan\theta$ for $0 < \theta \leq 2\pi$.

a. $\theta = \frac{\pi}{6}$ or $\frac{7\pi}{6}$
b. $\theta = \frac{\pi}{4}$ or $\frac{5\pi}{4}$
c. $\theta = \frac{3\pi}{4}$ or $\frac{7\pi}{4}$
d. There is no solution to the equation.

33. Three students attempt to put the equation $4x^2 - 8x + 9 = 20$ into the form $(x - p)^2 = q$ by completing the square. Which approach is correct?

a.

Student A

$$\begin{aligned} 4x^2 - 8x + 9 &= 20 \\ 4x^2 - 8x + 16 &= 27 \\ 4(x^2 - 2x + 4) &= 27 \\ (x - 2)^2 &= \frac{27}{4} \end{aligned}$$

b.

Student B

$$\begin{aligned} 4x^2 - 8x + 9 &= 20 \\ 4(x^2 - 2x) &= 11 \\ 4(x^2 - 2x + 1) &= 15 \\ (x - 1)^2 &= \frac{15}{4} \end{aligned}$$

c.

Student C

$$\begin{aligned} 4x^2 - 8x + 9 &= 20 \\ 4(x^2 - 2x) &= 11 \\ 4(x^2 - 2x + 1) &= 12 \\ (x - 1)^2 &= 3 \end{aligned}$$

d. More than one student has a correct approach.

34. The vertices of a polygon are $(-2, 3)$, $(1, -1)$, $(2, 6)$, and $(5, 2)$. Which of the following describes the polygon most specifically?

a. Parallelogram
b. Rhombus
c. Rectangle
d. Square

35. A teacher recorded the dates that students turned in projects in two different classes to compare trends and noticed that a large number of students were submitting their work late. She plotted charts to see how many students in each class were turning in late work and how late they were.

Which of the following is a true statement for these data?

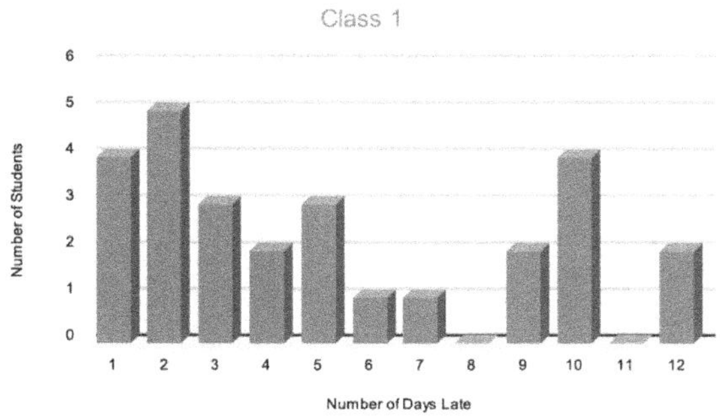

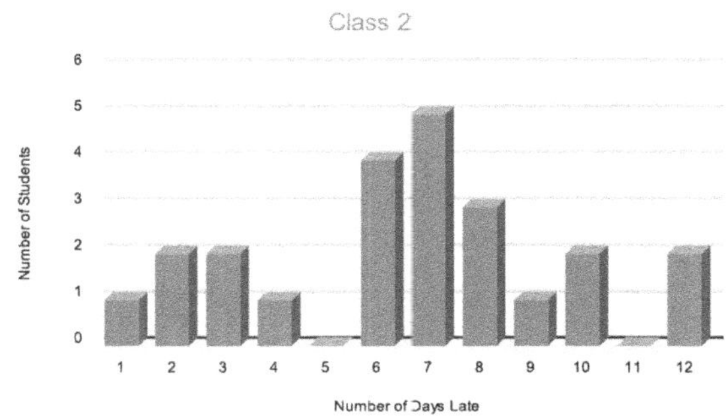

a. The median number of days late in Class 1 is 2 days less than in Class 2.
b. On average, students turned in work 4 days later in Class 2 than in Class 1 (round to nearest whole number).
c. The mode late time is higher for Class 2 than for Class 1.
d. The median late time for Class 1 is greater than the mean for Class 1.

36. A crane, 80 feet tall, moves a 200 lb. bucket suspended from a cable that weighs 3 lbs. per linear foot. You need to construct a function, F, that models the combined downward force (weight) of the bucket and the cable as the bucket rises and falls (changing the amount of cable hanging down). Ignore the height of the bucket itself. Which of the following functions constitute correct models? Select all correct solutions.

a. $F(y) = 320 - 3(y - 40)$, where y is the height (in feet) of the bucket above the ground.
b. $F(y) = 200 - 3(80 - y)$, where y is the height (in feet) of the bucket above the ground.
c. $F(y) = 200 - 3y$, where y is distance (in feet) of the bucket from the top of the crane.

37. You play a game as follows: you flip a fair coin up to five times, stopping the first time you flip tails. You win one dollar for every time you flip heads. Let X be the number of dollars you win, making X a random variable. Calculate $E(X)$, the expected value of this random variable (that is, your expected winnings from playing the game). Hint: If H represents flipping heads and T represents flipping tails, the sample space for this game is $\{T, HT, HHT, HHHT, HHHHT, HHHHH\}$.

a. $\frac{57}{64}$ dollars
b. $\frac{31}{32}$ dollars
c. 3 dollars
d. $\frac{5}{2}$ dollars

Praxis Practice Test #1

38. Solve: $\frac{x+2}{x+1} = \frac{x+9}{x+7} + \frac{2}{x+1}$.

a. $x = -3$
b. $x = -1$
c. $x = 1$
d. $x = 5$

39. If a, b, and c are multiples of 6 and $3a^2 + 2b^3 = c$, of the following integers, which is the largest integer that must be factor of c?

a. 6
b. 36
c. 72
d. 108

40. For ΔABC, what is the length of $\overline{BC}$?

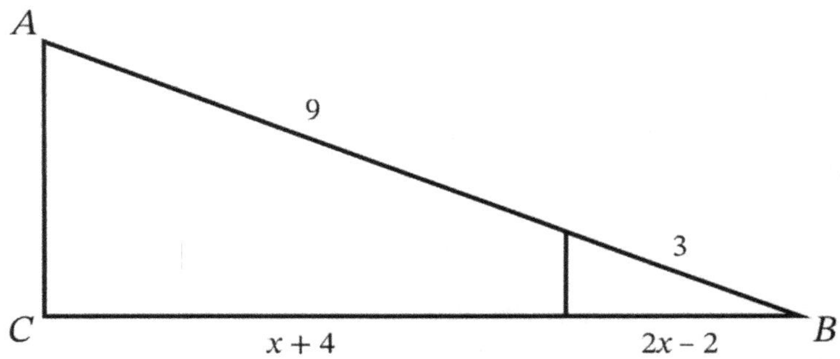

41. Given the partial table of values for $f(x)$ and $g(x)$, find $f(g(-4))$. (Assume that $f(x)$ and $g(x)$ are the simplest polynomials that fit the data.)

x	$f(x)$	$g(x)$
-2	8	1
-1	2	3
0	0	5
1	2	7
2	8	9

42. Three students solve the equation $y = a + 3ax$ for the variable x. They are not required to simplify their solutions. Which solutions are correct? Select all correct solutions.

a. $y = a + 3ax$
$y = 4ax$
$\frac{y}{4a} = x$

b. $y = a + 3ax$
$y - a = 3ax$

$$\frac{y-a}{3a} = x$$

c. $y = a + 3ax$

$$y = a(1 + 3x)$$

$$\frac{y}{a} = 1 + 3x$$

$$\frac{y}{a} - 1 = 3x$$

$$\frac{\frac{y}{a} - 1}{3} = x$$

43. For the right triangle below, where $a \neq b$, which of the following is a true statement of equality?

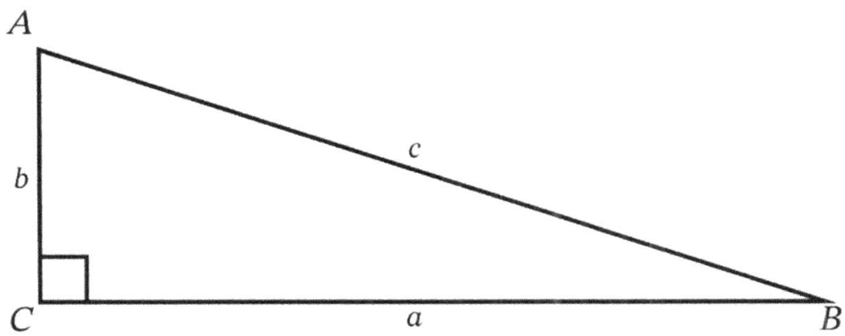

a. $\tan B = \frac{\sqrt{c^2-a^2}}{a}$

b. $\cos B = \frac{b\sqrt{a^2+b^2}}{a^2+b^2}$

c. $\sec B = \frac{\sqrt{a^2+b^2}}{b}$

d. $\csc B = \frac{a^2+b^2}{b}$

44. Solve the system of equations.

$$3x + 4y = 2$$
$$2x + 6y = -2$$

a. $\left(0, \frac{1}{2}\right)$

b. $\left(\frac{2}{5}, \frac{1}{5}\right)$

c. $(2, -1)$

d. $\left(-1, \frac{5}{4}\right)$

45. A colony of Escherichia coli is inoculated from a Petri dish into a test tube containing 50 mL of nutrient broth. The test tube is placed in a 37 °C incubator and agitator. After one hour, the number of bacteria in the test tube is determined to be 8×10^6. Given that the doubling time of Escherichia coli is 20 minutes with agitation at 37 °C, approximately how many bacteria should the test tube contain after eight hours of growth?

a. 2.56×10^8

Praxis Practice Test #1

b. 2.05×10^9
c. 1.7×10^{14}
d. 1.7×10^{13}

46. If $f(x) = 4x^3 - x^2 - 4x + 2$, which of the following statements is (are) true of its graph?

I. The point $\left(-\frac{1}{2}, 3\frac{1}{4}\right)$ is a relative maximum.

II. The graph of f is concave upward on the interval $(-\infty, \frac{1}{12})$.

a. I
b. II
c. I and II
d. Neither I nor II

47. A private tutor works with 11 five-year-olds, 8 six-year-olds, 9 seven-year-olds, 14 eight-year-olds, 11 nine-year-olds, and 13 ten-year-olds. If a student is randomly selected for a free session, what is the probability that he or she is no more than eight years old?

a. $\frac{7}{11}$
b. $\frac{14}{33}$
c. $\frac{4}{11}$
d. $\frac{3}{7}$

48. Which of the following is a subset of the set of rational numbers?

a. Integers
b. Real numbers
c. Imaginary numbers
d. Complex numbers

49. This question may have more than one correct answer. Select all correct answers. Three students attempt to solve the equation $6x^2 + 13x = 5$. All three get the correct answer that $x = \frac{1}{3}$ or $x = -\frac{5}{2}$, but they take three different approaches. Which of their approaches are mathematically correct? Select all correct approaches.

a.

Student A calculates the following.

$$6x^2 + 13x - 5 = 0$$

$$x = \frac{-13 \pm \sqrt{13^2 - 4(6)(-5)}}{2 \cdot 6} = \frac{-13 \pm \sqrt{289}}{12} = -\frac{13}{12} \pm \frac{17}{12}$$

$$x = -\frac{13}{12} + \frac{17}{12} = \frac{4}{12} = \frac{1}{3} \text{ or } x = -\frac{13}{12} - \frac{17}{12} = -\frac{30}{12} = -\frac{5}{2}$$

b.

Student B calculates the following.

$$6\left(x^2+\frac{13}{6}x+\frac{169}{144}\right)=5+\frac{169}{24}$$
$$6\left(x+\frac{13}{12}\right)^2=\frac{289}{24}$$
$$\left(x+\frac{13}{12}\right)^2=\frac{289}{144}$$
$$x+\frac{13}{12}=\pm\sqrt{\frac{289}{144}}$$
$$x=-\frac{13}{12}\pm\frac{17}{12}$$
$$x=-\frac{13}{12}-\frac{17}{12}=-\frac{30}{12}=-\frac{5}{2}\text{ or }x=-\frac{13}{12}+\frac{17}{12}=\frac{4}{12}=\frac{1}{3}$$

c.

Student C calculates the following.

$$6x^2+13x-5=0$$
$$(3x-1)(2x+5)=0$$
$$3x-1=0\text{ or }2x+5=0$$
$$x=\frac{1}{3}\text{ or }x=-\frac{5}{2}$$

Praxis Practice Test #1

50. You tell your students that in ΔABC, segment $\overline{AD}$ is a median. Further, the measure of $\angle BAD$ is 54° and the measure of segment $\overline{AD}$ is half the measure of side $\overline{BC}$. That is, $m\overline{AD} = \frac{1}{2} \cdot m\overline{BC}$. Based on this information, you ask your students to deduce the measure of $\angle CAD$.

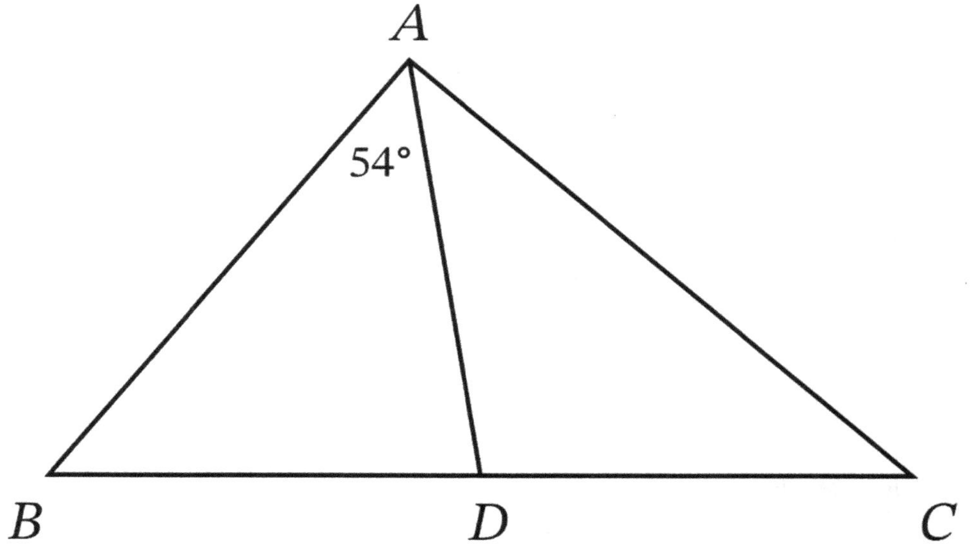

One student offers this argument: Since $\overline{AD}$ is a median, point D is the midpoint of side $\overline{BC}$. So $BD = AD = CD = \frac{1}{2} \cdot BC$, which makes ΔBAD and ΔCAD isosceles. Angles opposite equal sides are equal, so $m\angle B = 54°$. Then $m\angle ADC = m\angle B + m\angle BAD = 54° + 54° = 108°$, since an exterior angle equals the sum of the remote interior angles. Now $m\angle CAD = m\angle ACD$, which makes each of them half of $180° - m\angle ADC = 180° - 108° = 72°$. Hence, $m\angle CAD = \frac{1}{2} \cdot 72° = 36°$.

How do you evaluate this student's argument?

a. The argument is correct and reasonably simple.

b. The argument is correct but unnecessarily complicated. It is simpler to observe that $\angle BAC$ appears to be a right angle. Thus, $\angle CAD$ is the complement of $\angle BAD$ and has a measure of $90° - 54° = 36°$.

c. The argument is incorrect because a median bisects the angle it belongs to but not necessarily the opposite side. So $m\angle CAD = 54°$.

d. The argument is incorrect because a median is perpendicular to the opposite side of the triangle but does not necessarily bisect it. This makes ΔABD and ΔACD right triangles, but this provides inadequate information to determine the value of $m\angle CAD$.

51. This question may have more than one correct answer. Select all correct answers.

You give your students the following problem: On a computer system, users are assigned random usernames consisting of strings of six uppercase letters with repetitions allowed (e.g., RGMAQZ, TTOPTG). Administrators receive usernames that include the letter A at least once in the username. Under this scheme, how many administrator usernames are possible?

Three students take different approaches to answering this question. Which of their approaches are correct? Select all correct approaches.

a. Student A: Since we choose any letter of the alphabet for each of six positions, there are 26^6 usernames without restriction. Of these, 25^6 have no A's (put a non-A in each of 6 positions). Subtract these out to get that the number of usernames with at least one A is the difference $26^6 - 25^6$.

b. Student B: We need to place letters in all six positions in the username. First, guarantee there will be an A by choosing one of the six positions to receive an A (6 possibilities). Then, fill in the remaining five positions with any letter of the alphabet (26^5 possibilities). So, the number of possible administrator usernames is $6 \cdot 26^5$.

c. Student C: We can count the administrator usernames according to the number of A's in them. To count usernames with exactly one A, first choose one of the six positions in the word to get the A ($C(6,1)$ possibilities, where $C(n,r) = \frac{n!}{r!(n-r)!}$ is a binomial coefficient). Then choose letters besides A for the remaining five positions (25^5 possibilities). To get usernames with exactly two A's, choose two of the six positions to receive A's ($C(6,2)$ possibilities) and then choose letters besides A for the remaining four positions (25^4 possibilities). Continuing in this manner, we get a total of $C(6,1) \times 25^5 + C(6,2) \times 25^4 + C(6,3) \times 25^3 + C(6,4) \times 25^2 + C(6,5) \times 25^1 + C(6,6) \times 25^0$ administrator usernames.

52. Lisa is selling brownies and cupcakes at a bake sale. Brownies cost $2 each and cupcakes cost $3 each. Lisa must sell at least $150 worth of baked goods to break even on baking costs and the cost of renting the booth. If the x-axis represents the number of brownies sold and the y-axis represents the number of cupcakes sold, which of the following graphs shows the amount of baked goods Lisa must sell to break even or make a profit?

a.

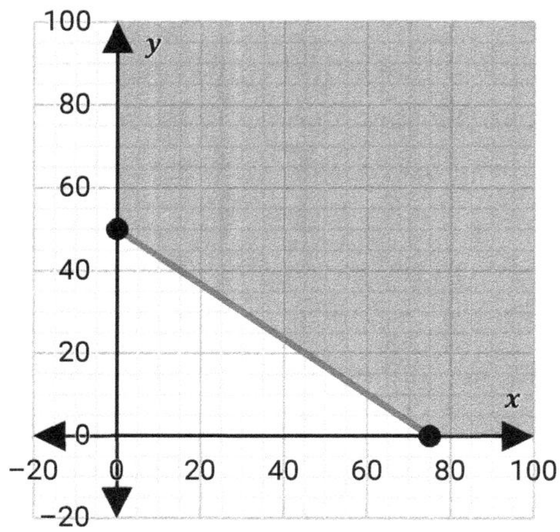

c.

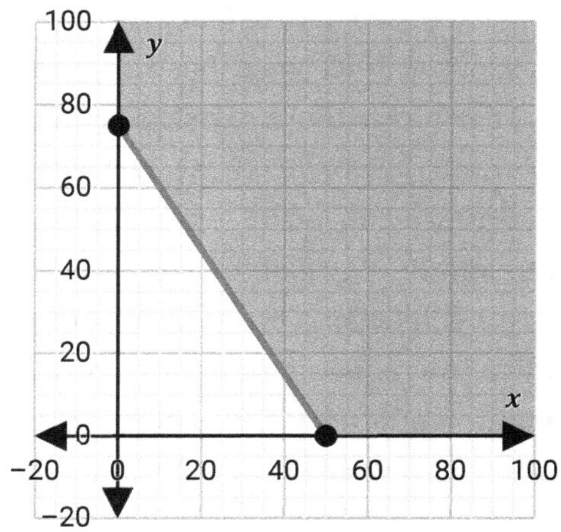

b.

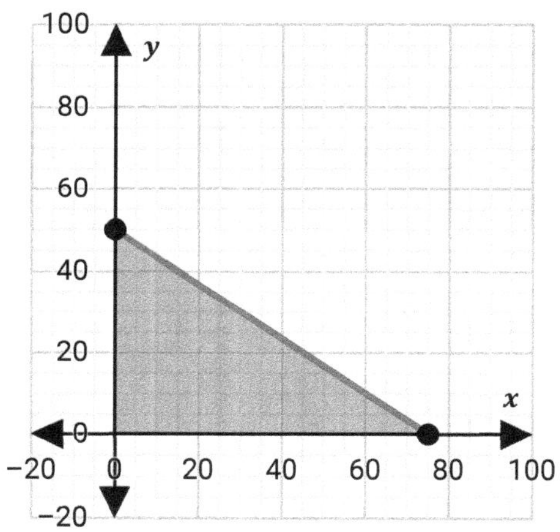

d.

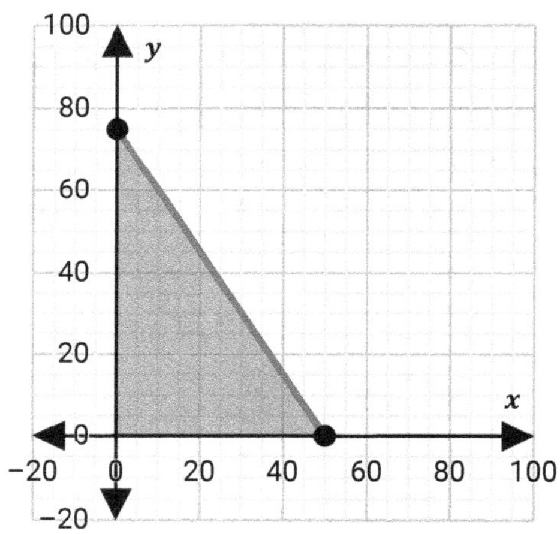

53. Given the relation $y - 2 = \pm\sqrt{x - 3}$, find the inverse relation and determine whether it is a function. If it is a function, express it in the form $y = f(x)$.

a. The inverse relation is not a function and is given by the equation $(y - 2)^2 = \pm(x - 3)$.
b. The inverse relation is the function $y = 2 \pm \sqrt{x - 3}$.
c. The inverse relation is the function $y = (x - 2)^2 + 3$.
d. The inverse relation is the function $y = x^2 + 7$.

54. You are writing a multiple-choice quiz with four questions. Each question has five answer choices, identified by letters *a, b, c, d,* and *e*. Each question has only one correct answer. Suppose you choose the letter for the correct answer to each question at random (i.e., each letter has an equal chance of being chosen). Use the fundamental counting principle to find the probability that none of the questions is assigned answer *c* as the correct answer.

a. 0.4096
b. 0.8
c. 0.2
d. 0.0384

55. The ratio of Jon's monthly expenses to his savings is 7: 1. If he is able to save $725 per month, what are his monthly expenses?

56. If $\sin\theta = \frac{\sqrt{3}}{2}$ when $\frac{\pi}{2} < \theta < \pi$, what is the value of θ?

a. $\frac{\pi}{6}$
b. $\frac{3\pi}{4}$
c. $\frac{2\pi}{3}$
d. $\frac{5\pi}{6}$

57. If the square of three times the sum of a and six is fourteen more than a, which of these is a possible value of a?

a. $\sqrt{2}$
b. $\frac{3}{2}$
c. 1
d. −5

58. A teacher asks students how many seconds are in a common year (a non-leap year). Three students set their solutions up in three different ways. Which, if any, will produce correct answers? Select all that apply.

a. Emily: $\frac{1\text{ year}}{365\text{ days}} \times \frac{24\text{ hours}}{1\text{ day}} \times \frac{60\text{ minutes}}{1\text{ hour}} \times \frac{60\text{ seconds}}{1\text{ minute}}$
b. Jonathon: $\frac{1\text{ year}}{1} \times \frac{365\text{ days}}{1\text{ year}} \times \frac{24\text{ hours}}{1\text{ day}} \times \frac{60\text{ minutes}}{1\text{ hour}} \times \frac{60\text{ seconds}}{1\text{ minute}}$
c. Sarah: $1\text{ year} \times \frac{60\text{ seconds}}{1\text{ minute}} \times \frac{1{,}440\text{ minutes}}{1\text{ day}} \times \frac{365\text{ days}}{1\text{ year}}$

59. Solve the equation $2 - \sqrt{x} = \sqrt{x - 20}$.

a. $x = 6$
b. $x = 36$
c. $x = 144$
d. No solution

Praxis Practice Test #1

60. The homework assignment says, "Find an equation of the line through the points $(12, 3)$ and $(0, -6)$. You do not need to simplify your answer." A student thinks of three possible approaches. Which produce correct equations? Select all correct approaches.

a. The slope of the line is $m = \frac{3-(-6)}{12-0} = \frac{9}{12} = \frac{3}{4}$. I can substitute the slope and the point $(12, 3)$ into the equation $y - y_1 = m(x - x_1)$. So, the line has the equation $y - 3 = \frac{3}{4}(x - 12)$.

b. The slope of the line is $m = \frac{-6-3}{0-12} = \frac{-9}{-12} = \frac{3}{4}$. Since the point $(0, -6)$ lies on the y-axis, I can substitute the slope and the value $b = -6$ into the equation $y = mx + b$. So, the line has the equation $y = \frac{3}{4}x - 6$.

c. The slope of the line is $m = \frac{-6-3}{0-12} = \frac{-9}{-12} = \frac{3}{4}$. I can substitute the slope and the point $(0, -6)$ into the equation $y - y_1 = m(x - x_1)$. So, the line has the equation $y + 6 = \frac{3}{4}(x - 0)$.

61. Find the area A of the finite region between the graphs of $y = x - 3$ and $y = x^2 - 2x - 7$.

a. 7

b. $\frac{56}{3}$

c. $\frac{125}{6}$

d. 12

62. Roxana walks x meters west and $x + 20$ meters south to get to her friend's house. On a neighborhood map which has a scale of 1 cm : 10 m, the direct distance between Roxana's house and her friend's house is 10 cm. How far did Roxana walk to her friend's house?

63. You draw a hexagon with a perimeter of 60 mm and an area of 250 mm^2. Then you make a photocopy of your drawing on a photocopier set to enlarge the image by 20%. That is, it produces a similar image with all lengths increased by 20%. What are the perimeter and area of the enlarged hexagon?

a. The perimeter is 72 mm, and the area is 300 mm^2.

b. The perimeter is 72 mm, and the area is 360 mm^2.

c. The perimeter is 67.2 mm, and the area is 300 mm^2.

d. The perimeter is 72 mm, and the area cannot be determined from the given information.

64. The velocity of a car which starts at position 0 at time 0 is given by the equation $v(t) = -t^2 + 8t$ for $0 \le t \le 8$. Find the position of the car when its acceleration is 0.

a. 64

b. $\frac{128}{3}$

c. 16

d. $\frac{56}{3}$

65. A pair of jeans is marked down 30% and placed on a clearance rack, on which is posted a sign reading, "Take an extra 15% off already reduced merchandise." What fraction of the original price is the final sale price of the jeans?

a. $\frac{61}{100}$
b. $\frac{119}{200}$
c. $\frac{4}{7}$
d. $\frac{3}{5}$

66. Evaluate $\lim_{x \to -3} \frac{x^3+3x^2-x-3}{x^2-9}$.

a. 0
b. $\frac{1}{3}$
c. $-\frac{4}{3}$
d. ∞

Answer Key and Explanations for Test #1

1. C: Ava and Jack hiked 8 miles from home and then back again, so they traveled a total of 16 miles. Therefore, statement III is true. They traveled these 16 miles in 7 hours, so their average speed was $\frac{16}{7}$ mph or approximately 2.3 mph. Therefore, statement I is false. Ava and Jack hiked 2 miles from point B to C in 1 hour, so they were hiking at a speed of 2 mph. They hiked 6 miles from D to E in 3 hours, so they hiked $\frac{6 \text{ miles}}{3 \text{ hours}} = \frac{2 \text{ miles}}{1 \text{ hour}} = 2$ mph. Therefore, their speed was the same during both these times, and statement II is also true.

2. D: By inspection of the graph, the function f has roots $x = -1$ and $x = 3$. We can factor g as $g(x) = -x^2 + 8x - 15 = -(x^2 - 8x + 15) = -(x - 3)(x - 5)$. To find the roots, set each factor equal to 0 and solve for x. The roots are $x = 3$ and $x = 5$. Thus functions f and g share the root $x = 3$ but not a second root. The y-intercept of f can be found by determining the point where the graph crosses the y-axis, which is at the point $(0, -3)$. The y-intercept of g can be found by determining $g(0)$, which is –15, so the y-intercept is at the point $(0,15)$. Therefore, the functions f and g have different y-intercepts.

3. B: Choice B does not simulate random selection because the probability of different outcomes is not equal. For instance, there is only one combination of dice outcomes that can produce the value 2 or 12, whereas there are several combinations that produce the value of 7.

4. B: To graph a piecewise function, graph each part of the function for the given domain values. In this case, graph the function $4 - x^2$ for the x-values $x < 0$, graph the function $4 - 2x$ for the values $0 \leq x < 2$, and graph the function $7 - x$ for the values $x \geq 2$. By inspection of the graph, we see that there is no discontinuity at $x = 0$, because the two functions on either side of $x = 0$ meet at 4. There is a discontinuity at $x = 2$, because the function to the left of $x = 2$ approaches 0, while the function to the right of $x = 2$ is equal to 5 at the point $x = 2$. Also, the maximum value of the function is $f(2) = 5$.

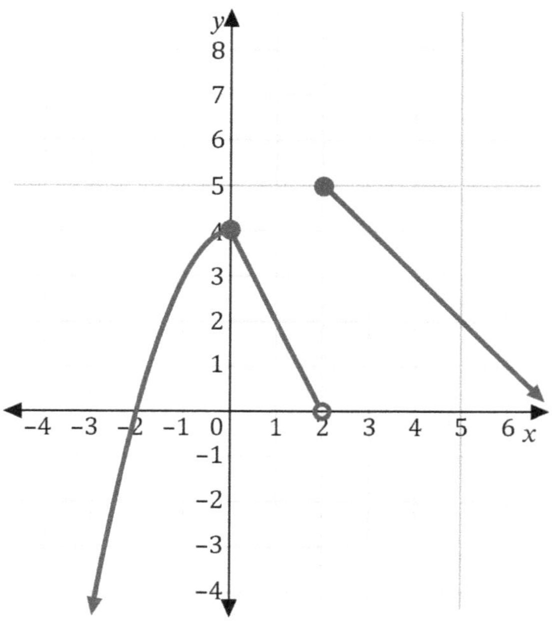

5. A: Since $f(1) = \frac{1}{2}, g(1) = 0, \frac{1}{2} > 0$, and the graphs never cross, we know that $f(x) > g(x)$ throughout the interval $[1,4]$. Thus, the area between the curves over this interval is the definite integral from $x = 1$ to $x = 4$ of $f(x) - g(x)$, the difference of the larger curve and the smaller curve. This gives us:

$$\begin{aligned}
A &= \int_1^4 \frac{1}{2}x^2 - \left(\sqrt{x} - \frac{1}{x^2}\right) dx \\
&= \int_1^4 \left(\frac{1}{2}x^2 - x^{\frac{1}{2}} + x^{-2}\right) dx = \frac{1}{6}x^3 - \frac{2}{3}x^{\frac{3}{2}} - x^{-1}\Big|_1^4 \\
&= \left(\frac{1}{6}(4)^3 - \frac{2}{3}(4)^{\frac{3}{2}} - (4)^{-1}\right) - \left(\frac{1}{6}(1)^3 - \frac{2}{3}(1)^{\frac{3}{2}} - (1)^{-1}\right) \\
&= \frac{1}{6} \cdot 64 - \frac{2}{3} \cdot 8 - \frac{1}{4} - \frac{1}{6} + \frac{2}{3} + 1 \\
&= \frac{32}{3} - \frac{16}{3} - \frac{1}{4} - \frac{1}{6} + \frac{2}{3} + 1 \\
&= \frac{16}{3} - \frac{3}{12} - \frac{2}{12} + \frac{2}{3} + 1 \\
&= \frac{18}{3} - \frac{5}{12} + 1 \\
&= 7 - \frac{5}{12} \\
&= 6\frac{7}{12}
\end{aligned}$$

6. B: The center of the sphere is shared by the center of the cube, and each of the corners of the cube touches the surface of the sphere. Therefore, the diameter of the sphere is the line that passes through the center of the cube and connects one corner of the cube to the opposite corner on the opposite face. Notice in the illustration below that the diameter d of the sphere can be represented as the hypotenuse of a right triangle with a short leg measuring 5 units. Since the volume of the cube is 125 cubic units, each of its sides measure $\sqrt[3]{125} = 5$ units. The long leg of the triangle is the diagonal of the base of the cube. Its length can be found using the Pythagorean theorem:

$$5^2 + 5^2 = x^2$$
$$x = \sqrt{50} = 5\sqrt{2}$$

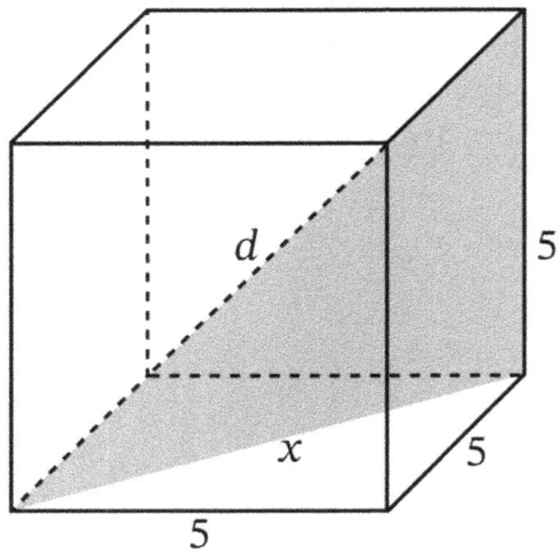

Use the Pythagorean theorem again to find d, the diameter of the sphere: $d^2 = \left(5\sqrt{2}\right)^2 + 5^2$, which then becomes $d = \sqrt{75} = 5\sqrt{3}$.

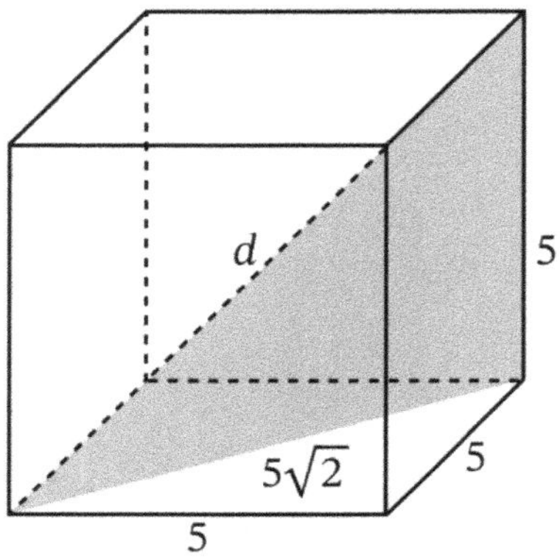

To find the volume of the sphere, use the formula $V = \frac{4}{3}\pi r^3$. Since the radius r of the sphere is half the diameter, $r = \frac{5\sqrt{3}}{2}$, and $V = \frac{4}{3}\pi\left(\frac{5\sqrt{3}}{2}\right)^3 = \frac{4}{3}\pi\left(\frac{375\sqrt{3}}{8}\right) = \frac{125\pi\sqrt{3}}{2}$ cubic units.

7. C: We can find the area of the white part by viewing the target as five separate circles of various sizes, one inside the other. We ignore the outer circle since we are only interested in the white parts, and add the areas of the two white circles and subtract the areas of the two interior shaded circles. The formula for area of a circle is πr^2, so the area of the largest white circle is $\pi(8m)^2$. We subtract the shaded circle inside it, add the white circle inside that one, and then subtract the smallest shaded circle.

$$\begin{aligned} A &= \pi(8m)^2 - \pi(6m)^2 + \pi(4m)^2 - \pi(2m)^2 \\ &= 64\pi m^2 - 36\pi m^2 + 16\pi m^2 - 4\pi m^2 \\ &= (64 - 36 + 16 - 4)\pi m^2 \\ &= 40\pi m^2 \end{aligned}$$

8. B: Use the law of sines to find the distance d between the bird in the tree and the worm:

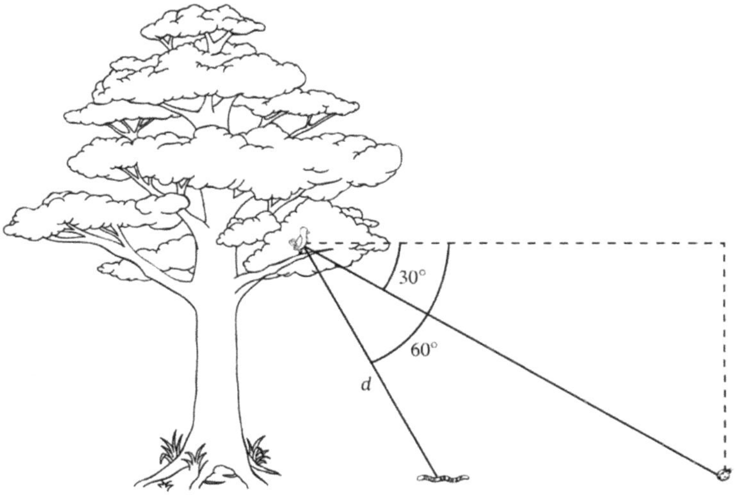

$$\frac{30}{\sin 30°} = \frac{d}{\sin 30°}$$
$$d = \frac{30 \sin 30°}{\sin 30°}$$
$$d = 30 \text{ ft}$$

9. B: Since it has been established that $m\angle a + m\angle b = m\angle c + m\angle d$ and $m\angle a = m\angle d$, we can conclude that $m\angle b = m\angle c$ because of the subtraction property of equality, which states that subtracting equal amounts from each side of a balanced equation will keep it balanced.

10. A: Because Step 6 established that $\angle b \cong \angle c$, we can conclude that the sides opposite these congruent angles are congruent, or $\overline{XY} \cong \overline{ZY}$.

11. 48: Rate in miles per hour can be expressed as, mph $= \frac{\text{distance in miles}}{\text{time in hours}}$. So, Zeke's driving speed on the way to Atlanta and home from Atlanta in mph can be expressed as $\frac{d}{3}$ and $\frac{d}{2}$, respectively, where d is the distance between Zeke's house and his destination. Since Zeke drove 20 mph faster on his way home, (speed home) − (speed to store) = 20. Substitute Zeke's speeds and solve for d.

$$\frac{d}{2} - \frac{d}{3} = 20$$
$$6\left(\frac{d}{2} - \frac{d}{3} = 20\right)$$
$$3d - 2d = 120$$
$$d = 120$$

Since the distance between Zeke's house and the store in Atlanta is 120 miles, Zeke drove a total distance of 240 miles in five hours. Therefore, his average speed was $\frac{240 \text{ miles}}{5 \text{ hours}} = 48$ mph.

Answer Key and Explanations #1

12. D: To draw a box-and-whisker plot from the data, find the median, quartiles, and upper and lower limits.

		Key
3	6 799	
4	2|3889	
5	0|11158|	2|123 = 21, 22, 23
6	0 0123	

The median is $\frac{50+51}{2} = 50.5$, the lower quartile is $\frac{42+43}{2} = 42.5$, and the upper quartile is $\frac{58+60}{2} = 59$. The box of the box-and-whisker plot goes through the quartiles, and a line through the box represents the median of the data. The whiskers extend from the box to the lower and upper limits, unless there are any outliers in the set. In this case, there are no outliers, so the box-and-whisker plot in choice A correctly represents the data set.

To draw a pie chart, find the percentage of data contained in each of the ranges shown. There are four out of twenty numbers between 30 and 39, inclusive, so the percentage shown in the pie chart for that range of data is $\frac{4}{20} \times 100\% = 20\%$; there are five values between 40 to 49, inclusive, so the percentage of data for that sector is $\frac{5}{20} \times 100\% = 25\%$; $\frac{6}{20} \times 100\% = 30\%$ of the data is within the range of 50–59, and $\frac{5}{20} \times 100\% = 25\%$ is within the range of 60–69. The pie chart shows the correct percentage of data in each category.

To draw a cumulative frequency histogram, find the cumulative frequency of the data.

Range	Frequency	Cumulative frequency
30–39	4	4
40–49	5	9
50–59	6	15
60–69	5	20

The histogram shows the correct cumulative frequencies. Therefore, all of the graphs represent the data set.

13. 14: Use a Venn diagram to help organize the given information. Start by filling in the space where the three circles intersect: Jenny tutored three students in all three areas. Use that information to fill in the spaces where two circles intersect. For example, she tutored 4 students in chemistry and for the ACT, and 3 of those were students she tutored in all three areas, so 1 student was tutored in chemistry and for the ACT but not for math. Once the diagram is completed, add the number of students who were tutored in all areas to the number of students tutored in only two of the three areas to the number of students tutored in only one area. The total number of students tutored was $3 + 2 + 2 + 1 + 3 + 2 + 1 = 14$.

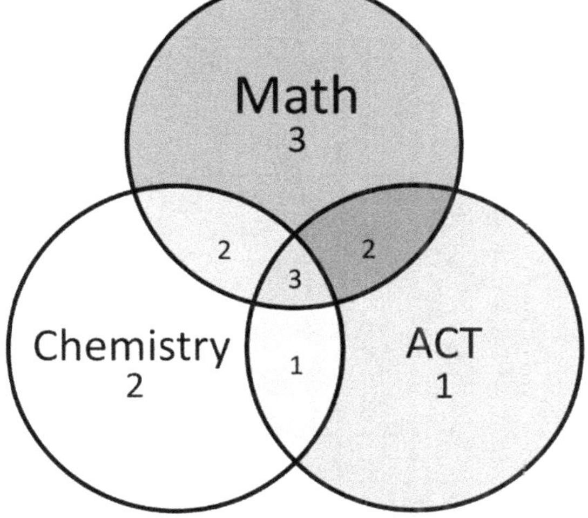

14. A, B, C: Student A uses a form of the Euclidean Algorithm for finding the greatest common factor of two numbers. Student B uses the fact that the greatest common factor of numbers is the product of their common prime factors, each raised to the lowest power it appears to. Student C uses the definition of greatest common factor, by listing the factors of both numbers and finding the largest factor they have in common.

15. C: A height of 5'11", or 71 inches, is one standard deviation above the mean. Since approximately 68% of the data is within one standard deviation of the mean, about 32% (100% − 68%) of the data is outside of one standard deviation within the mean. Normally distributed data is symmetric about the mean, which means that about 16% of the data lies below one standard deviation below the mean and about 16% of data lies above one standard deviation above the mean. Therefore, approximately 16% of individuals are 5'11" or taller, while approximately 84% of the population is shorter than 5'11". Multiply to find 84% of 200, $0.84 \times 200 = 168$. So, about 168 people from the selected group are shorter than 5'11".

16. D: There are many ways to solve quadratic equations in the form $ax^2 + bx + c = 0$. However, some methods, such as graphing and factoring, are not useful for equations with irrational or complex roots. Solve this equation by using the quadratic formula, $x = \frac{-b \pm \sqrt{b^2 - 4ac}}{2a}$. Set the given

Answer Key and Explanations #1

equation equal to zero, so $7x^2 + 6x + 2 = 0$. Substitute the values $a = 7$, $b = 6$, and $c = 2$ into the quadratic formula.

$$x = \frac{-b \pm \sqrt{b^2 - 4ac}}{2a}$$

$$x = \frac{-6 \pm \sqrt{6^2 - 4(7)(2)}}{2(7)}$$

$$x = \frac{-6 \pm \sqrt{36 - 56}}{14}$$

$$x = \frac{-6 \pm \sqrt{-20}}{14}$$

$$x = \frac{-6 \pm 2i\sqrt{5}}{14}$$

$$x = \frac{-3 \pm i\sqrt{5}}{7}$$

17. C: Evaluate the integral by adding 1 to the exponent of each term and dividing the coefficient by the new exponent. Don't forget to add the constant of integration to the end.

$$\int 6x^2 + \frac{1}{2}x - 3\,dx = \frac{6}{2+1}x^{(2+1)} + \frac{\frac{1}{2}}{1+1}x^{(1+1)} - \frac{3}{0+1}x^{(0+1)} + C$$
$$= 2x^3 + \frac{1}{4}x^2 - 3x + C$$

18. C: Choice A represents convenience sampling since it is limited to one age. The pool of 8th grade students may not represent the opinions of all youth in the district. Choice B is biased because participants are self-selected rather than randomly selected. It may be that customers who have a strong opinion are more likely to respond than those who are more neutral, and this would give a skewed perspective of opinions. In choice C, families are randomly selected, so the sampling technique is not biased. Choice D, like choice B, is biased because participants are self-selected rather than randomly selected.

19. B: The range is the spread of the data. It can be calculated for each test by subtracting the lowest test score from the highest, or it can be determined visually from the graph. The difference between the highest and lowest test scores on the Midterm is $91 - 22 = 69$ points. The range for each of the other classes is much smaller.

20. D: For the Final, the third quartile is 87. Since 75% of the data in a set is below the third quartile, there is a 75% chance, or a probability of 0.75, that a test score chosen at random from the Final scored below 87.

21. A: Student A correctly uses the rule $\log_a x^y = y \log_a x$ to convert $\log_2 5^{3x-1}$ into $(3x - 1)\log_2 5$ and combines this rule with the rules $\log_a xy = \log_a x + \log_a y$ and $\log_a a^x = x$ to calculate $\log_2(32 \cdot 3) = \log_2 32 + \log_2 3 = \log_2 2^5 + \log_2 3 = 5 + \log_2 3$.

Student B tries to apply the rule $\log\left(\frac{a}{b}\right) = \log a - \log b$ but confuses $\log\left(\frac{a}{b}\right)$ with $\frac{\log a}{\log b}$, incorrectly rewriting $\frac{\log 96}{\log 5}$ as $\log 96 - \log 5$.

Student C incorrectly "cancels the logarithms" turning $\frac{\ln 96}{\ln 5}$ into $\frac{96}{5}$.

22. B: Student A is wrong because the GCF of the coefficients of a quadratic function does not determine the range. For instance, the function $f(x) = x^2 + 5$ has coefficients with a GCF of 1 but a range of $[5, \infty)$.

Student B is correct because the y-value of the vertex of parabola that opens upward is the minimum value of the function.

Student C is wrong because the axis of symmetry is the x-value of the vertex, not the y-value.

23. B: This problem can be solved using proportions. Determine the number of feet between the two points by setting up the following proportion.

$$\frac{1 \text{ in}}{10 \text{ ft}} = \frac{24 \text{ in}}{x \text{ ft}}$$

Cross multiply to solve for x.

$$1 \times x = 24 \times 10$$
$$x = 240$$

Therefore, the two points are 240 feet apart. From here, convert 240 feet to a distance in yards. To do so, multiply by the conversion factor $\frac{1 \text{ yd}}{3 \text{ ft}}$.

$$240 \text{ ft} \times \frac{1 \text{ yd}}{3 \text{ ft}} = 80 \text{ yd}$$

The two points are 80 yards apart.

24. B: A simple test of each option at $x = -5$, $x = 0$, and $x = 2$ demonstrates that only choice B works:

x	**Observed**	y_a	y_b	y_c	y_d
-5	$\sim\mathbf{\frac{1}{4}}$	-2	$\frac{1}{6}$	$-\frac{7}{6}$	$\frac{1}{9}$
0	$\sim-\mathbf{\frac{1}{4}}$	$\frac{1}{2}$	$-\frac{1}{4}$	$-\frac{3}{4}$	$-\frac{1}{6}$
2	$\sim\mathbf{\frac{1}{4}}$	$\frac{1}{3}$	$\frac{1}{6}$	$\frac{7}{6}$	$\frac{1}{9}$

25. D: The cross-section is a hexagon.

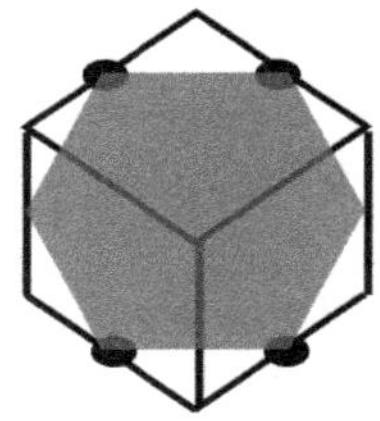

26. A: The composition function $(g \circ f)(x)$ can be rewritten as $g(f(x))$. This means that anywhere there is an x in the function $g(x)$, it should be replaced with $2x + 4$. After substituting, simplify the expression.

$$\begin{aligned} g(2x+4) &= (2x+4)^2 - 3(2x+4) + 2 \\ &= 4x^2 + 16x + 16 - 6x - 12 + 2 \\ &= 4x^2 + 10x + 6 \end{aligned}$$

27. B: The recursive definition of the sequence gives the first term of the series, $a_1 = 2$. The definition also defines each term in the series as the sum of the previous term and -3. Therefore, the second term in the series is $2 + (-3) = -1$, the third term in the series is $-1 + (-3) = -4$, and so on.

n	a_n
1	2
2	-1
3	-4

The relationship between n and a_n is linear, so the equation of the sequence can be found using the explicit formula for an arithmetic sequence, $a_n = a_1 + d(n-1)$, where a_n is the value of any term n, a_1 is the value of the first term, and d is the common difference between each term. The value of a_n decreases by 3 each time the value of n increases by 1, so the common difference (d) is -3.

Substitute all the known values into the equation for the explicit formula for an arithmetic sequence, and simplify the formula.

$$\begin{aligned} a_n &= 2 + (-3)(n-1) \\ a_n &= 2 - 3n + 3 \\ a_n &= -3n + 5 \end{aligned}$$

28. B: To find the second derivative of the function, take the derivative of the first derivative of the function. Remember, to take a derivative, multiply the exponent of each term by the term's coefficient and then decrease the exponent by 1.

$$\begin{aligned} f(x) &= -3x^3 - x^2 + 2x - 4 \\ f'(x) &= -9x^2 - 2x + 2 \\ f''(x) &= -18x - 2 \end{aligned}$$

29. A: Beginning by adding 7 to both sides of the inequality is an easier, more common, and recommended approach; however, dividing both sides by 3 is also mathematically correct because it produces an equivalent inequality and makes progress in isolating the variable. Dividing by 3, however, requires application of the distributive property on the left side to get the inequality $x - \frac{7}{3} < \frac{20}{3}$. This yields the solution $x < \frac{27}{3}$, which simplifies to $x < 9$.

30. B: A standard result from high school geometry states that the perpendicular bisector of a chord passes through the center of the circle. Thus, two distinct perpendicular bisectors of chords intersect at the center of the circle. Therefore, this construction works except in the case that both chords share the same perpendicular bisector, which happens precisely when the chords are parallel, the special case in which the student's construction fails.

31. C: There are three ways in which two women from the same department can be selected: two women can be selected from Biology, two women can be selected from Chemistry, or two women can be selected from Physics. Since the events of choosing one woman and then another are both independent events, multiply the two probabilities together to get the probability of choosing two women from the same department.

Biology	Chemistry	Physics
$\frac{26}{129} \times \frac{25}{128} = \frac{650}{16{,}512}$	$\frac{31}{129} \times \frac{30}{128} = \frac{930}{16{,}512}$	$\frac{20}{129} \times \frac{19}{128} = \frac{380}{16{,}512}$

Since any of these is a distinct possible outcome, the probability that two women will be selected from the same department is the sum of these outcomes.

$$\frac{650}{16{,}512} + \frac{930}{16{,}512} + \frac{380}{16{,}512} = \frac{1{,}960}{16{,}512} \approx 0.119 = 11.9\%$$

32. B: The trigonometric identity $\sec^2\theta = \tan^2\theta + 1$ can be used to rewrite the equation $\sec^2\theta = 2\tan\theta$ as $\tan^2\theta + 1 = 2\tan\theta$, which can then be rearranged into the form $\tan^2\theta - 2\tan\theta + 1 = 0$. Solve by factoring and using the zero-product property.

$$\begin{aligned} \tan^2\theta - 2\tan\theta + 1 &= 0 \\ (\tan\theta - 1)^2 &= 0 \\ \tan\theta - 1 &= 0 \\ \tan\theta &= 1 \end{aligned}$$

Since $\tan\theta = 1$ when $\sin\theta = \cos\theta$, for $0 < \theta \leq 2\pi$, $\theta = \frac{\pi}{4}$ or $\frac{5\pi}{4}$.

33. B: Student A goes wrong by incorrectly treating $x^2 - 2x + 4$ as the perfect square $(x - 2)^2$.

Student B's approach is correct. First, we subtract 9 from both sides of the equation and factor 4 out of the terms on the lefthand side to get $4(x^2 - 2x) = 11$. Then we add 1 inside the parentheses to complete the square. Because of the leading factor of 4, this has the effect of adding 4 on the left side of the equation. So we also add 4 to the right side of the equation, getting $4(x^2 - 2x + 1) = 15$. Now we divide by 4 on both sides to get $(x - 1)^2 = \frac{15}{4}$.

Student C goes wrong by adding 4 to the left side of the equation $4(x^2 - 2x) = 11$ when he or she adds 1 inside the parentheses, but only adds 1 to the right side of the equation.

34. D: Since all of the answer choices are parallelograms, determine whether the parallelogram is also a rhombus or a rectangle, or both. One way to do this is by examining the parallelogram's diagonals. If the parallelogram's diagonals are perpendicular, then the parallelogram is a rhombus. If the parallelogram's diagonals are congruent, then the parallelogram is a rectangle. If a parallelogram is both a rhombus and a rectangle, then it is a square. To determine whether the diagonals are perpendicular, find the slopes of the diagonals of the quadrilateral.

Diagonal 1: $(-2, 3)$ and $(5, 2)$

$$m = \frac{y_2 - y_1}{x_2 - x_1} = \frac{2 - 3}{5 - (-2)} = \frac{-1}{7} = -\frac{1}{7}$$

Diagonal 2: $(1, -1)$ and $(2, 6)$

$$m = \frac{y_2 - y_1}{x_2 - x_1} = \frac{6 - (-1)}{3 - 2} = \frac{7}{1} = 7$$

The diagonals have opposite inverse slopes and are therefore perpendicular. Thus, the parallelogram is a rhombus. To determine whether the diagonals are congruent, find the lengths of the diagonals of the quadrilateral using the distance formula.

Diagonal 1: $(-2, 3)$ and $(5, 2)$

$$d = \sqrt{(x_2 - x_1)^2 + (y_2 - y_1)^2}$$
$$d = \sqrt{\left(5 - (-2)\right)^2 + (2 - 3)^2}$$
$$d = \sqrt{(7)^2 + (-1)^2}$$
$$d = \sqrt{49 + 1}$$
$$d = \sqrt{50} = 5\sqrt{2}$$

Diagonal 2: $(1, -1)$ and $(2, 6)$

$$d = \sqrt{(x_2 - x_1)^2 + (y_2 - y_1)^2}$$
$$d = \sqrt{(2 - 1)^2 + (6 - (-1))^2}$$
$$d = \sqrt{(1)^2 + (7)^2}$$
$$d = \sqrt{1 + 49}$$
$$d = \sqrt{50} = 5\sqrt{2}$$

The diagonals are congruent, so the parallelogram is a rectangle. Since the parallelogram is a rhombus and a rectangle, it is also a square.

35. C: In Class 1, 27 students submitted late work, so the median is between the 13th and 14th piece of homework. Both the 13th and 14th fall on Day 4, so the median for Class 1 is 4. In Class 2, 23 students submitted late work, so the median is at the 12th piece of homework. This falls on Day 7. So, because $7 - 4 = 3$, the median number of days late in Class 1 is 3 less than the median number of days late in Class 2.

To find the mean late time for each class, find the sum of the late times and divide by the number of students turning in late work in each class. If the frequency of a one-day late time is 4, the number 1 is added four times (1×4), if frequency of a two-day late time is 5, the number 2 is added five times (2×5), and so on. So, to find the average turn-in time, divide the sum of the products of each late time and its corresponding frequency by the number of late turn-ins in each class. For Class 1, the mean late time is 5.22 days, which rounds to approximately 5 days.

$$\text{Mean} = \frac{4(1) + 5(2) + 3(3) + 2(4) + 3(5) + 1(6) + 1(7) + 2(9) + 4(10) + 2(12)}{4 + 5 + 3 + 2 + 3 + 1 + 1 + 2 + 4 + 2} = \frac{141}{27} \approx 5.22$$

For Class 2, the mean late time is 6.57, which rounds to approximately 7 days.

$$\text{Mean} = \frac{1(1) + 2(2) + 2(3) + 1(4) + 4(6) + 5(7) + 3(8) + 1(9) + 2(10) + 2(12)}{1 + 2 + 2 + 1 + 4 + 5 + 3 + 1 + 2 + 2} = \frac{151}{23} \approx 6.57$$

So, because $7 - 5 = 2$, the students in Class 2 turned in work 2 days later than in Class 1.

The mode late time is the date for which the frequency is the highest. For Class 1, the mode late time is 2 days, and for Class 2, the mode late time is 7 days.

The median late time for Class 1 is 4 and the mean is 4.96, so the median late time is less than the mean.

36. A: In correct answer choice A, the variable y is the height of the bucket above the ground, making $y - 40$ the position (negative if $y < 40$) of the bucket above the point halfway to the top of the crane. At this halfway point, the cable weighs $3 \cdot 40 = 120$ lbs. which, added to the weight of the bucket, gives a total weight of 320 lbs. Each foot above halfway means one foot less of cable hanging down and thus subtracts 3 lbs. from this weight so that at position y, the total is $320 - 3(y - 40)$ lbs.

In incorrect answer choice B, the variable y is the height of the bucket above the ground, making $80 - y$ the length of cable hanging down to support the bucket. The weight of this cable is $3(80 - y)$ lbs., which must be added to, not subtracted from, the weight of the bucket, giving a total weight of $200 + 3(80 - y)$ lbs. instead of the $200 - 3(80 - y)$ lbs. in choice B. When the bucket is at ground level ($y = 0$), this incorrect function makes the total weight $F(0) = 200 - 3(80 - 0) = -40$ lbs., which is wrong.

In incorrect answer choice C, the variable y is the distance of the bucket from the top of the crane. In other words, y is the length of cable hanging down to support the bucket. The weight of this cable is $3y$ lbs., which must be added to, not subtracted from, the weight of the bucket, giving a total weight of $200 + 3y$ lbs. instead of the $200 - 3y$ lbs. in choice C. When the bucket is at ground level ($y = 80$), this incorrect function makes the total weight $F(80) = 200 - 3 \cdot 80 = -40$ lbs., which is wrong.

37. B: The range of random variable X is $\{0, 1, 2, 3, 4, 5\}$. To calculate the probability of getting the possible values of X keep in mind the coin flips are independent with the probability of getting heads (and, similarly, tails) being $\frac{1}{2}$. Thus, $P(0) = P(\{T\}) = \frac{1}{2}$, $P(1) = P(\{HT\}) = \left(\frac{1}{2}\right)^2 = \frac{1}{4}$, $P(2) = P(\{HHT\}) = \left(\frac{1}{2}\right)^3 = \frac{1}{8}$, $P(3) = P(\{HHHT\}) = \left(\frac{1}{2}\right)^4 = \frac{1}{16}$, $P(4) = P(\{HHHHT\}) = \left(\frac{1}{2}\right)^5 = \frac{1}{32}$, and $P(5) = P(\{HHHHH\}) = \left(\frac{1}{2}\right)^5 = \frac{1}{32}$. So, using the definition of expected value, we get:

$$\begin{aligned} E(X) &= \sum_{x=0}^{5} xP(x) = 0 \cdot P(0) + 1 \cdot P(1) + 2 \cdot P(2) + 3 \cdot P(3) + 4 \cdot P(4) + 5 \cdot P(5) \\ &= 0 \cdot \frac{1}{2} + 1 \cdot \frac{1}{4} + 2 \cdot \frac{1}{8} + 3 \cdot \frac{1}{16} + 4 \cdot \frac{1}{32} + 5 \cdot \frac{1}{32} \\ &= 0 + \frac{1}{4} + \frac{1}{4} + \frac{3}{16} + \frac{1}{8} + \frac{5}{32} \\ &= \frac{8}{32} + \frac{8}{32} + \frac{6}{32} + \frac{4}{32} + \frac{5}{32} = \frac{31}{32} \end{aligned}$$

38. A: Start by determining which values of x are not possible solutions. For this equation, x cannot equal –1 or –7 because these would result in denominator values of 0.

Then, solve the equation as normal.

$\frac{x+2}{x+1} = \frac{x+9}{x+7} + \frac{2}{x+1}$	
$\frac{x}{x+1} = \frac{x+9}{x+7}$	Subtract $\frac{2}{x+1}$ from both sides of the equation.
$(x)(x+7) = (x+9)(x+1)$	Cross multiply to eliminate the fractions.
$x^2 + 7x = x^2 + 10x + 9$	Multiply the polynomials.
$7x = 10x + 9$	Subtract x^2 from both sides of the equation
$-3x = 9$	Subtract $10x$ from both sides of the equation.
$x = -3$	Divide both sides of the equation by –3.

Since –3 is not one of the values x cannot be equal to, our answer is $x = -3$.

39. D: Since a and b are multiples of 6, each can be expressed as the product of 6 and an integer. So, if we write $a = 6x$ and $b = 6y$ then, $3(6x)^2 + 2(6y)^3 = c$. Simplify and factor.

$$3(6x)^2 + 2(6y)^3 = c$$
$$108x^2 + 432y^3 = c$$
$$108(x^2 + 4y^3) = c$$

Since c is the product of 108 and some other integer, 108 must be a factor of c.

40. 8: ΔABC is similar to the smaller triangle with which it shares vertex B.

$$\overline{BC} = (2x-2) + (x+4) = 3x+2$$
$$\overline{AB} = 3 + 9 = 12$$

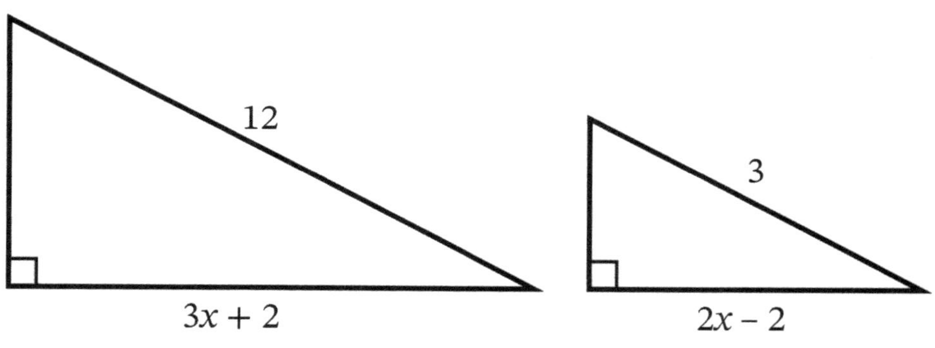

Set up a proportion and solve for x:

$$\frac{3x+2}{12}=\frac{2x-2}{3}$$
$$9x+6=24x-24$$
$$30=15x$$
$$x=2$$

So, $BC=3x+2=3(2)+2=8$.

41. 18: Use the table of values to first write equations for $f(x)$ and $g(x)$: $f(x)=2x^2$ and $g(x)=2x+5$. Then, use those equations to find $f(g(-4))$. Start by finding $g(-4)$.

$$g(-4)=2(-4)+5=-3$$

From here, substitute $g(-4)$ into the function $f(x)$ to find $f(g(-4))$. Since $g(-4)=-3$, plug in -3 anywhere an x is found in $f(x)$.

$$f(-3)=2(-3)^2=18$$

So, $f(g(-4))=18$.

42. B, C: Student A makes a mistake at the first step, violating order of operations by adding the a to $3a$ which is joined to x by multiplication (or, alternatively, by trying to combine unlike terms as though they were like terms). The correct way to combine these terms is by factoring out the a, getting $a+3ax=a(1+3x)$ as student C does.

Student B takes a correct approach, first subtracting a from both sides of the equation and then dividing both sides of the equation by the coefficient $3a$ to isolate the x.

Student C also takes a correct approach, first factoring out the common factor of a on the right-hand side of the equation, then dividing both sides of the equation by the factor a. Next, they subtract 1 from both sides of the equation, and finally divide both sides of the equation by the coefficient 3 to isolate the x

43. A: Determine the veracity of each option given that the hypotenuse of a right triangle is equal to the square root of the sum of the squares of the legs or $c=\sqrt{a^2+b^2}$:

A	$\tan B=?\frac{\sqrt{c^2-a^2}}{a}$	$\tan B=\frac{\text{opposite}}{\text{adjacent}}=\frac{b}{a}=\frac{\sqrt{c^2-a^2}}{a}$
B	$\cos B=?\frac{b\sqrt{a^2+b^2}}{a^2+b^2}$	$\cos B=\frac{\text{adjacent}}{\text{hypotenuse}}=\frac{a}{\sqrt{a^2+b^2}}=\frac{a}{\sqrt{a^2+b^2}}\times\frac{\sqrt{a^2+b^2}}{\sqrt{a^2+b^2}}=\frac{a\sqrt{a^2+b^2}}{a^2+b^2}\neq\frac{b\sqrt{a^2+b^2}}{a^2+b^2}$
C	$\sec B=?\frac{\sqrt{a^2+b^2}}{b}$	$\sec B=\frac{\text{hypotenuse}}{\text{adjacent}}=\frac{\sqrt{a^2+b^2}}{a}\neq\frac{\sqrt{a^2+b^2}}{b}$
D	$\csc B=?\frac{a^2+b^2}{b}$	$\csc B=\frac{\text{hypotenuse}}{\text{opposite}}=\frac{\sqrt{a^2+b^2}}{b}\neq\frac{a^2+b^2}{b}$

44. C: A system of linear equations can be solved by using matrices or by using the graphing, substitution, or elimination (also called linear combination) method. The elimination method is shown here:

$$3x + 4y = 2$$
$$2x + 6y = -2$$

In order to eliminate x by linear combination, multiply the top equation by 2 and the bottom equation by –3 so that the coefficients of the x-terms will be additive inverses.

$$2(3x + 4y = 2)$$
$$-3(2x + 6y = -2)$$

Then, add the two equations and solve for y.

$$6x + 8y = 4$$
$$\underline{-6x - 18y = 6}$$
$$-10y = 10$$
$$y = -1$$

Substitute –1 for y in either of the given equations and solve for x.

$$3x + 4y = 2$$
$$3x + 4(-1) = 2$$
$$3x - 4 = 2$$
$$3x = 6$$
$$x = 2$$

The solution to the system of equations is $(2, -1)$.

45. D: Bacterial growth is exponential. Let x be the number of times the population doubles, a be the number of bacteria in the colony originally transferred into the broth, and y be the number of bacteria in the broth after x doubling times. After 1 hour, the population would have doubled three times:

$$a(2)^x = y$$
$$a(2^3) = 8 \times 10^6$$
$$8a = 8 \times 10^6$$
$$a = 10^6$$

So, the number of bacteria originally transferred into the petri dish was 10^6. The equation for determining the number of bacteria is $y = (2^x) \times 10^6$. Since the bacteria double every twenty minutes, they go through three doubling times every hour. So, when the bacteria are allowed to grow for eight hours, they will have gone through $8 \times 3 = 24$ doubling times. When $x = 24$, $y = (2^{24}) \times 10^6 = 16{,}777{,}216 \times 10^6$, which is approximately 1.7×10^{13}.

46. A: The critical points of the graph occur when $f'(x) = 0$.

$f(x) = 4x^3 - x^2 - 4x + 2$	
$f'(x) = 12x^2 - 2x - 4$	Take the derivative of $f(x)$.
$f'(x) = 2(6x^2 - x - 2)$	Factor a 2 out of each term in $f'(x)$.
$f'(x) = 2(3x - 2)(2x + 1)$	Factor the new polynomial in $f'(x)$.
$0 = 2(3x - 2)(2x + 1)$	Set $f'(x)$ equal to 0.
$3x - 2 = 0$, $x = \frac{2}{3}$	Solve the first factor to find the value of x that makes it 0.
$2x + 1 = 0$, $x = -\frac{1}{2}$	Solve the second factor to find the value of x that makes it 0.

The critical points of the graph are at $x = \frac{2}{3}$ and $x = -\frac{1}{2}$.

To determine if these points are relative maximums or minimums, test points on each end of the intervals surrounding them. Start by testing a point less than $-\frac{1}{2}$. For example, solve for the value of $f'(x)$ when $x = -1$.

$$f'(-1) = 12(-1)^2 - 2(-1) - 4 = 12 + 2 - 4 = 10$$

Now, test a point greater than $-\frac{1}{2}$ and less than $\frac{2}{3}$, like 0.

$$f'(0) = 12(0)^2 - 2(0) - 4 = 0 - 0 - 4 = -4$$

Therefore, when $x < -\frac{1}{2}$, $f'(x) > 0$, and when $-\frac{1}{2} < x < \frac{2}{3}$, $f'(x) < 0$, so the point where $x = -\frac{1}{2}$ is a relative maximum of the function. Now, test a point greater than $\frac{2}{3}$, like 1.

$$f'(1) = 12(1)^2 - 2(1) - 4 = 12 - 2 - 4 = 6$$

When $x > \frac{2}{3}$, $f'(x) > 0$, so the point where $x = \frac{2}{3}$ is a relative minimum. To find the corresponding y-values for these points, substitute each value of x into the original function and solve for y.

If $f''(x) > 0$ for all x in an interval, the graph of the function is concave upward on that interval, and if $f''(x) < 0$ for all x in an interval, the graph of the function is concave downward on that interval. Find the second derivative of the function and determine the intervals in which $f''(x)$ is less than zero and greater than zero.

$$f''(x) = 24x - 2$$

$24x - 2 < 0$	$24x - 2 > 0$
$x < \frac{1}{12}$	$x > \frac{1}{12}$

The graph of f is concave downward on the interval $\left(-\infty, \frac{1}{12}\right)$ and concave upward on the interval $\left(\frac{1}{12}, \infty\right)$.

47. A: The probability of an event is the number of possible occurrences of that event divided by the number of all possible outcomes. A student who is eight years old or less can be five, six, seven, or eight years old.

$$\frac{\text{number of } 5-, 6-, 7-, \text{ and } 8-\text{year}-\text{old students}}{\text{total number of students}} = \frac{11+8+9+14}{11+8+9+14+11+13} = \frac{42}{66} = \frac{7}{11}$$

Therefore, the probability that he or she is no more than eight years old is $\frac{7}{11}$.

48. A: The set of integers is a subset of the set of rational numbers, which is a subset of the set real numbers, which is a subset of the set of complex numbers.

49. A, B, C: Student A uses the quadratic formula. Student B uses the completing the square method. Student C uses factoring. These are the three common ways to find the exact solution of a quadratic equation.

50. A: Choice B is incorrect because we are not given that $\angle BAC$ is a right angle and we cannot assume it is right because it looks like a right angle. It is possible to prove that it is right, but this requires an argument very much like the one the student makes. Choices C and D are incorrect because the definition of median is a line segment joining a vertex of a triangle to the midpoint of the opposite side. The student gave an argument that is correct and reasonably simple, so choice A is correct.

51. A, C: Student A recognizes that the set of all usernames is the disjoint union of usernames with A and those without A, and from this correctly derives the equation "usernames with A" $+ 25^6 = 26^6$ and concludes "usernames with A" $= 26^6 - 25^6$.

Student B's plan looks promising but overcounts administrator usernames with more than one A. For instance, a username that begins with two A's gets counted twice—once when the first position is chosen as the "guaranteed" A and then A happens to be chosen as the second letter, and once when the second position is the "guaranteed" A and then A happens to be chosen as the first letter.

Student C recognizes that the set of usernames with A's is the disjoint union of the subset of usernames with exactly one A, the subset of those with exactly two A's, etc. Thus, we can get the total by adding up the sizes of these six subsets, which Student C calculates correctly.

52. A: The equation for this scenario is $2b + 3c \geq 150$, where b is the number of brownies that must be sold and c is the number of cupcakes that must be sold. Solve for the intercepts of the inequality by momentarily changing the inequality sign to an equal sign, substituting 0 for b and solving for c, and then substituting 0 for c and solving for b. The intercepts of this linear inequality are $b = 75$ and $c = 50$. The x-axis represents the number of brownies sold, so the x-intercept is $(75, 0)$. The y-axis represents the number of cupcakes sold, so the y-intercept is $(0, 50)$. The solid line through the two intercepts, (75,0) and (0,50), represents the minimum number of each type of baked good that must be sold to offset production costs. Since you are breaking even or making a profit, shade above the line.

53. C: To get the inverse relation, we swap the variables, replacing y with x and vice versa. This gives us the equation $x - 2 = \pm\sqrt{y - 3}$, which we solve for y:

$$\pm\sqrt{y-3} = x - 2$$
$$\left(\pm\sqrt{y-3}\right)^2 = (x-2)^2$$
$$y - 3 = (x-2)^2$$
$$y = (x-2)^2 + 3$$

The variable x in this final equation can take on any real value, and each choice of x will produce a single value of y. Thus, this relation is a function, written in the form $y = f(x)$. Alternatively, we might note that the quadratic equation $y = (x-2)^2 + 3$ is in vertex form, indicating that its graph is a parabola opening upward with a vertex of (2,3). It passes the vertical line test and is, thus, a function. To verify our work, we can graph this function and the original equation $y - 2 = \pm\sqrt{x - 3}$, which is a parabola opening to the right with a vertex of (3,2), to find that they are—as they should be—mirror images of each other across the line $y = x$.

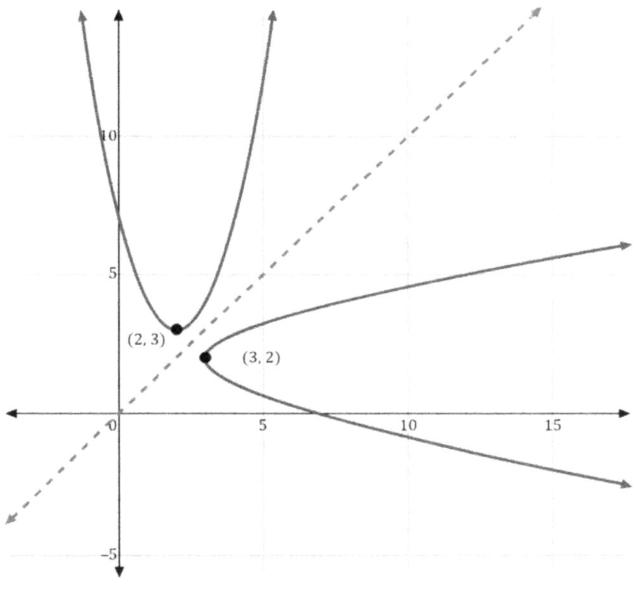

54. A: Since there are five possible choices for the correct answer to the first problem, five for the second, five for the third, and five for the fourth, by the Fundamental Counting Principle, the total number of possible ways to assign the letters of the correct answers is $5 \cdot 5 \cdot 5 \cdot 5 = 5^4$. But if we exclude the letter c from consideration, there are four choices for the correct answer to the first problem, four for the second, etc. Thus, there are $4 \cdot 4 \cdot 4 \cdot 4 = 4^4$ ways to assign the letters with c never being correct. Thus, the probability of randomly assigning the letters so that c is never correct is:

$$\frac{\#\text{ of assignments without } c}{\#\text{ of unrestricted assignments}} = \frac{4^4}{5^4} = \frac{256}{625} = 0.4096$$

55. 5,075: The ratio of expenses to income is 7: 1. Since we know he saves \$725 per month, we need to multiply this value by 7 to find out how much he spends on expenses each month. $\$725 \times 7 = \$5,075$. So, Jon spends \$5,075 on expenses each month.

Answer Key and Explanations #1

56. C: On the unit circle, $\sin\theta = \frac{\sqrt{3}}{2}$ when $\theta = \frac{\pi}{3}$ and when $\theta = \frac{2\pi}{3}$. Since only $\frac{2\pi}{3}$ is in the given range of $\frac{\pi}{2} < \theta < \pi, \theta = \frac{2\pi}{3}$.

57. D: "The square of three times the sum of a and six is fourteen more than a" is represented by the equation $[3(a+6)]^2 = a + 14$. Solve for a.

$$\begin{aligned} [3a+18]^2 &= a + 14 \\ 9a^2 + 108a + 324 &= a + 14 \\ 9a^2 + 107a + 310 &= 0 \\ (a+5)(9a+62) &= 0 \\ a &= -5, -\frac{62}{9} \end{aligned}$$

So, –5 is a possible value of a.

58. B, C: To convert a quantity with units (in this case, 1 year), we multiply it by conversion factors (namely, fractions that equal 1) in such a fashion that all units cancel except those we want in our answer (in this case, seconds). From the time facts, 1 (common) year = 365 days, 1 day = 24 hours, 1 hour = 60 minutes, and 1 minute = 60 seconds, we get the conversion factors $\frac{365 \text{ days}}{1 \text{ year}}$, $\frac{24 \text{ hours}}{1 \text{ day}}$, $\frac{60 \text{ minutes}}{1 \text{ hour}}$, and $\frac{60 \text{ seconds}}{1 \text{ minute}}$.

Emily sets the problem up incorrectly. She multiplies only conversion factors without including the actual quantity, 1 year, that needs conversion. She also combines the factors in such a fashion that the units do not cancel correctly, leaving years and seconds in the numerator and two factors of days in the denominator.

Jonathon sets the problem up correctly, beginning with the quantity 1 year (written as the fraction $\frac{1 \text{ year}}{1}$) and multiplying by conversion factors in order from longer times to shorter times (years to days, days to hours, etc.). He does this in such a fashion that all units cancel except for seconds.

Sarah sets the problem up correctly, beginning with the quantity 1 year and multiplying by conversion factors in order from shorter times to longer ones in such a fashion that all units cancel except for seconds. This student converts minutes directly to days, bypassing hours, by using the time fact 1,440 minutes = 1 day.

59. D: To solve this equation, start by squaring both sides, then isolate the variable.

$$\begin{aligned} 2 - \sqrt{x} &= \sqrt{x - 20} \\ \left(2 - \sqrt{x}\right)^2 &= \left(\sqrt{x - 20}\right)^2 \\ \left(2 - \sqrt{x}\right)\left(2 - \sqrt{x}\right) &= x - 20 \\ 4 - 4\sqrt{x} + x &= x - 20 \\ -4\sqrt{x} &= -24 \\ \sqrt{x} &= 6 \\ x &= 36 \end{aligned}$$

When solving radical equations, check for extraneous solutions by substituting the found value into the original equation.

$$2-\sqrt{x}=\sqrt{x-20}$$
$$2-\sqrt{36}=\sqrt{36-20}$$
$$2-6=\sqrt{16}$$
$$-4\neq 4$$

Since the solution results in an untrue statement, there is no solution. Alternatively, consider the graphs of the left and right side of the given equation. Notice the graphs $y=2-\sqrt{x}$ and $y=\sqrt{x-20}$ do not intersect, which confirms there is no solution.

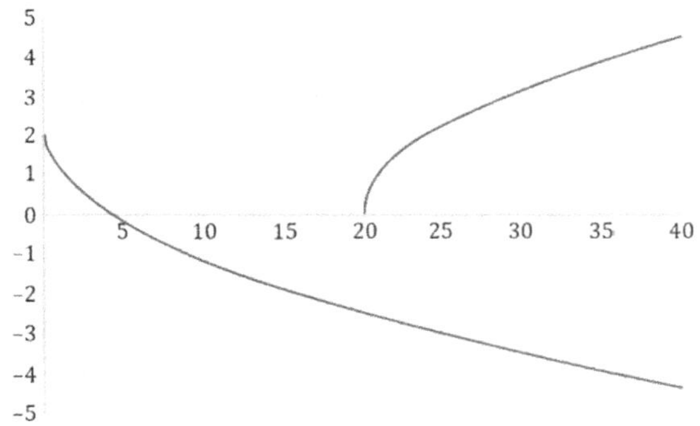

60. A, B, C: In the slope formula, $m=\frac{y_2-y_1}{x_2-x_1}$, the ordering of the points—that is, which is (x_1,y_1) and which is (x_2,y_2)—is irrelevant. So, all three approaches correctly calculate the slope as $m=\frac{3}{4}$.

Approach A correctly substitutes the slope of the line, $m=\frac{3}{4}$, and the coordinates of a point on the line, $(x_1,y_1)=(12,3)$, into point-slope form of a linear equation, $y-y_1=m(x-x_1)$, to get an equation of the line.

Approach B correctly substitutes the slope of the line, $m=\frac{3}{4}$, and the y-intercept of the line, $b=-6$, into the slope-intercept form of a linear equation, $y=mx+b$.

Approach C is the same as approach A, except that it substitutes the coordinates of the other point on the line, $(0,-6)$, into point-slope form. Since the choice of points on the line is irrelevant, this produces a different correct equation of the same line.

61. C: Find the points of intersection of the two graphs:

$$x^2-2x-7=x-3$$
$$x^2-3x-4=0$$
$$(x+1)(x-4)=0$$
$$x=-1 \qquad x=4$$

So, the area is between the graphs on the interval $[-1,4]$. To determine which of the graphs is the upper or lower bound, pick a test point in the interval. Using $x = 0$ as the test point.

$$\begin{array}{cc} y = x^2 - 2x - 7 & y = x - 3 \\ y = 0^2 - 2(0) - 7 & y = 0 - 3 \\ y = -7 & y = -3 \end{array}$$

The finite region is bound at the top by the line $y = x - 3$ and at the bottom by $y = x^2 - 2x - 7$. The height of the region at point x is defined by $[(x - 3) - (x^2 - 2x - 7)]$, so the area is:

$$\begin{aligned} A &= \int_{-1}^{4} [(x-3) - (x^2 - 2x - 7)]dx \\ &= \int_{-1}^{4} (-x^2 + 3x + 4)\, dx \\ &= \left[-\frac{1}{3}x^3 + \frac{3}{2}x^2 + 4x\right]_{-1}^{4} \\ &= \left[-\frac{1}{3}(4)^3 + \frac{3}{2}(4)^2 + 4(4)\right] - \left[-\frac{1}{3}(-1)^3 + \frac{3}{2}(-1)^2 + 4(-1)\right] \\ &= \left[-\frac{64}{3} + 24 + 16\right] - \left[\frac{1}{3} + \frac{3}{2} - 4\right] \\ &= \frac{56}{3} - \left(-\frac{13}{6}\right) \\ &= \frac{125}{6} \end{aligned}$$

62. 140: If the distance between the two houses is 10 cm on the map, then the actual distance between the houses is 100 m. To find x, use the Pythagorean theorem.

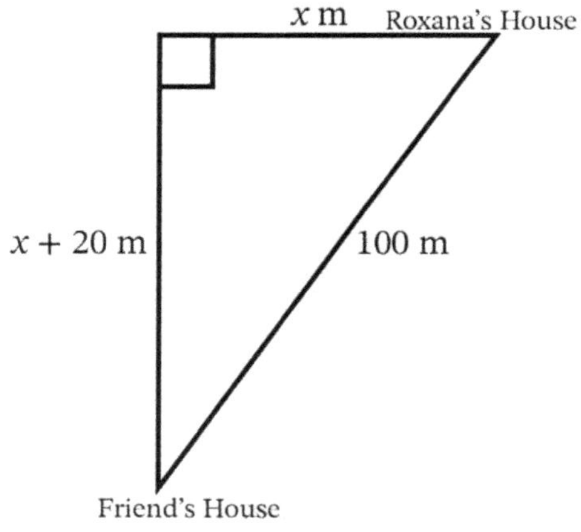

$$\begin{aligned} x^2 + (x+20)^2 &= (100)^2 \\ x^2 + x^2 + 40x + 400 &= 10{,}000 \\ 2x^2 + 40x - 9{,}600 &= 0 \\ 2(x^2 + 20x - 4{,}800) &= 0 \\ 2(x - 60)(x + 80) &= 0 \\ x = 60 \quad x &= -80 \end{aligned}$$

Since x represents a distance, it cannot equal –80. Since $x = 60$, $x + 20 = 80$. Roxana walks a total of $60 + 80 = 140$ meters to get to her friend's house.

63. B: To increase a value by 20%, multiply the value by 1.2. Multiplying the lengths in a closed figure by a given factor (to produce a similar figure) multiplies the perimeter of the figure by the same factor. So, multiplying all lengths by 1.2 (increasing them 20%) increases the perimeter by a factor of 1.2. The new perimeter is $1.2 \cdot 60 \text{ mm} = 72 \text{ mm}$. On the other hand, multiplying the lengths in a closed figure by a given factor multiplies the area of the figure by the square of that factor. So, multiplying all lengths by 1.2 increases the area by a factor of 1.2^2. Thus, the new area is $1.2^2 \cdot 250 \text{ mm}^2 = 360 \text{ mm}^2$.

64. B: The acceleration a of an object at time t is the derivative of the velocity v of the object with respect to time t. Velocity of the object is the derivative of the position s of the object with respect to time t. So, given the velocity of an object at time t, $s(t)$ can be found by taking the integral of the $v(t)$, and $a(t)$ can be found by taking the derivative of $v(t)$.

$$\begin{aligned} s(t) &= \int v(t)dt \\ &= \int(-t^2 + 8t)dt \\ &= -\frac{1}{3}t^3 + 4t^2 + C \end{aligned}$$

Since the position of the car at time 0 is 0:

$$\begin{aligned} 0 &= s(0) \\ 0 &= 4(0)^2 - \frac{1}{3}(0)^3 + C \\ 0 &= 0 - 0 + C \\ 0 &= C \end{aligned}$$

Therefore, $s(t) = -\frac{1}{3}t^3 + 4t^2$.

The acceleration at time t is: $a(t) = v'(t) = -2t + 8$.

Find the time at which the acceleration is equal to 0.

$$\begin{aligned} 0 &= 8 - 2t \\ t &= 4 \end{aligned}$$

Then, find $s(4)$ to find the position of the car when the velocity is 0.

$$s(t) = -\frac{1}{3}t^3 + 4t^2$$
$$s(4) = -\frac{1}{3}(4)^3 + 4(4)^2$$
$$s(4) = -\frac{64}{3} + 64$$
$$s(4) = \frac{128}{3}$$

Therefore, the position of the car when its acceleration is 0 is $\frac{128}{3}$.

65. B: When the jeans are marked down by 30%, the cost is 70% of its original price. Thus, the reduced price of the dress can be written as $\frac{70}{100}x$, or $\frac{7}{10}x$, where x is the original price. When discounted an extra 15%, the jeans cost 85% of the reduced price, or $\frac{85}{100}\left(\frac{7}{10}x\right)$. This can be simplified to $\frac{17}{20}\left(\frac{7}{10}x\right)$. Multiply the fractions to get $\frac{119}{200}x$. So, the final price is $\frac{119}{200}$ of the original price.

66. C: Evaluate $\frac{x^3+3x^2-x-3}{x^2-9}$ at $x = -3$.

$$\frac{(-3)^3 + 3(-3)^2 - (-3) - 3}{(-3)^2 - 9} = \frac{-27 + 3(9) + 3 - 3}{9 - 9} = \frac{-27 + 27 + 3 - 3}{9 - 9} = \frac{0}{0}$$

Since evaluating $\frac{x^3+3x^2-x-3}{x^2-9}$ at $x = -3$ produces a fraction with a zero denominator, simplify the polynomial expression before evaluating the limit.

Simplify the polynomial expression:

$$\frac{x^3 + 3x^2 - x - 3}{x^2 - 9}$$

$$\frac{x^2(x+3) - 1(x+3)}{(x+3)(x-3)}$$ Factor out $(x + 3)$ from the first two terms and the 3rd and 4th terms in the numerator, and factor the denominator.

$$\frac{(x+3)(x^2-1)}{(x+3)(x-3)}$$ Factor out $(x + 3)$ completely in the numerator.

$$\frac{(x+1)(x-1)}{x-3}$$ Cancel the common $(x + 3)$ terms in the numerator and denominator.

Now, evaluate the limit.

$$\lim_{x\to-3}\frac{(x+1)(x-1)}{x-3}$$

$$\frac{(-3+1)(-3-1)}{-3-3}$$ Substitute –3 into the expression for x.

$$\frac{8}{-6}$$ Simplify the numerator and denominator.

$$-\frac{4}{3}$$ Simplify the fraction.

Praxis Practice Test #2

1. Examine the triangles below:

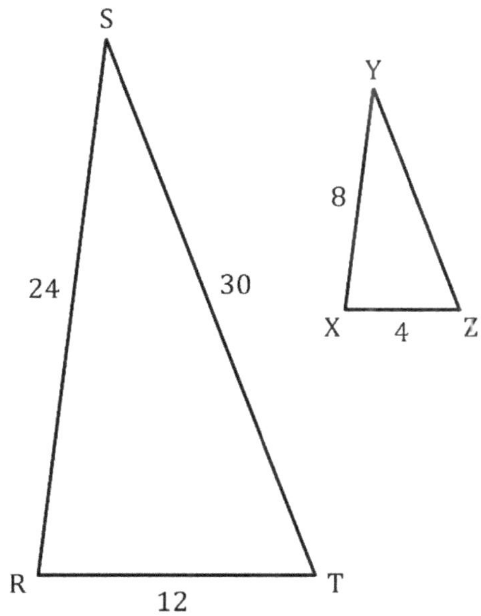

In order for ΔRST to be similar to ΔXYZ, what must the length of $\overline{YZ}$ be?

a. 10
b. 14
c. 15
d. 22

2. You are teaching a class of 10 students. During the semester you give them four exams. On which exam do the scores best seem to fit a normal distribution?

Exam #1	**37**	**48**	**90**	**80**	**1**	**88**	**98**	**83**	**99**	**52**
Exam #2	**20**	**78**	**4**	**37**	**5**	**68**	**23**	**9**	**99**	**30**
Exam #3	**83**	**11**	**18**	**14**	**25**	**80**	**88**	**72**	**27**	**79**
Exam #4	**57**	**48**	**47**	**51**	**62**	**49**	**56**	**72**	**61**	**33**

a. Exam 1
b. Exam 2
c. Exam 3
d. Exam 4

3. Which function is neither odd nor even?

a. $f(x) = x + \sin x$
b. $g(x) = x + \cos x$
c. $h(x) = x \sin x$
d. $k(x) = x \cos x$

Refer to the following for question 4:

The box-and-whisker plot displays a district's standardized test scores by subject.

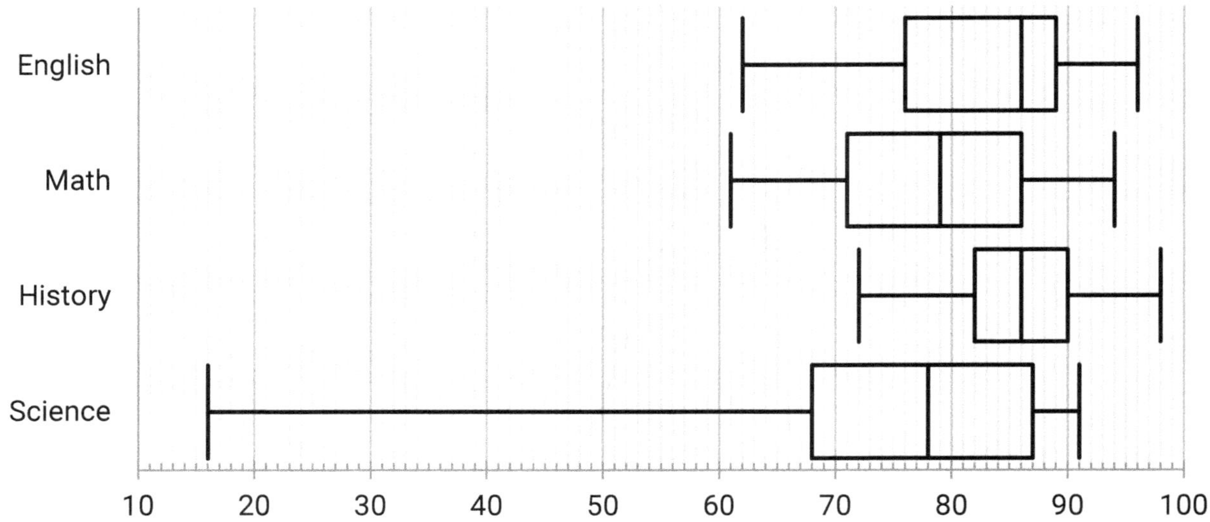

4. Which of the following statements is NOT true of the data?

a. The median better reflects student performance in Science than the mean.
b. The mean test score for English and history is the same.
c. The median of the History scores is the same as the 3rd quartile score for Math.
d. The median test score is above the mean for English.

5. This question may have more than one correct answer. Select all correct answers.

Your class attempts to solve the following rational equation.

$$\frac{x^2 - x - 6}{x - 3} = 1$$

Three students offer the following solutions. Which solutions are mathematically correct? Select all correct solutions.

a.

Student A:

$$(x-3)\frac{x^2 - x - 6}{x - 3} = 1(x-3)$$
$$x^2 - x - 6 = x - 3$$
$$x^2 - 2x - 3 = 0$$
$$(x-3)(x+1) = 0$$
$$x = 3 \text{ or } x = -1$$

b.

Student B:

$$\frac{(x-3)(x+2)}{x-3} = 1$$
$$x + 2 = 1, (\text{with } x \neq 3)$$
$$x = -1$$

c.

Student C:

$$\frac{x^2 - x - 6}{x-3} - 1 = 0$$
$$\frac{x^2 - x - 6}{x-3} - \frac{x-3}{x-3} = 0$$
$$\frac{x^2 - 2x - 3}{x-3} = 0$$
$$x^2 - 2x - 3 = 0$$
$$(x-3)(x+1) = 0$$
$$x = 3 \text{ or } x = -1$$

Refer to the following for questions 6–7:

This data shows the number of students at a high school who are enrolled in various electives.

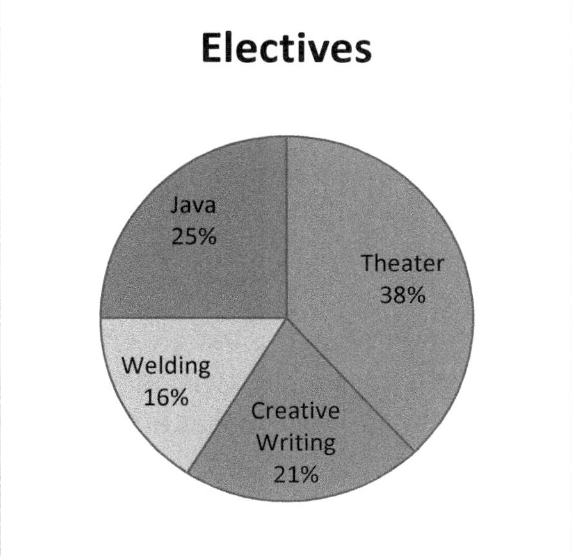

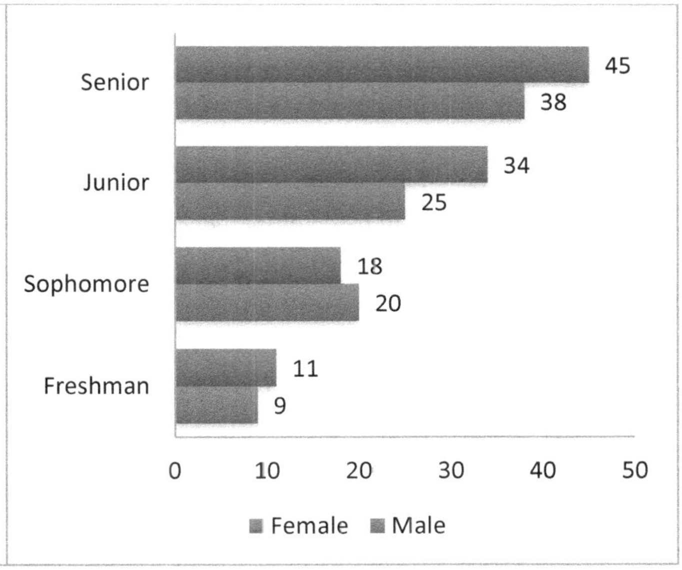

6. Which of these is the greatest quantity?

a. The number of students taking welding
b. The number of male sophomores taking an elective
c. The difference in the number of male and female students taking an elective
d. The difference in the number of juniors and seniors taking an elective

7. Compare the two quantities.

Quantity A	Quantity B
The percentage of students taking Java or Creative Writing	The percentage of students in electives who are male

a. Quantity A is greater.
b. Quantity B is greater.
c. The two quantities are the same.
d. The relationship cannot be determined from the given information.

8. Which of these is the least biased sampling technique?

a. A PE teacher records the height of the entire girls' volleyball team to show that the average height of high school girls today is greater than it was 50 years ago.
b. A woman surveys her entire neighborhood for opinions on the new business coming to town.
c. A news website requests feedback via an anonymous online form.
d. A journalist calls the first 50 names on a randomly generated list of all voters in the county to request opinions on a politician.

9. If $f(x) = \frac{1}{4}x^2 - 3$, find the slope of the line tangent to graph of $f(x)$ at $x = 2$.

10. This question may have more than one correct answer. Select all correct answers.

You give your students this problem: "A small business has sales of \$316,447 in 2025, which is 15% more than 2024. The sales in 2024 were 5% more than 2023. What were the sales in 2023? Round your answer to the nearest dollar." Three students give the following answers. Which answers are mathematically correct? Choose all correct answers.

a. Student A: "This means that the 2025 sales were 115% of the 2024 sales, so I divide by 115% to get the 2024 sales: \$316,447 ÷ 1.15 = \$275,171.30. The 2024 sales were 105% of the 2023 sales, so I divide by 105% to get the 2023 sales: \$275,171.30 ÷ 1.05 = \$262,067.90, which rounds to \$262,068."
b. Student B: "I find 15% of the 2025 sales and subtract it to get the 2024 sales: \$316,447 − 0.15(\$316,447) = \$268,979.95, which I round to \$268,980. Then I subtract another 5% to get the 2023 sales: \$268,980 − 0.05(\$316,447) = \$253,157.65, which rounds to \$253,158."
c. Student C: "I add the 5% and 15% to get a total 20% increase. I calculate 20% of 2025 sales, getting 0.2(\$316,447) = \$63,289.40. I subtract this from the 2025 sales to get \$316,447 − \$63,289.40 = \$253,157.60, which rounds to \$253,258 sales in 2023."

11. You ask a student to find the axis of symmetry of the parabola whose equation is $y = x^2 - 11x + 24$. The student rewrites the equation in factored form $y = (x-3)(x-8)$, notes that the function has zeros at $x = 3$ and $x = 8$, and concludes that the axis of symmetry is the average of the two x-intercepts: $x = \frac{3+8}{2} = \frac{11}{2}$. How do you critique this solution?

a. The answer is wrong, and the student's approach is unsound.
b. The answer is correct. The student's approach yields the correct answer when the quadratic function has integer zeros, but it may yield wrong answers in other cases.
c. The answer happens to be correct, but the student's approach is unsound.
d. The answer is correct. The student's approach always yields the correct answer but may be difficult to carry out when the quadratic function has fractional, irrational, or complex zeros.

12. Which of these is the equation graphed below?

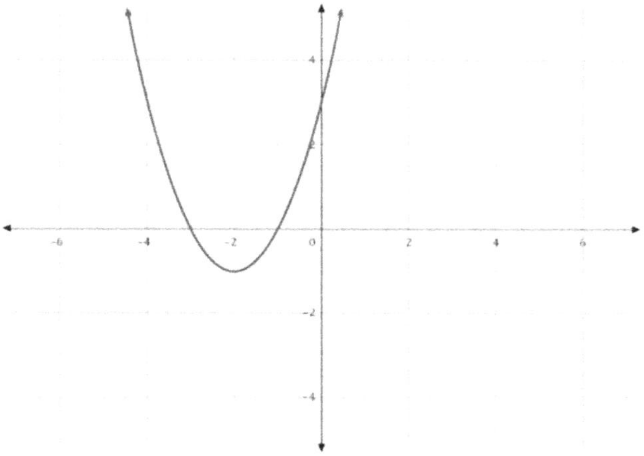

a. $y = x^2 + 4x + 3$
b. $y = x^2 - 2x - 1$
c. $y = 2x^2 + 4x + 1$
d. $y = \frac{1}{2}x^2 - 2x + 3$

13. Let $f(x) = 3 - 5x$ and $g(x) = \frac{1}{5}x + \frac{3}{5}$, where the domain and range of both functions is the set of all real numbers. Find the composition $(f \circ g)(x)$ and use it to determine whether g is the inverse function of f.

a. From $(f \circ g)(x) = -x$, we conclude $g \neq f^{-1}$.
b. From $(f \circ g)(x) = -x$, we conclude $g = f^{-1}$.
c. From $(f \circ g)(x) = x$, we conclude $g = f^{-1}$.
d. From $(f \circ g)(x) = x$, we conclude $g \neq f^{-1}$.

14. Solve $\sqrt{2x} - 3 = \sqrt{2x - 15}$.

a. $x = 0$
b. $x = 4$
c. $x = 8$
d. $x = 10$

15. Find the area A of the finite region between the graphs of $y = -x + 2$ and $y = x^2 - 4$.

a. 18

b. $\frac{125}{6}$

c. $\frac{45}{2}$

d. 25

16. This question may have more than one correct answer. Select all the correct answers.

In $\triangle ABC$, side $\overline{AC}$ has length $AC = 4$, the median $\overline{BD}$ has length $BD = 2$, and the measure of $\angle BDA$ is $m\angle BDA = 72°$. Which of the following statements about $\triangle ABC$ are correct? Select all that are correct.

a. $BC = 2$

b. $BA = 2$

c. $m\angle A = 54°$

17. Given the figure and the following information, find DE to the nearest tenth.

$\overline{AD}$ is an altitude of ΔABC

$\overline{DE}$ is an altitude of triangle ΔADC

$\overline{BD} \cong \overline{DC}$

$AB = 17$; $AD = 8$

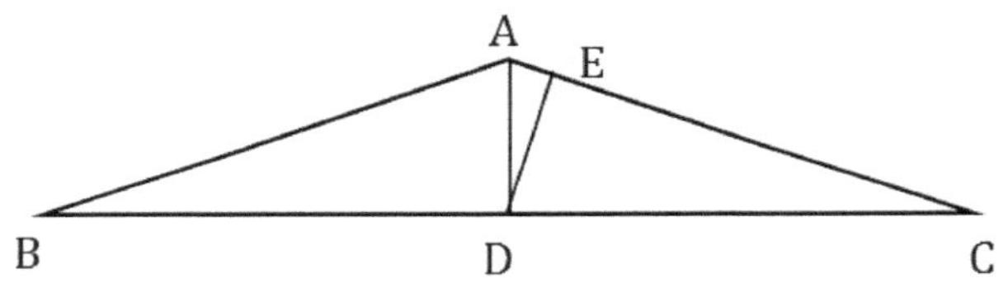

a. 5.8

b. 6.2

c. 6.9

d. 7.1

18. Expand the complex linear polynomial $(2 + 3i)[(7 - i)z - (3 + 5i)]$ using the distributive property and then simplify the result. Note that i is the imaginary unit and z is a complex variable.

a. $(11 + 19i)z - (21 + 19i)$

b. $(17 - 23i)z + (9 - i)$

c. $(17 + 19i)z + (9 - 19i)$

d. $(17 + 23i)z + (9 + i)$

19. Which of these relationships represents y as a function of x?

a. $x = y^4$

c. $y = |x - 2|$

b.

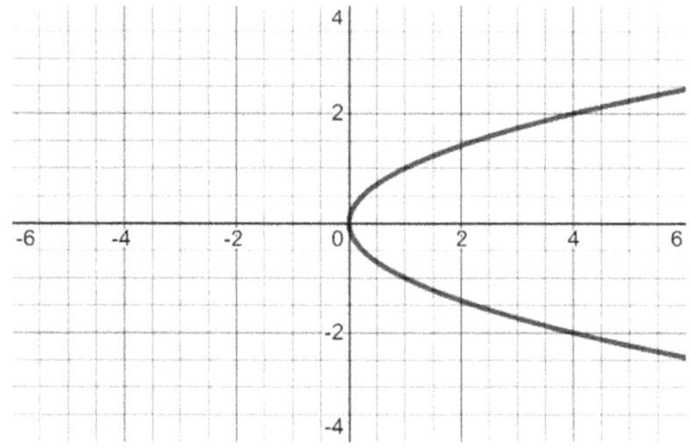

d.

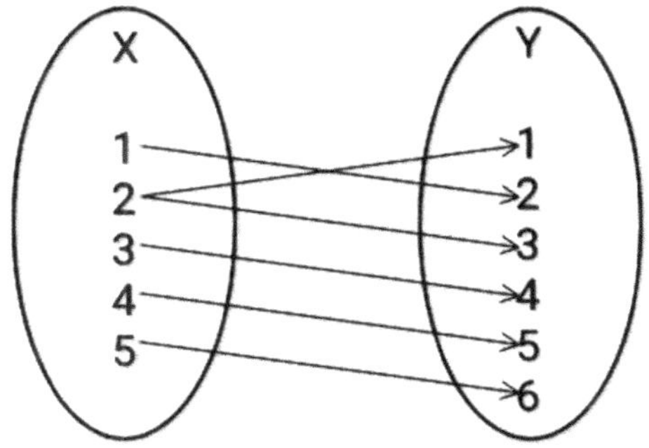

20. A greenhouse is built in the shape of a square pyramid. Each of the sides of the square base measures x ft, and the height is y feet. If the height were increased by 4 ft, how much more interior space would the greenhouse have?

a. $\frac{x^2y+4x^2}{3}$ ft^3
b. $\frac{y(x^2+4)}{3}$ ft^3
c. $\frac{4x^2}{3}$ ft^3
d. 64 ft^3

Refer to the following for question 21:

For a science fair project, Paul wanted to find the probability of rolling an even number with a weighted die versus a fair die. He did fifty tests, rolling the die three times for each test and recording the number of even numbers that came up. The chart below shows his results.

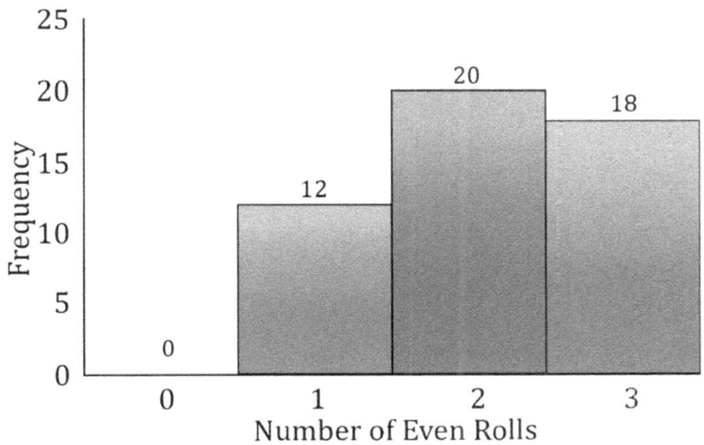

21. Given these experimental data, which of these approximates $P(\text{even})$ for a single roll of this die?

a. 0.468
b. 0.5
c. 0.625
d. 0.707

22. If $f(x)$ and $g(x)$ are inverse functions, which of these is the value of x when $f(g(x)) = \frac{1}{8}$?

a. -8
b. $\frac{1}{8}$
c. 8
d. $-\frac{1}{8}$

23. Use the operation table to determine $(w * y) * (x * x)$.

$*$	w	x	y	z
w	w	x	y	z
x	z	w	x	y
y	y	z	w	x
z	x	y	z	w

a. w
b. x
c. y
d. z

24. You pose the following question to your statistics students:

"A team of engineers performs an experiment on a large, slow Ferris wheel by dropping small weights from one of the cars each minute after boarding and measuring the time, in seconds, that it takes each weight to reach the ground. The scatter plot of their data appears here. How would you describe the relationship between these variables?"

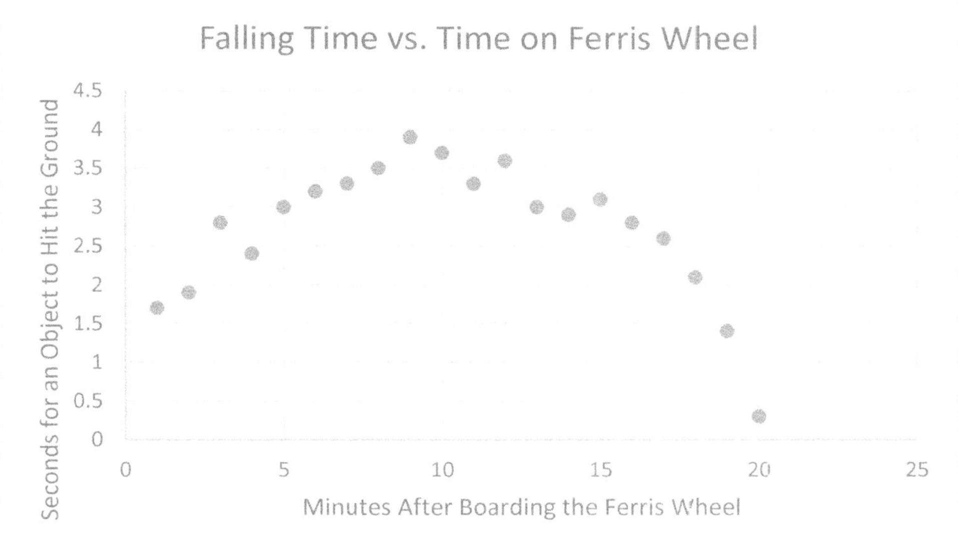

After some discussion, the class divides into three groups, each with a different opinion. Group A says that these variables have a negative correlation. Group B says that these variables have no correlation. Group C says that these variables have a positive correlation for the first ten minutes and a negative linear correlation for the second ten minutes. What do you say to the class?

a. "Group A is correct."
b. "Group B is correct."
c. "Group C is correct."
d. "All three groups' observations are true, and they reflect a more complex relationship between the variables."

25. Aaron goes on a run every morning down the straight country road that he lives on. The graph below shows Aaron's distance from home at times throughout his morning run. Which of the following statements is (are) true?

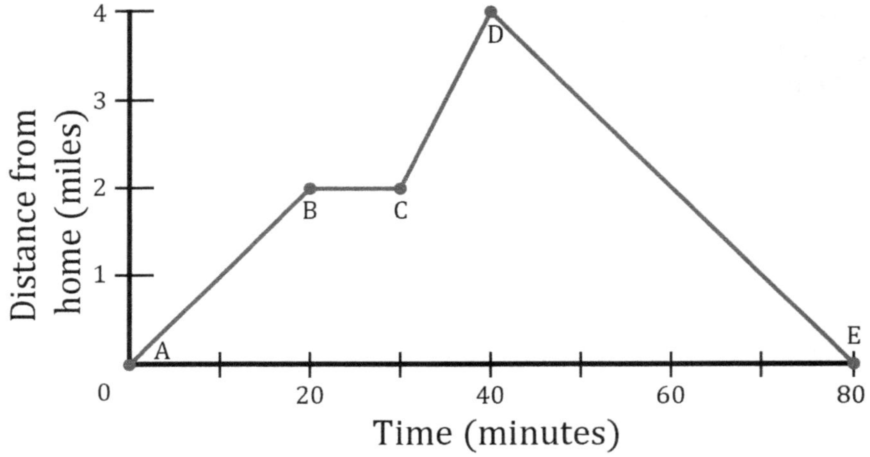

I. Aaron's average running speed was 6 mph.
II. Aaron's running speed from A to B was the same as from D to E.
III. Aaron ran a total distance of four miles.

a. I only
b. II only
c. I and II
d. I, II, and III

26. How many diagonals are in a dodecagon?

27. The ratio of employee wages and benefits to all other operational costs of a business is 2: 3. If a business's total operating expenses are $130,000 per month, how much money does the company spend on employee wages and benefits?

a. $43,333.33
b. $86,666.67
c. $52,000.00
d. $78,000.00

28. This question may have more than one correct answer. Select all correct answers.

In a statistics class of 21 students, you ask each student to flip a coin 10 times and record the number of heads on the whiteboard. Here are the results, sorted from smallest to largest:

3, 3, 3, 4, 4, 4, 4, 5, 5, 5, 6, 6, 6, 6, 6, 6, 7, 7, 7, 7, 9

Next, you divide the class into three groups and ask each group to produce a graphical display of this data. Which of the following displays accurately represents the data? Select all displays that accurately represent the data, regardless of whether they are aesthetically pleasing or the ideal format for communicating the data.

a. Group A:

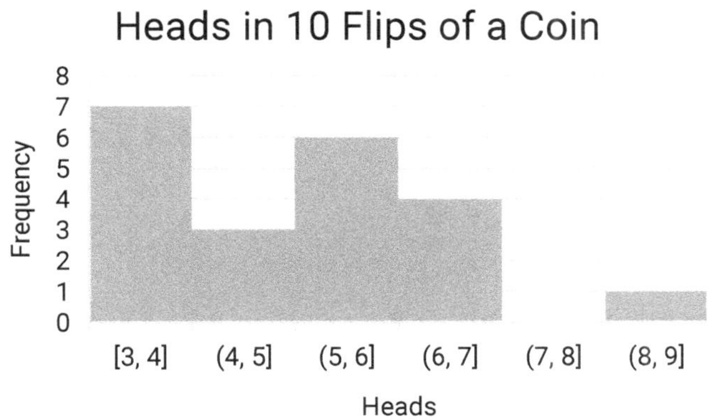

b. Group B:

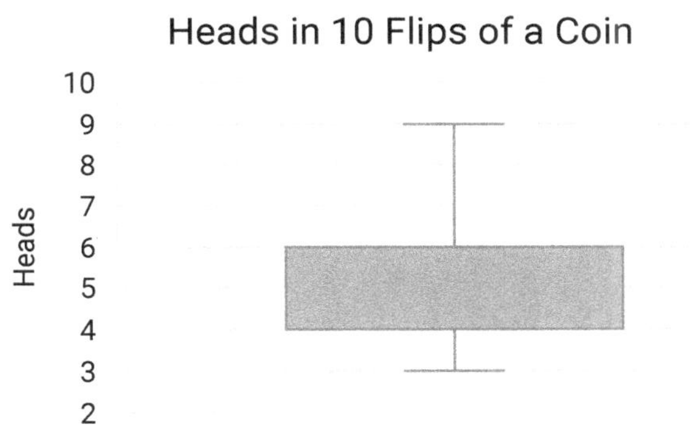

c. Group C:

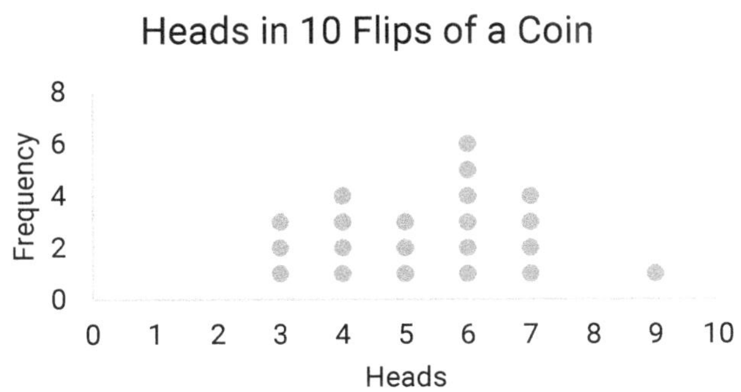

29. This question may have more than one correct answer. Select all correct answers.

You ask your students to find a sequence of one or more rigid transformations of the plane that will transform the solid black G in the 2nd quadrant to the gray G in the 4th quadrant. Three students give the following answers. Which of these answers produce the desired transformation? Select all sequences that transform the black G onto the gray G.

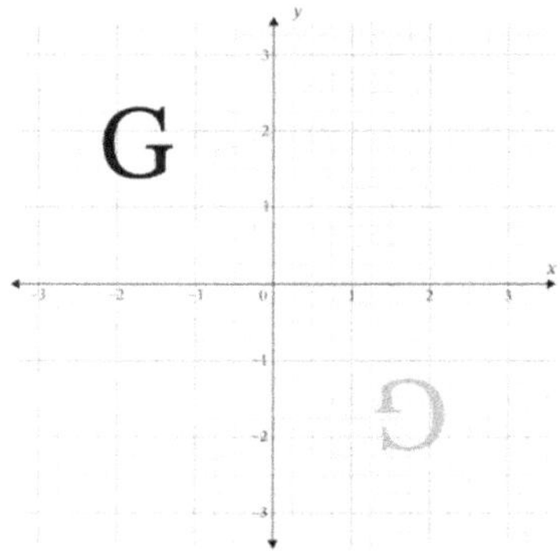

a. Student A: Simply rotate the plane 180° clockwise around the origin.
b. Student B: First reflect the plane across the y-axis. Then, reflect across the x-axis.
c. Student C: First, rotate the plane 90° counterclockwise around the origin. Then, reflect across the y-axis.

30. Which of these defines the recursive sequence $a_1 = -2, a_{n+1} = a_n + \frac{1}{2}$ explicitly?

a. $a_n = \frac{1}{2}n - 2$
b. $a_n = -2n + \frac{1}{2}$
c. $a_n = \frac{1}{2}n - \frac{5}{2}$
d. $a_n = -2n + \frac{5}{2}$

31. Below is the graph of a single polynomial function $p(x)$. (Parts of it extend outside the "viewing window.") Based on what you can see, which of the following could be $p(x)$?

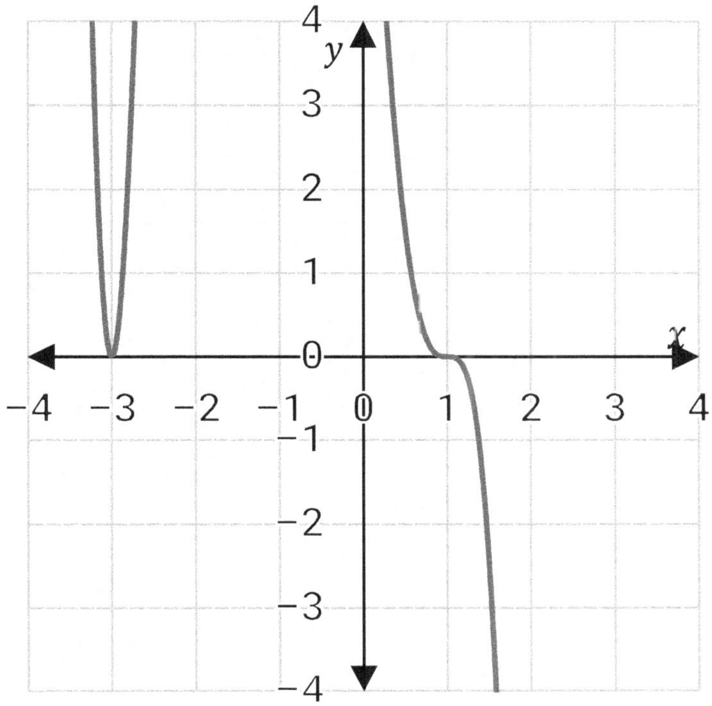

a. $p(x) = -(x-3)^4(x+1)$
b. $p(x) = -(x-3)^2(x+1)^3$
c. $p(x) = -(x+3)^2(x-1)^3$
d. $p(x) = -(x+3)^4(x-1)$

32. What is the radius of the circle defined by the equation $x^2 + y^2 + 2x - 10y + 18 = 0$?

a. $2\sqrt{2}$
b. 4
c. $2\sqrt{5}$
d. $3\sqrt{2}$

33. Solve the system of equations: $\begin{cases} -4x + 3y = 8 \\ 6x - 2y = -7 \end{cases}$.

a. $\left(\frac{1}{2}, \frac{10}{3}\right)$
b. $\left(-1, \frac{3}{2}\right)$
c. $\left(2, -\frac{5}{2}\right)$
d. $\left(-\frac{1}{2}, 2\right)$

34. Which of the graphs shown represents $f(x) = -2|-x+4| - 1$?

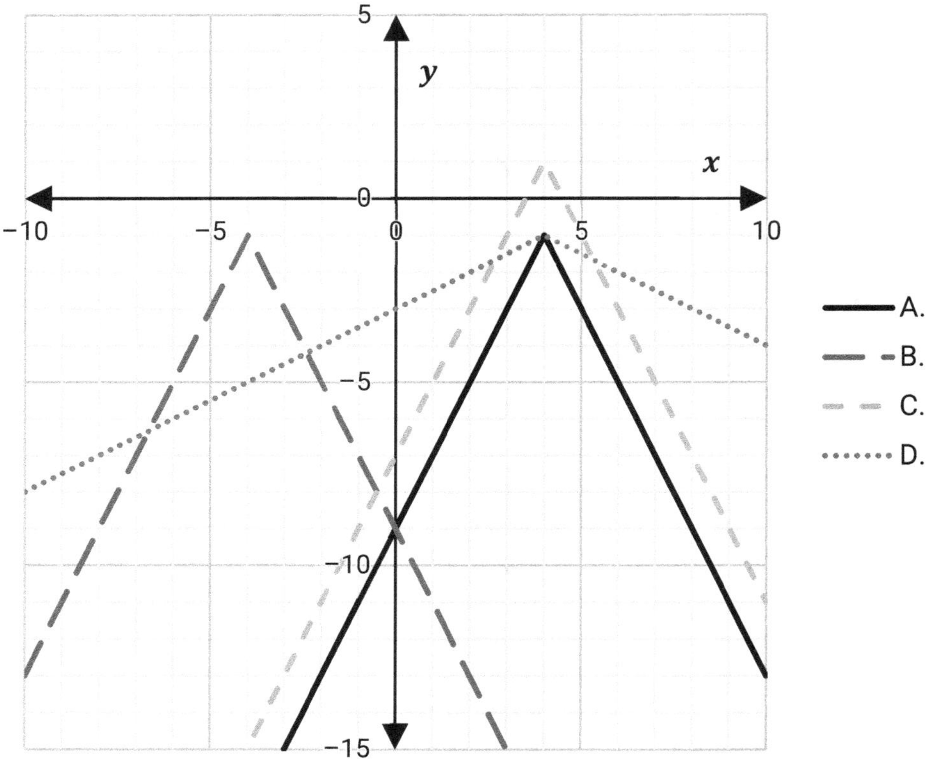

a. 
b. 
c. 
d. 

35. Which of the following expressions is equal to $\cos\theta \cot\theta$?

a. $\sin\theta$
b. $\sec\theta \tan\theta$
c. $\csc\theta - \sin\theta$
d. $\sec\theta - \sin\theta$

Refer to the following for question 36:

The box-and-whisker plot displays a district's standardized test scores by subject.

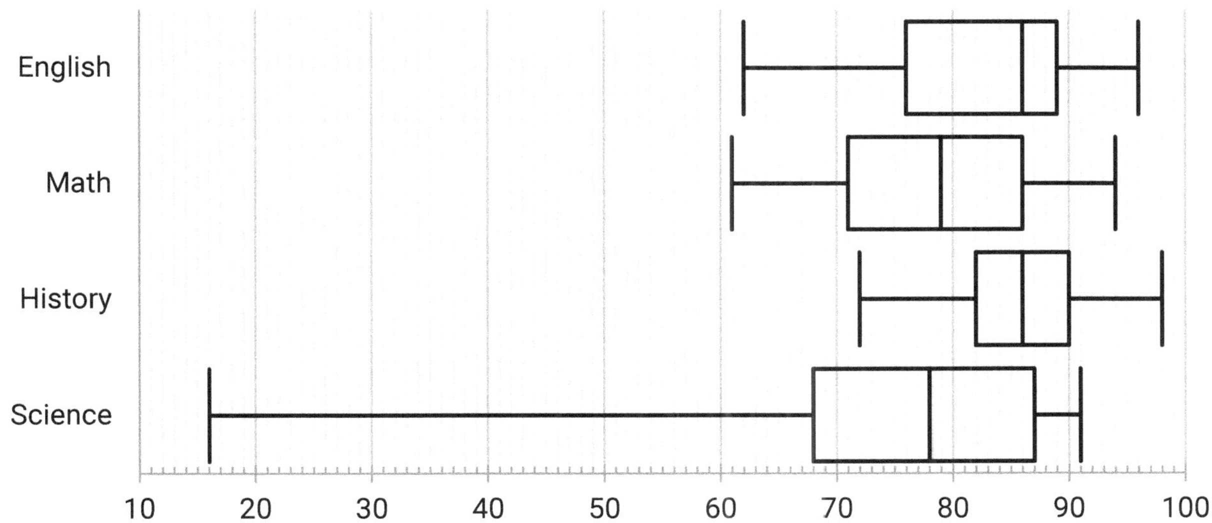

36. Which subject has the smallest range of test scores?

a. English
b. Math
c. History
d. Science

37. How many different seven-digit telephone numbers can be created in which no digit repeats and in which zero cannot be the first digit?

a. 5,040
b. 35,280
c. 544,320
d. 3,265,920

38. Identical rugs are offered for sale at two local shops and one online retailer, designated Stores A, B, and C, respectively. The rug's regular sales price is $296 at Store A, $220 at Store B, and $198.00 at Store C. Stores A and B collect 8% in sales tax on any after-discount price, while Store C collects no tax but charges a $35 shipping fee. A buyer has a 30% off coupon for Store A and a $10 off coupon for Store B. Which of these lists the stores in order of lowest to highest final sales price after all discounts, taxes, and fees are applied?

a. Store A, Store B, Store C
b. Store B, Store C, Store A
c. Store C, Store A, Store B
d. Store C, Store B, Store A

39. A random sample of 90 students at an elementary school were asked these three questions:

Do you like carrots?
Do you like broccoli?
Do you like cauliflower?

The results of the survey are shown below. If these data are representative of the population of students at the school, which of these is most probable?

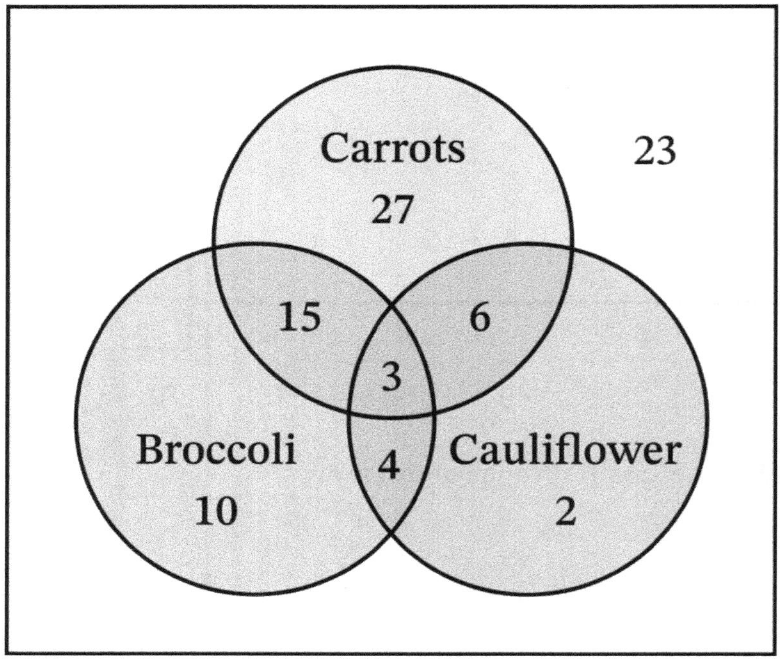

a. A student chosen at random likes broccoli.
b. Given a student chosen at random likes carrots, they will also like at least one other vegetable.
c. Given a student chosen at random likes cauliflower and broccoli, they will also like carrots.
d. A student chosen at random does not like carrots, broccoli, or cauliflower.

40. You ask your students to find the maximum value of the function $y = 3 - \sqrt{-x^2 + 4x + 21}$. One student offers the following solution:

"First, I rearrange this equation into standard form.

$$y - 3 = -\sqrt{-x^2 + 4x + 21}$$
$$(y - 3)^2 = -x^2 + 4x + 21$$
$$x^2 - 4x + (y - 3)^2 = 21$$
$$\left(x^2 - 4x + 4\right) + (y - 3)^2 = 21 + 4$$
$$(x - 2)^2 + (y - 3)^2 = 25$$

This is a circle of radius 5 centered at the point $(2, 3)$. So, its maximum value is 5 units above its center's y-value, namely 8."

What do you say to this student?

a. "The center of the circle is $(-2, -3)$, so the maximum is 2 not 8."

b. "Squaring both sides of the equation requires caution because it turns the graph of the original function, a semicircle, into a whole circle. In this case, however, the semicircle is the upper half of the full circle, so they share the same maximum, 8."

c. "Your excellent solution is entirely correct."

d. "Squaring both sides of the equation requires caution because it turns the graph of the original function, a semicircle, into a whole circle. In this case, the semicircle is the lower half of the whole circle, so its maximum is not 8, but 3, which it attains at both ends of the semicircle."

41. Determine which of the four labeled regions solves the system:

$$\begin{cases} 4x - 2y \geq -4 \\ 3x + 2y \leq 2 \end{cases}$$

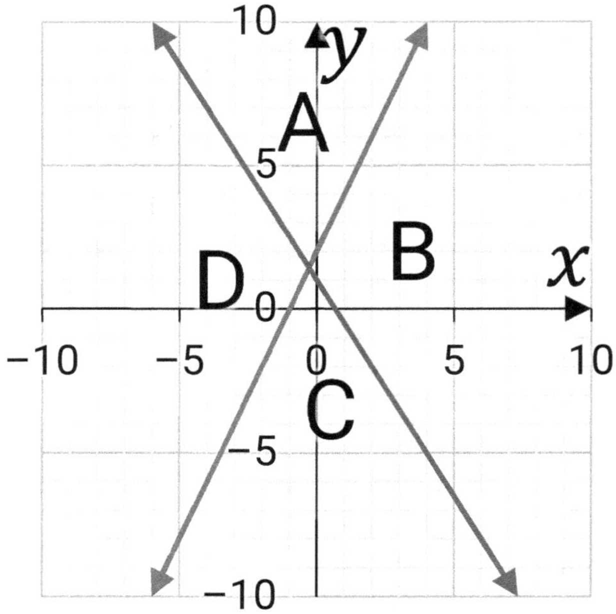

a. Region A
b. Region B
c. Region C
d. Region D

42. Right triangle $\triangle RST$ has altitude $\overline{SU} \perp \overline{RT}$. Line segment $\overline{UV} \perp \overline{ST}$, as shown. Also, line segments $\overline{SU}$, $\overline{UV}$, and $\overline{RT}$ have lengths $SU = a$, $UV = b$, and $RT = c$, respectively, as shown. Use similar triangles to find the length, RS, of side $\overline{RS}$.

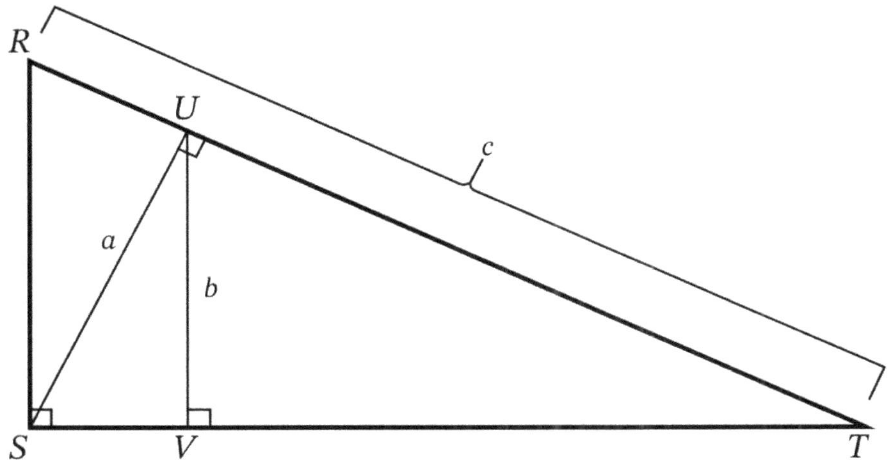

a. $RS = \frac{bc}{a}$

b. $RS = \frac{c\sqrt{a^2+b^2}}{a}$

c. $RS = \frac{c\sqrt{a^2-b^2}}{a}$

d. $RS = \frac{ac}{\sqrt{a^2-b^2}}$

43. You give your students the following assignment: "A robotic car travels along a straight track for nine seconds. During the first three seconds, it moves at 10 feet per second away from its starting point. During the next three seconds, it moves in the same direction at 15 feet per second. During the final three seconds, it returns to its starting point at a constant speed. Sketch a rough graph of the total distance traveled by the car as a function of time." One student produces the graph displayed here. What do you say to the student?

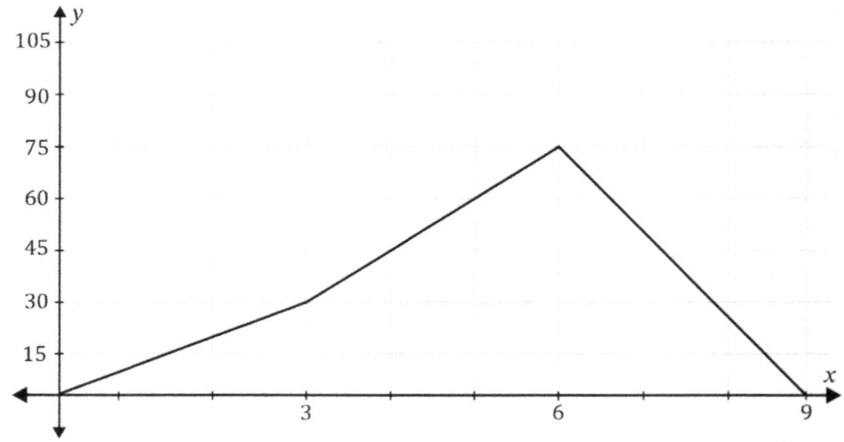

a. "You have mistakenly graphed the velocity of the car instead of the distance it traveled."
b. "Excellent work! This is exactly right."
c. "You have mistakenly graphed the speed of the car instead of the distance it traveled."
d. "This is close to the correct graph, but you have graphed the distance of the car from its starting point rather than the distance it traveled."

44. Solve the equation $\frac{x-1}{x+2} = \frac{6}{x+2} - \frac{x-3}{x-2}$.

a. $x = 0, 2$
b. $x = 1, 4$
c. $x = -3, 1$
d. $x = 1, 3$

45. The high school honors program is offering three special electives: Biophysics, Intro to Law, and Latin. Seventeen students enrolled in Biophysics, twenty-two in Intro to Law, and nineteen in Latin. Two students are taking both Biophysics and Intro to Law, seven are taking both Intro to Law and Latin, and four are taking both Biophysics and Latin. One student is enrolled in all three electives. How many students are taking special electives?

46. This question may have more than one correct answer. Select all the correct answers.

You manufacture widgets and gadgets. To make a widget requires 5 gears and 20 levers. To make a gadget requires 20 gears and 10 levers. You have 650 gears and 1,200 levers available. If you make $8 profit on each widget and $32 profit on each gadget, which of the following pairs of production levels can be ruled out as possible solutions that maximize profit? Select all answers that CANNOT maximize profit.

a. 30 widgets and 25 gadgets
b. 55 widgets and 20 gadgets
c. 5 widgets and 30 gadgets

47. Solve $x^4 + 9 = 10x^2$.

a. $x = \{1, 3\}$
b. $x = \{-3i, -i, i, 3i\}$
c. $x = \{i, 3i\}$
d. $x = \{-3, -1, 1, 3\}$

48. This question may have more than one correct answer. Select all the correct answers.

A trigonometry test asks students to simplify the expression 2 sin 38° sin 52°. Students A, B, and C give the three answers below. Which of these simplifications produce correct results? Select all that are correct.

a. $2\sin 38° \sin 52° = 2\cos 52° \sin 52° = \sin(2 \times 52°) = \sin 104°$
b. $2\sin 38° \sin 52° = 2\sin(38° + 52°) = 2\sin 90° = 2 \times 1 = 2$
c. $2\sin 38° \sin 52° = 2\sin 38° \cos 38° = \sin(2 \times 38°) = \sin 76°$

49. To the nearest hundredth, what is the area in square units under the curve $f(x) = \frac{1}{x}$ on the interval $[1, 2]$?

50. Divide the least common multiple (lcm) of 5,760 and 4,800 by the least common multiple of 3,600 and 960. That is, find $lcm(5,760,\ 4,800) \div lcm(3,600,\ 960)$.

a. 4
b. 2
c. 8
d. 3

51. This question may have more than one correct answer. Select all correct answers.

A calculus test has the following problem involving finding a one-sided limit.

$$\lim_{x\to 3^-}\sqrt{\frac{3-x}{3+x}}$$

Three students submit the following answers. Which answers are both correct and correctly reasoned? Select all that apply.

a. Student A: $\lim_{x\to 3^-}\sqrt{\frac{3-x}{3+x}} = \sqrt{\lim_{x\to 3^-}\frac{3-x}{3+x}} = \sqrt{\frac{\lim_{x\to 3^-}(3-x)}{\lim_{x\to 3^-}(3+x)}}$, which does not exist since $\lim_{x\to 3^-}(3-x) = 3-3 = 0$

b. Student B: $\lim_{x\to 3^-}\sqrt{\frac{3-x}{3+x}} = \sqrt{\lim_{x\to 3^-}\frac{3-x}{3+x}} = \sqrt{\frac{\lim_{x\to 3^-}(3-x)}{\lim_{x\to 3^-}(3+x)}} = 0$, since $\lim_{x\to 3^-}(3+x) = 3+(-3) = 0$

c. Student C: $\lim_{x\to 3^-}\sqrt{\frac{3-x}{3+x}} = \sqrt{\lim_{x\to 3^-}\frac{3-x}{3+x}} = \sqrt{\frac{\lim_{x\to 3^-}(3-x)}{\lim_{x\to 3^-}(3+x)}} = 0$, since $\lim_{x\to 3^-}(3-x) = 3-3 = 0$

52. A target is made of concentric circles, as shown below. The width of each band is a and the center is a circle with radius a. Which of the following gives the area A of the shaded region as a function of a?

a. $A = 12\pi a^2$
b. $A = 15\pi a^2$
c. $A = 18\pi a^2$
d. $A = 25\pi a^2$

Refer to the following for question 53:

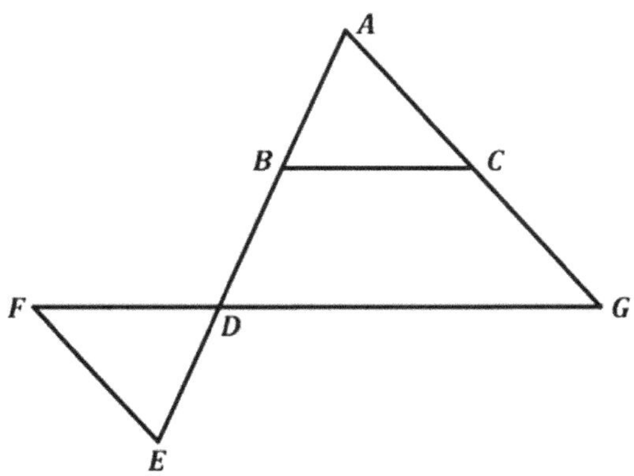

Statement	Reason
1. $\overline{BC} \parallel \overline{FG}$	Given
2.	
3. $\overline{FD} \cong \overline{BC}$	Given
4. $\overline{AB} \cong \overline{DE}$	Given
5. $\Delta ABC \cong \Delta EDF$	________
6. ________	
7. $\overline{FE} \parallel \overline{AG}$	

■

Given: $\overline{BC} \parallel \overline{FG}; \overline{FD} \cong \overline{BC}; \overline{AB} \cong \overline{DE}$

Prove: $\overline{FE} \parallel \overline{AG}$

53. Which of the following justifies step 5 in the proof?

a. AAS
b. SSS
c. ASA
d. SAS

54. Which of these demonstrates the relationship between the sets of prime numbers, real numbers, natural numbers, complex numbers, rational numbers, and integers?

$\mathbb{P}$–Prime; $\mathbb{R}$–Real; $\mathbb{N}$–Natural; $\mathbb{C}$–Complex; $\mathbb{Q}$–Rational; $\mathbb{Z}$–Integer

a. $\mathbb{P} \subseteq \mathbb{Q} \subseteq \mathbb{R} \subseteq \mathbb{Z} \subseteq \mathbb{C} \subseteq \mathbb{N}$
b. $\mathbb{P} \subseteq \mathbb{N} \subseteq \mathbb{Z} \subseteq \mathbb{Q} \subseteq \mathbb{R} \subseteq \mathbb{C}$
c. $\mathbb{C} \subseteq \mathbb{R} \subseteq \mathbb{Q} \subseteq \mathbb{Z} \subseteq \mathbb{N} \subseteq \mathbb{P}$
d. None of these

55. A right square pyramid has a height of h meters. Each edge of its base also has a length of h meters. If the lateral surface area of the pyramid (in square meters) equals the volume of the pyramid (in cubic meters)—that is, if the numbers are equal even though the units are different—what is the height of the pyramid?

a. $3\sqrt{5}$ meters
b. $3\sqrt{3}$ meters
c. 6 meters
d. $\frac{3\sqrt{5}}{4}$ meters

56. Suppose the path of a baseball hit straight up from three feet above the ground is modeled by the first quadrant graph of the function $h(t) = -16t^2 + 50t + 3$, where t is the flight time of the ball in seconds and h is the height of the ball in feet. What is the velocity of the ball two seconds after it is hit?

a. 39 ft/s upward
b. 19.5 ft/s upward

c. 19.5 ft/s downward
d. 14 ft/s downward

57. This question may have more than one correct answer. Select all the correct answers.

You invest $374 at 3% annual interest, compounded continuously. After t years, the dollar amount of your investment is $A = 374e^{0.03t}$. Which of the following calculations correctly finds the time it will take for your investment to double in value? Select all that are correct. (Recall that $\log x$, also called the common logarithm, means $\log_{10} x$, the logarithm of x with base 10. And $\ln x$, also called the natural logarithm, means $\log_e x$, the logarithm of x with base e.)

a.

$$\begin{aligned} 374e^{0.03t} &= 748 \\ \ln(374e^{0.03t}) &= \ln 748 \\ \ln(374)\ln(e^{0.03t}) &= \ln 748 \\ 0.03t &= \frac{\ln 748}{\ln 374} \\ t &= \frac{\ln 748}{0.03 \ln 374} \end{aligned}$$

b.

$$\begin{aligned} 374e^{0.03t} &= 748 \\ e^{0.03t} &= 2 \\ 0.03t &= \ln 2 \\ t &= \frac{\ln 2}{0.03} \end{aligned}$$

c.

$$\begin{aligned} 374e^{0.03t} &= 748 \\ \log(374e^{0.03t}) &= \log 748 \\ \log 374 + \log e^{0.03t} &= \log 748 \\ \log e^{0.03t} &= \log 748 - \log 374 \\ 0.03t \log e &= \log 748 - \log 374 \\ t &= \frac{\log 748 - \log 374}{0.03 \log e} \end{aligned}$$

58. Which of these would best illustrate the percentage of a budget allocated to various departments?

a. Pie chart
b. Line graph
c. Box-and-whisker plot
d. Venn diagram

59. At a particular moment one afternoon, the sun is at an angle of elevation of 46°. At that moment, an old flagpole is leaning (directly away from the sun) at 11° from vertical. The pole casts a shadow 96 feet long. Use the law of sines to determine the length of the flagpole to the nearest tenth of a foot.

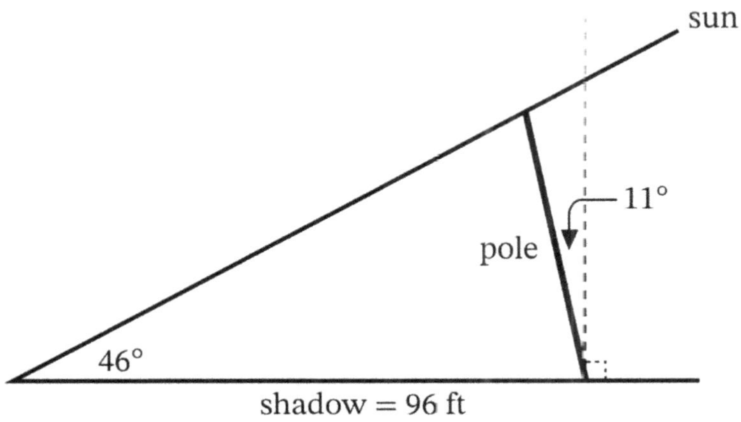

a. 84.3 feet
b. 82.9 feet
c. 79.4 feet
d. 109.3 feet

60. Evaluate $\int(3x^2 + 2x - 1)\,dx$.

a. $x^3 + x^2 - x + C$
b. $6x^2 + 2$
c. $\frac{3}{2}x^3 + 2x^2 - x + C$
d. $6x^2 + 2 + C$

61. Gillian is deciding between two data plans for her cellphone. Plan A provides 2.5 GB of data for a flat rate of $20/month and charges $15 per GB for any extra use. Plan B provides unlimited data for $50/month. What amount of data would Gillian have to use in a month for both plans to cost the same amount?

62. This question may have more than one correct answer. Select all the correct answers.

We are given the function $f(x) = (x + 3)(x^2 - 3x)$ and told to rearrange it algebraically into a form that reveals important features of the function or its graph. Which of the following procedures lead to correct conclusions? Select all that are correct.

a. We complete the square in the second factor to get $f(x) = (x + 3)\left(x^2 - 3x + \frac{9}{4} - \frac{9}{4}\right) = (x + 3)\left(x^2 - 3x + \frac{9}{4}\right) - (x + 3)\frac{9}{4} = (x + 3)\left(x - \frac{3}{2}\right)^2 - \frac{9}{4}(x + 3)$. From this form, we conclude that f has a local minimum at $x = \frac{3}{2}$ and that the graph of f has the line $x = \frac{3}{2}$ as an axis of symmetry.

b. We expand the formula to get $f(x) = x^3 - 3x^2 + 3x^2 - 9x = x^3 - 9x$. From this general form, we conclude that the graph of f has a y-intercept of $y = 0$ (since the constant term is 0) and that the graph is symmetric with respect to the origin (since $x^3 - 9x$ has only odd-degree terms, making f an odd function).

c. We factor x out of the second factor to get $f(x) = (x + 3)(x - 3)x$. From this factored form, we conclude that f has simple zeroes (x-intercepts) at $x = -3, x = 3$, and $x = 0$. Also, since $x = 0$ is a zero, the graph has a y-intercept of $y = 0$.

63. Use integration by parts to find the area under the curve $y = x\sin(2x)$ over the interval $\left[0, \frac{\pi}{2}\right]$.

a. $\frac{\pi}{4}$

b. $\frac{\pi}{2}$

c. $\frac{\pi}{4} + \frac{1}{4}$

d. $\frac{\pi}{2} + \frac{1}{2}$

64. Ben is purchasing notebooks and boxes of pencils for a class he is teaching. Notebooks cost \$1 each and pencils cost \$3 per box. If the x-axis represents boxes of pencils, the y-axis represents notebooks, and Ben can spend a maximum of \$30, which of the following graphs shows the number of notebooks and pencils Ben can purchase?

a. 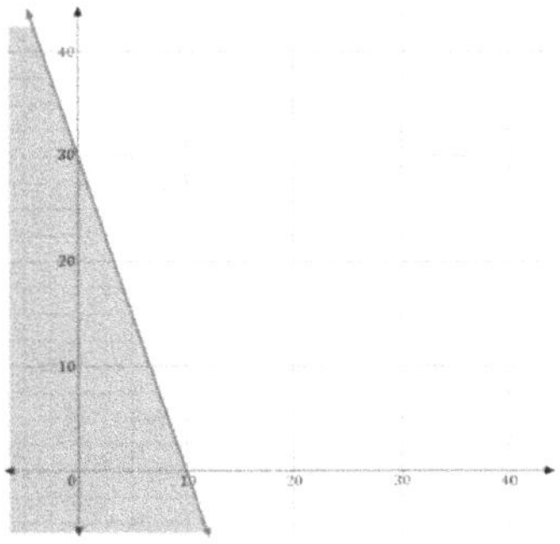

c. 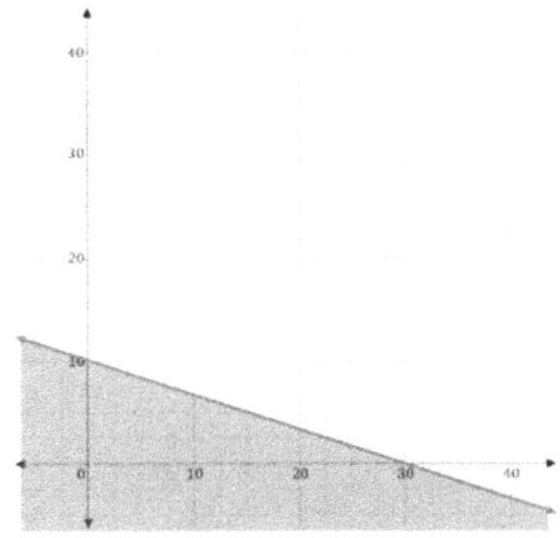

b. 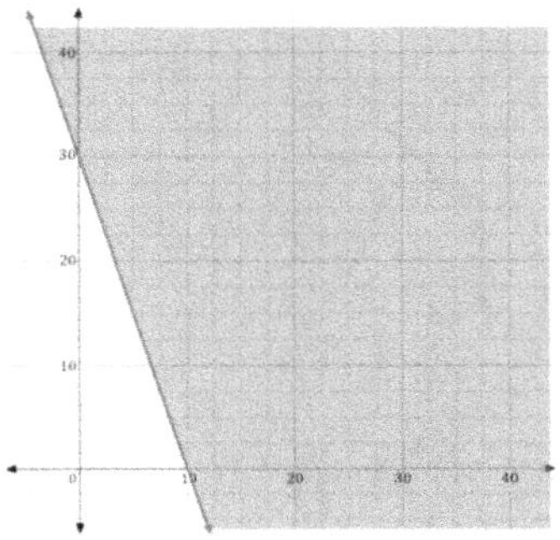

d. 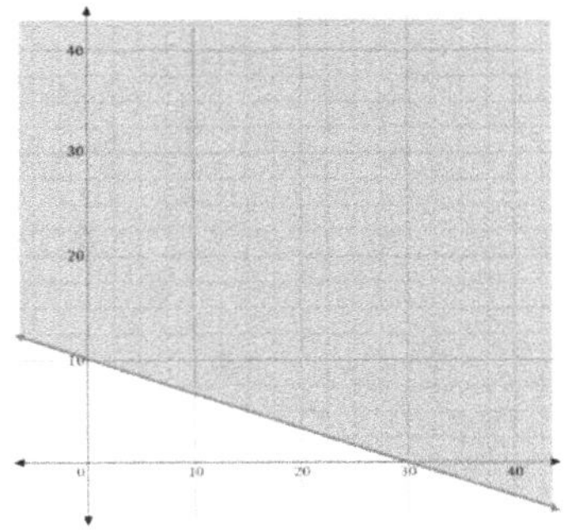

65. An economist surveys a random sample of 10 homes in two different neighborhoods of a city to find out what their annual household incomes are. Here are the results, in thousands of dollars:

Neighborhood A: 26, 26, 27, 28, 32, 34, 37, 42, 48, 400
Neighborhood B: 36, 38, 39, 45, 48, 52, 53, 62, 68, 79

Which of the following statements is the most appropriate comment on this data (pending further investigation)?

a. Neighborhood A appears to be more prosperous than neighborhood B because its *mean* income is higher.
b. Neighborhood B appears to be more prosperous than neighborhood A because its *mean* income is higher.
c. Neighborhood A appears to be more prosperous than neighborhood B because its *median* income is higher.
d. Neighborhood B appears to be more prosperous than neighborhood A because its *median* income is higher.

66. The function $P(\theta)$ is the "wrapping function"—the function that "wraps" the number line around the unit circle. Its value is the point where the terminal side of an angle of radian measure θ, in standard position, intersects the unit circle (see diagram). With this definition in mind, choose the answer that correctly explains why the graph of $y = \cos x$ (not shown) is symmetric with respect to the y-axis.

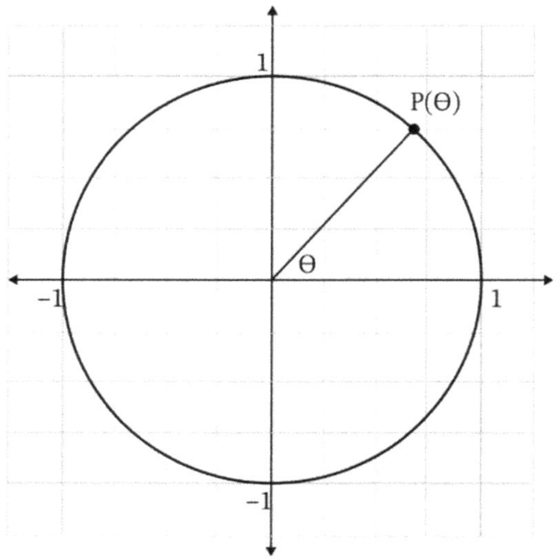

a. Since the points $P(\theta)$ and $P(-\theta)$ are symmetric with respect to the origin, they have opposite x- and y-coordinates. Consequently, $\cos(-\theta) = -\cos(\theta)$ for all real numbers θ. Thus, whenever the point $(x, \cos(x))$ appears on the graph of $y = \cos x$, so does the point $(-x, \cos(-x)) = (-x, -\cos(x))$.

b. Since the points $P(\theta)$ and $P(-\theta)$ are symmetric with respect to the x-axis, they have opposite y-coordinates. Consequently, whenever the point $(x, \cos(x))$ appears on the graph of $y = \cos x$, so does the point $(x, -\cos(x))$.

c. Since the points $P(\theta)$ and $P(-\theta)$ are symmetric with respect to the x-axis, they have the same x-coordinate. Consequently, $\cos(\theta) = \cos(-\theta)$ for all real numbers θ. Thus, whenever the point $(x, \cos(x))$ appears on the graph, so does the point $(-x, \cos(-x)) = (-x, \cos(x))$.

d. The graph of the equation $y = \cos x$ is not symmetric with respect to the y-axis.

Answer Key and Explanations for Test #2

1. A: If two triangles are similar, then all pairs of corresponding sides are proportional. For ΔRST to be similar to ΔXYZ, we need $\frac{RS}{XY} = \frac{RT}{XZ} = \frac{ST}{YZ}$. Substituting the numbers from the figure in for those values gives $\frac{24}{8} = \frac{12}{4} = \frac{30}{YZ}$. Simplifying the fractions results in $\frac{3}{1} = \frac{3}{1} = \frac{30}{YZ}$. Therefore, in order for the triangles to be similar, we need $\frac{3}{1} = \frac{30}{YZ}$. After cross-multiplying the terms, it becomes $3(YZ) = 30(1)$, so $3(YZ) = 30$. Divide both sides by 3 to get $YZ = 10$.

2. D: To understand the distribution of the data, we need a way to display the "shape" of it using a tool such as a histogram or a stem-and-leaf plot. Here are stem-and-leaf plots for the four exams:

Exam #1

Stem	Leaf
0	1
1	
2	
3	7
4	8
5	2
6	
7	
8	0 3 8
9	0 8 9

Exam #2

Stem	Leaf
0	4 5 9
1	
2	0 3
3	0 7
4	
5	
6	8
7	8
8	
9	9

Exam #3

Stem	Leaf
0	
1	1 4 8
2	5 7
3	
4	
5	
6	
7	2 9
8	0 3 8
9	

Exam #4

Stem	Leaf
0	
1	
2	
3	3
4	7 8 9
5	1 6 7
6	1 2
7	2
8	
9	

The distribution of Exam #1 appears negatively skewed, and that of Exam #2 appears positively skewed. The distribution of Exam #3 appears bimodal, with no data near the center. Only in Exam #4 is the data roughly mound-shaped, clustering around a central value with tails going off roughly symmetrically on both sides. Thus, Exam #4 looks most like a normal distribution.

3. B: To see that function g is neither odd nor even, we observe that, for example, $g(\pi) = \pi + \cos\pi = \pi - 1$, and $g(-\pi) = -\pi + \cos(-\pi) = -\pi - 1$. Since $g(-\pi) \neq -g(\pi)$, function g is not odd, and because $g(-\pi) \neq g(\pi)$, function g is not even.

In contrast, recalling that sine is an odd function and cosine is an even function, we see that:

$$f(-x) = -x + \sin(-x) = -x - \sin x = -(x + \sin x) = -f(x)$$
$$h(-x) = -x\,sin(-x) = -x(-\sin x) = x\,sin\,x = h(x)$$
$$k(-x) = -x\cos(-x) = -x\cos x = -k(x)$$

This means functions f and k are odd, and function h is even.

4. B: The line through the center of the box represents the median. The median test score for English and History is the same, not the mean test score, which makes choice B the correct answer. Note that for Science, the median is a better representation of the data than the mean. There is at least one outlier (a point that lies outside of two standard deviations from the mean), which brings down the average test score. In cases such as this, the mean is not the best measure of central tendency, so choice A is not the correct answer. The median of the History scores is 87, and the 3rd quartile scores for Math is also 87, so choice C is not the answer. The median of the English scores is 86, and based on the spread of data, the mean is lower than the median, so choice D is also not correct.

5. B: Student B cancels the common factor of $x - 3$ and correctly notes that the new equation is equivalent to the old under the condition that $x \neq 3$. Students A and C, who otherwise follow correct procedures, neglect to add the condition that $x \neq 3$ where it is called for. Thus, they get the extraneous solution $x = 3$, which they then fail to exclude by checking their solutions.

6. A: The total number of students taking electives is 200, as seen in the chart on the right, and 16% of these are taking welding, so the number of students taking welding is $0.16 \times 200 = 32$. The number of male sophomores taking an elective is 20. The difference in the number of male and female students taking an elective is $108 - 92 = 16$. The difference in the number of juniors and seniors taking an elective is $83 - 59 = 24$. The greatest quantity of these is 32, the number of students taking welding.

7. C: The percentage of students taking Java or Creative Writing is $25\% + 21\% = 46\%$. The percentage of students in electives who are male is $\frac{92}{200} \times 100\% = 46\%$. So, the two quantities are equal.

8. D: In choice A, the teacher uses convenience sampling, which is a biased technique. A group of athletes is likely not representative of all high school girls. Choice B also represents convenience sampling, as all surveys were taken from the same neighborhood. While this likely included a variety of ages, career paths, etc., it most likely represented a narrow demographic since homes in a neighborhood tend to be similar in value. Choice C is biased because the participants are self-selected rather than randomly selected. It may be that people who have a strong opinion are more likely to respond than those who are more neutral, and this would give a skewed perspective of opinions. Choice D represents random selection since names are organized by computer generation rather than any particular demographic.

9. 1: The slope of the line tangent to the graph of a function f at $x = a$ is $f'(a)$. Find the derivative of the given function.

$$f(x) = \frac{1}{4}x^2 - 3$$
$$f'(x) = 2\left(\frac{1}{4}\right)x^{(2-1)} - 0$$
$$f'(x) = \frac{1}{2}x$$

So, the slope at $x = 2$ is $f'(2) = \frac{1}{2}(2) = 1$.

10. A: Let S_{23}, S_{24} and S_{25} be the sales in 2023, 2024, and 2025 respectively. Sales in 2025 are 15% more than 2024, so $S_{25} = 1.15S_{24}$, which gives us $S_{24} = S_{25} \div 1.15$. Sales in 2024 are 5% more than 2023, so $S_{24} = 1.05S_{23}$, which gives us $S_{23} = S_{24} \div 1.05$. This is exactly the procedure that Student A follows. Students B and C, on the other hand, erroneously calculate the 5% and 15% increases relative to the 2025 sales rather than 2023 and 2024 sales, respectively.

11. D: The axis of symmetry of the general quadratic trinomial $y = ax^2 + bx + c$ is $x = -\frac{b}{2a}$. The quadratic formula tells us that the zeros of this trinomial, whether real or complex, are symmetric on either side of this axis of symmetry.

$$x = \frac{-b \pm \sqrt{b^2 - 4ac}}{2a} = \frac{-b}{2a} \pm \frac{\sqrt{b^2 - 4ac}}{2a}$$

Thus, if we average the two zeros, we get the following expression.

$$\frac{\left(\frac{-b}{2a} + \frac{\sqrt{b^2 - 4ac}}{2a}\right) + \left(\frac{-b}{2a} - \frac{\sqrt{b^2 - 4ac}}{2a}\right)}{2} = \frac{\frac{-b}{2a} + \frac{-b}{2a}}{2} = -\frac{b}{2a}$$

This is the axis of symmetry. So, the student's method always works, but it is hard to use unless the trinomial is easy to factor. If the trinomial has fractional, irrational, or complex zeros, our only way of factoring it may be the quadratic formula, which gives us the axis of symmetry, $x = -\frac{b}{2a}$, directly.

12. A: The vertex form of a quadratic equation is $y = a(x - h)^2 + k$, where $x = h$ is the parabola's axis of symmetry and (h, k) is the parabola's vertex. The vertex of the graph is $(-2, -1)$, so the equation can be written as $y = a(x + 2)^2 - 1$. The parabola passes through point (0,3), so $3 = a(0 + 2)^2 - 1$. Solve for a.

$$\begin{aligned} 3 &= a(0 + 2)^2 - 1 \\ 3 &= a(2)^2 - 1 \\ 3 &= 4a - 1 \\ 4 &= 4a \\ 1 &= a \end{aligned}$$

So, the vertex form of the parabola is $y = (x + 2)^2 - 1$. Write the equation in standard form, $y = ax^2 + bx + c$.

$$\begin{aligned} y &= (x + 2)^2 - 1 \\ y &= (x + 2)(x + 2) - 1 \\ y &= (x^2 + 4x + 4) - 1 \\ y &= x^2 + 4x + 3 \end{aligned}$$

13. A: For f and g to be inverse functions, the range of f must be the domain of g and vice versa, and it must be true that $(f \circ g)(x) = x$ for every x in the domain of g and that $(g \circ f)(x) = x$ for every x in the domain of f. We calculate:

$$(f \circ g)(x) = f(g(x)) = f\left(\frac{1}{5}x + \frac{3}{5}\right) = 3 - 5\left(\frac{1}{5}x + \frac{3}{5}\right) = 3 - x - 3 = -x$$

Answer Key and Explanations #2

Since $(f \circ g)(x) = -x \neq x$, we conclude that $g \neq f^{-1}$.

14. C: When solving radical equations, check for extraneous solutions.

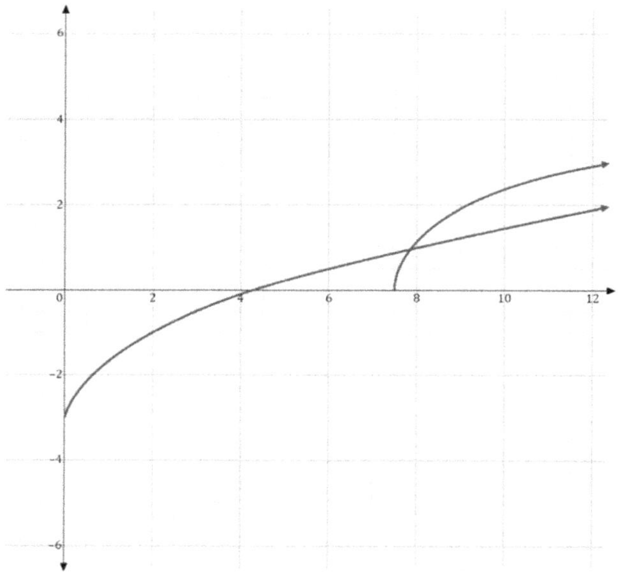

$$\sqrt{2x} - 3 = \sqrt{2x - 15}$$
$$\left(\sqrt{2x} - 3\right)^2 = \left(\sqrt{2x - 15}\right)^2$$
$$2x - 6\sqrt{2x} + 9 = 2x - 15$$
$$-6\sqrt{2x} + 9 = -15$$
$$-6\sqrt{2x} = -24$$
$$\sqrt{2x} = 4$$
$$2x = 16$$
$$x = 8$$

Check: $$\sqrt{2(8)} - 3 = \sqrt{2(8) - 15}$$
$$\sqrt{16} - 3 = \sqrt{16 - 15}$$
$$4 - 3 = \sqrt{1}$$
$$1 = 1$$

Since the solution checks, it is a valid solution. Notice that the graphs $y = \sqrt{2x} - 3$ and $y = \sqrt{2x - 15}$ intersect, which confirms there is a solution.

15. B: Find the points of intersection of the two graphs:

$$x^2 - 4 = -x + 2$$
$$x^2 + x - 6 = 0$$
$$(x + 3)(x - 2) = 0$$
$$x = -3, x = 2$$

So, the area is between the graphs on $[-3,2]$. To determine which of the graphs is the upper or lower bound, pick a test point in the interval. Using $x = 0$ as the test point.

$$y = x^2 - 4 \qquad y = -x + 2$$
$$y = (0)^2 - 4 \qquad y = -(0) + 2$$
$$y = -4 \qquad y = 2$$

The finite region is bound at the top by the line $y = -x + 2$ and at the bottom by $y = x^2 - 4$, and the height of the region at point x is defined by $[(-x + 2) - (x^2 - 4)]$, so the area of the region is:

$$\begin{aligned}
A &= \int_{-3}^{2} [(-x + 2) - (x^2 - 4)]dx \\
&= \int_{-3}^{2} (-x^2 - x + 6)\, dx \\
&= \left[-\frac{1}{3}x^3 - \frac{1}{2}x^2 + 6x\right]_{-3}^{2} \\
&= \left[-\frac{1}{3}(2)^3 - \frac{1}{2}(2)^2 + 6(2)\right] - \left[-\frac{1}{3}(-3)^3 - \frac{1}{2}(-3)^2 + 6(-3)\right] \\
&= \left[-\frac{8}{3} - 2 + 12\right] - \left[9 - \frac{9}{2} - 18\right] \\
&= \frac{22}{3} - \left(-\frac{27}{2}\right) \\
&= \frac{125}{6}
\end{aligned}$$

16. C: We can draw a diagram like the one below, which shows $\triangle ABC$ with $AC = 4$, median $\overline{BD}$ having length 2, and $m\angle BDA = 72°$. Angles $\angle BDA$ and $\angle BDC$ are a linear pair and are thus supplementary, so $m\angle BDC = 180° - 72° = 108°$. Because $\overline{BD}$ is a median, the point D is the midpoint of side $\overline{AC}$, making the length of $\overline{AD}$ half that of side $\overline{AC}$, namely $AD = 2$. This makes $\triangle ABD$ an isosceles triangle whose base angles, $\angle A$ and $\angle ABD$, are congruent. Similarly, $\triangle CBD$ is an isosceles triangle whose base angles, $\angle C$ and $\angle CBD$, are congruent.

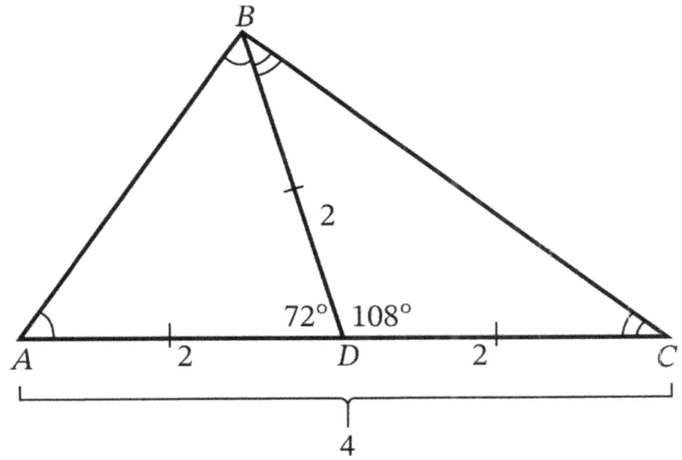

Choice A is incorrect. If side $\overline{BC}$ were 2 units long, then $\triangle CBD$ would be equilateral, requiring $m\angle BDC = 60°$ instead of 108°.

Choice B is similarly incorrect. If side $\overline{BA}$ were 2 units long, then $\triangle ABD$ would be equilateral, requiring $m\angle BDA = 60°$ instead of 72°.

Choice C is correct. Since the angles of $\triangle ABD$ must add up to 180°, by the equality of the base angles:

$$m\angle A + m\angle ABD + 72° = 180°$$
$$m\angle A + m\angle A + 72° = 180°$$
$$2 \times m\angle A = 108°$$
$$m\angle A = 54°$$

17. D: ΔADC is a right triangle with legs measuring 8 and 15 and a hypotenuse of 17. (8–15–17 is a Pythagorean triple.) ΔADC and ΔDEC are both right triangles which share vertex C. By the AA similarity theorem, $\Delta ADC \sim \Delta DEC$. Therefore, a proportion can be written and solved to find DE.

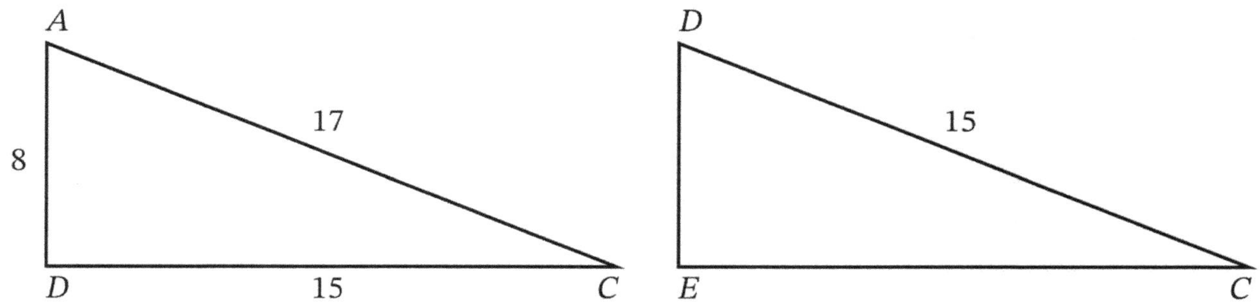

$$\frac{8}{DE} = \frac{17}{15}$$
$$17DE = 120$$
$$DE \approx 7.1$$

18. C: Distribute the factor $(2 + 3i)$, and then multiply the complex numbers using the FOIL (first, outer, inner, last) method, recalling that $i^2 = -1$:

$$(2 + 3i)[(7 - i)z - (3 + 5i)] = (2 + 3i)(7 - i)z - (2 + 3i)(3 + 5i)$$
$$= (14 - 2i + 21i - 3i^2)z - (6 + 10i + 9i + 15i^2)$$
$$= \big(14 - 2i + 21i - 3(-1)\big)z - \big(6 + 10i + 9i + 15(-1)\big)$$
$$= (17 + 19i)z - (-9 + 19i)$$
$$= (17 + 19i)z + (9 - 19i)$$

19. C: Choice C is an absolute value function. A function is a relationship in which for every element of the domain (x), there is exactly one element of the range (y). Graphically, a relationship between x and y can be identified as a function if the graph passes the vertical line test.

The first relation is a quartic parabola on its side, which fails the vertical line test for functions. Choice B also fails the vertical line test and is therefore not a function. The relation in choice D pairs two elements of the range with one of the elements of the domain, so it is also not a function.

20. C: To find the volume of square pyramid use the formula $A = \frac{1}{3}lwh$, where l is the length of the base, w is the width of the base, and h is the height of the pyramid. The volume of the greenhouse is

$\frac{x^2y}{3}$. If the height were increased by 4 ft, the volume would be $\frac{x^2(y+4)}{3} = \frac{x^2y+4x^2}{3}$. Notice this is the volume of the greenhouse, $\frac{x^2y}{3}$, increased by $\frac{4x^2}{3}$.

21. D: To find the probability, we count the number of evens that were rolled, divided by the number of total rolls.

$$\frac{0 + 12 + 20(2) + 18(3)}{50(3)} = \frac{106}{150} = \frac{53}{75} \approx 0.707$$

Therefore, the approximate probability of a single roll being even is 0.707.

22. B: By definition, when $f(x)$ and $g(x)$ are inverse functions, $f(g(x)) = g(f(x)) = x$. So, $f\left(g\left(\frac{1}{8}\right)\right) = x = \frac{1}{8}$.

23. C: First, use the table to determine the values of $(w * y)$ and $(x * x)$:

$*$	**w**	**x**	**y**	**z**
w	w	x	y	z
x	z	w	x	y
y	y	z	w	x
z	x	y	z	w

Move horizontally and then vertically. So, $(w * y) = y$ and $(x * x) = w$, that means $(w * y) * (x * x) = y * w$, which is equal to y.

24. D: Group A's comment is true: these variables appear to have an overall downward trend from left to right, indicating a negative correlation. Group B's comment is true: the correlation between these variables is close to zero, indicating little or no correlation. More to the point, by *correlation* we usually mean *linear* correlation, and this data does not seem to have a strong linear relationship. Group C's comment is true: the data for the first ten minutes has a clear upward trend, and the data for the second ten minutes has a clear downward trend. The bigger story, however, is that the variables appear to have a strong nonlinear relationship, more in the shape of a parabola or semicircle than a line. It is important to make note of a "correlation" (in the broad sense) like this, even though it is not a linear correlation of the kind that usually attracts our attention in an introductory statistics course.

25. C: Aaron ran four miles from home and then back again, so he ran a total of eight miles. Therefore, statement III is false. Statements I and II, however, are both true. Since Aaron ran eight miles in eighty minutes, he ran an average of 1 mile every 10 minutes, or 6 miles every 60 minutes, which is 6 mph. He ran two miles from point (A) to (B) in 20 minutes, which is a rate of 1 mile per 10 minutes, and four miles from (D) to (E) in 40 minutes, which is also a rate of 1 mile per 10 minutes, so his running speed between both sets of points was the same.

26. 54: To find the number of diagonals in a polygon, use the formula $\frac{n(n-3)}{2}$, where n is the number of sides on the polygon. A dodecagon has 12 sides, so substitute 12 for n and simplify.

$$\frac{12(12-3)}{2}=\frac{12(9)}{2}=\frac{108}{2}=54$$

Therefore, a dodecagon has 54 diagonals.

27. C: When you have a ratio, you can find the fraction that each part of the ratio is of the whole by dividing a part by the sum of the parts. Since the ratio of wages and benefits to other costs is 2: 3, the amount of money spent on wages and benefits is $\frac{2}{2+3}=\frac{2}{5}$ of total operational costs.

$$\frac{2}{5}\times \$130{,}000=\$52{,}000$$

28. A, B, C: All three answers, A, B, and C are accurate representations of the data. Group A has a histogram with the horizontal axis labeled to indicate that each bin, except the first, excludes the left endpoint and includes the right endpoint. Group B has a box-and-whiskers plot, or boxplot, in which it might be helpful to note that they are following the convention, typical in introductory statistics courses, of including the median of the whole set of data in calculating the medians of both the upper and lower halves of the data. Group C has a simple dot plot.

29. A, B: The diagrams below show the results of following the students' instructions, with the result after the first transformation being shown as a hollow G. The hollow G for student A, who gives only one step, shows the result halfway through the 180° rotation. That is, after rotating 90° clockwise around the origin.

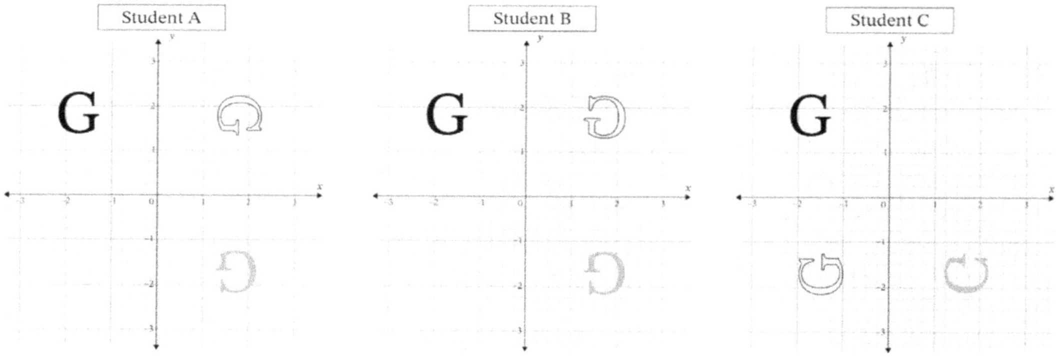

30. C: The recursive definition of the sequence gives the first term of the series, $a_1=-2$. The definition also defines each term in the series as the sum of the previous term and $\frac{1}{2}$. Therefore, the second term in the series is $-2+\frac{1}{2}=-\frac{3}{2}$, the third term in the series is $-\frac{3}{2}+\frac{1}{2}=-1$, and so on.

n	a_n
1	-2
2	$-\frac{3}{2}$
3	-1

The relationship between n and a_n is linear, so the equation of the sequence can be found in the same way as the equation of a line. The value of a_n increases by $\frac{1}{2}$ each time the value of n increases by 1.

n	$\frac{n}{2}$	a_n
1	$\frac{1}{2}$	-2
2	1	$-\frac{3}{2}$
3	$\frac{3}{2}$	-1

Since the difference between $\frac{1}{2}n$ and a_n is $-\frac{5}{2}$. So, $a_n = \frac{1}{2}n - \frac{5}{2}$.

n	$\frac{n}{2} - \frac{5}{2}$	a_n
1	$\frac{1}{2} - \frac{5}{2}$	-2
2	$1 - \frac{5}{2}$	$-\frac{3}{2}$
3	$\frac{3}{2} - \frac{5}{2}$	-1

31. C: The zero at $x = -3$ indicates that the polynomial has a factor of $x + 3$. And the way the curve "rebounds" at that zero—touching the x-axis without crossing it—indicates that $x = -3$ is a zero with even multiplicity. That is, in $p(x)$, the factor $x + 3$ is raised to an even power.

The zero at $x = 1$ indicates that the polynomial has a factor of $x - 1$; and the way the curve "flattens out" as it crosses the x-axis at $x = 1$ indicates that $x = 1$ is a zero with odd multiplicity strictly greater than 1. That is, in $p(x)$, the factor $x - 1$ is raised to an odd power that is 3 or higher. (The power cannot be 1 because a polynomial factor with degree 1 is simply a line, which would not cause any flattening effect.) The polynomial $p(x) = -(x + 3)^2(x - 1)^3$ is the only answer choice with factors that meet these constraints.

32. A: The equation of the circle is given in general form. When the equation is written in the standard form $(x - h)^2 + (y - k)^2 = r^2$, where (h, k) is the center of the circle and r is the radius of the circle, the radius can be determined by taking the positive square root of the right side. Putting the equation into standard form requires completing the square for x and y:

$$x^2 + y^2 + 2x - 10y = -18$$
$$(x^2 + 2x + 1) + (y^2 - 10y + 25) = -18 + 1 + 25$$
$$(x + 1)^2 + (y - 5)^2 = 8$$

Answer Key and Explanations #2

Since $r^2 = 8$, and since r must be a positive number, $r = \sqrt{8} = 2\sqrt{2}$.

33. D: A system of linear equations can be solved by using matrices or by using the graphing, substitution, or elimination method. The elimination method is shown here.

$$\begin{cases} -4x + 3y = 8 \\ 6x - 2y = -7 \end{cases}$$

To eliminate y by the elimination method, multiply the top equation by 2 and the bottom equation by 3 so that the coefficients of the y-terms will be additive inverses.

$$2(-4x + 3y) = (8)2 \qquad\qquad 3(6x - 2y) = (-7)(3)$$
$$-8x + 6y = 16 \qquad\qquad 18x - 6y = -21$$

Then, add the two equations and solve for x.

$$\begin{array}{rrrl} & -8x & +6y & = \ \ 16 \\ + & 18x & -6y & = -21 \\ \hline & & 10x & = -5 \\ & & x & = -\frac{1}{2} \end{array}$$

Substitute $-\frac{1}{2}$ for x in either of the given equations and solve for y.

$$\begin{aligned} -4x + 3y &= 8 \\ -4\left(-\frac{1}{2}\right) + 3y &= 8 \\ 2 + 3y &= 8 \\ 3y &= 6 \\ y &= 2 \end{aligned}$$

The solution to the system of equations is $\left(-\frac{1}{2}, 2\right)$.

34. A: One way to determine which is the graph of $f(x) = -2|-x + 4| - 1$ is to find $f(x)$ for a few values of x. For example, when $x = 0$, $f(0) = -2|0 + 4| - 1 = -9$. Graphs A and B pass through $(0, -9)$, but graphs C and D do not. When $x = 4$, $f(4) = -2|-4 + 4| - 1 = -1$. Graph A passes through $(4, -1)$, but graph B does not. Graph A is the correct graph.

35. C: Use trigonometric identities to simplify.

$\cos\theta\cot\theta$	Write out the original expression.
$\cos\theta\times\frac{\cos\theta}{\sin\theta}$	Convert $\cot\theta$ to its identity $\frac{\cos\theta}{\sin\theta}$.
$\frac{\cos^2\theta}{\sin\theta}$	Multiply the two terms.
$\frac{1-\sin^2\theta}{\sin\theta}$	Convert $\cos^2\theta$ to its identity $1-\sin^2\theta$.
$\frac{1}{\sin\theta}-\frac{\sin^2\theta}{\sin\theta}$	Split the fraction into two separate fractions.
$\frac{1}{\sin\theta}-\sin\theta$	Simplify the second term by dividing.
$\csc\theta-\sin\theta$	Convert $\frac{1}{\sin\theta}$ to $\csc\theta$.

36. C: The range is the spread of the data. It can be calculated for each subject by subtracting the lowest test score from the highest, or it can be determined visually from the graph. The difference between the highest and lowest test scores in History is $98-72=26$ points. The range for each of the other subjects is greater.

37. C: There are nine ways to assign the first digit since it can be any of the numbers 1-9. There are nine ways to assign the second digit since it can be any digit 0-9 except for the digit assigned in place 1. There are eight ways to assign the third number since there are ten digits, two of which have already been assigned. There are seven ways to assign the fourth number, six ways to assign the fifth, five ways to assign the sixth, and four ways to assign the seventh. Multiply to find the total number of combinations.

$$9\times9\times8\times7\times6\times5\times4=544{,}320.$$

Another way to approach the problem is to notice that the arrangement of nine digits in the last six places is a sequence without repetition, or a permutation. (Note: this may be called a partial permutation since all of the elements of the set need not be used.) The number of possible sequences of a fixed length r of elements taken from a given set of size n is permutation ${}_n\mathrm{P}_r=\frac{n!}{(n-r)!}$. So, the number of ways to arrange the last six digits is ${}_9\mathrm{P}_6=\frac{9!}{(9-6)!}=\frac{9!}{3!}=60{,}480$. Multiply this number by nine since there are nine possibilities for the first digit of the phone number: ${}_9\mathrm{P}_6\times9=544{,}320$.

38. A: Calculate the final sales price of the rug at each store. Please note that the explanation below uses 1.08 instead of 0.08 for tax in order to avoid having to perform addition later. If 0.08 were to be used, only the amount of tax charged would result instead of the total cost of the product with tax included. Thus, 1.08 is used to make sure 108% of the product price is calculated.

$$1.08(0.7\times\$296)=\$223.78\text{ at Store A}$$
$$1.08(\$220-\$10)=\$226.80\text{ at Store B}$$
$$\$198+\$35=\$233\text{ at Store C}$$

39. B: Determine the probability of each choice.

For choice A, this is the total number of students in the broccoli circle of the Venn diagram divided by the total number of students surveyed:

$$\frac{10+4+3+15}{90} = \frac{32}{90} \approx 35.6\%$$

For choice B, this is the total number of students in the carrots circle and also in at least one other circle divided by the total number in the carrots circle:

$$\frac{15+3+6}{15+3+6+27} = \frac{24}{51} \approx 47.1\%$$

For choice C, this is the number of students in the intersection of all three circles divided by the total number in the overlap of the broccoli and cauliflower circles:

$$\frac{3}{3+4} = \frac{3}{7} \approx 42.9\%$$

For choice D, this is the number of students outside of all the circles divided by the total number of students surveyed:

$$\frac{23}{90} \approx 25.6\%$$

Since choice B has the highest percentage, it is the choice that is most probable.

40. D: The standard equation of the circle with center (2,3) and radius 5 is indeed $(x-2)^2 + (y-3)^2 = 5^2$, but a circle fails the vertical line test and cannot be the graph of the function. Working backwards, we see that if the student's first equation were $y - 3 = \pm\sqrt{-x^2+4x+21}$, its graph would be the whole circle. As it is, the graph of $y - 3 = -\sqrt{-x^2+4x+21}$ is the lower half of the circle, so its maximum is level with the center of the circle, namely $y = 3$.

41. C: Graph the solutions of the individual inequalities and find the region where they overlap.

To graph the inequality $4x - 2y \geq -4$, we first graph the corresponding linear equation $4x - 2y = -4$. The rest of the solution is either the region above this line or the region below. To determine which, choose a convenient test point, say (0,0), and substitute it into the inequality: $4(0) - 2(0) \geq -4$. This statement is true, so the solution lies on the side that includes the test point, namely the side below.

Repeat this procedure for $3x + 2y \leq 2$, graphing the corresponding line $3x + 2y = 2$. Check a test point to see that the solution lies below the line.

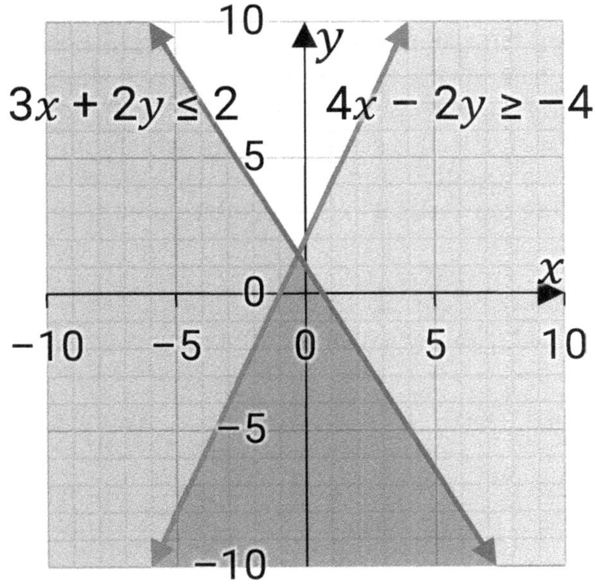

The solution of the system consists of the points common to the solutions of the two individual inequalities. The graph of this solution is the darker-shaded region where the two shaded half-planes overlap. This is region C.

42. C: $\triangle RST$ and $\triangle SUT$ are right triangles with $\angle T$ in common, so $\angle R$ and $\angle USV$ are both complements of $\angle T$. Complements of the same angle are congruent, so $\angle R \cong \angle USV$. Thus, $\triangle RST$ and $\triangle SVU$ are right triangles with congruent acute angles $\angle R \cong \angle USV$. By the AA (angle-angle) similarity criterion, $\triangle RST \sim \triangle SVU$. Corresponding sides of similar triangles are proportional, so $\frac{RS}{RT} = \frac{SV}{SU}$. Solving for RS, we get $RS = \frac{RT \times SV}{SU}$. It is given that $SU = a$ and $RT = c$; and by the Pythagorean theorem, $SV = \sqrt{a^2 - b^2}$. This gives us $RS = \frac{RT \times SV}{SU} = \frac{c\sqrt{a^2-b^2}}{a}$.

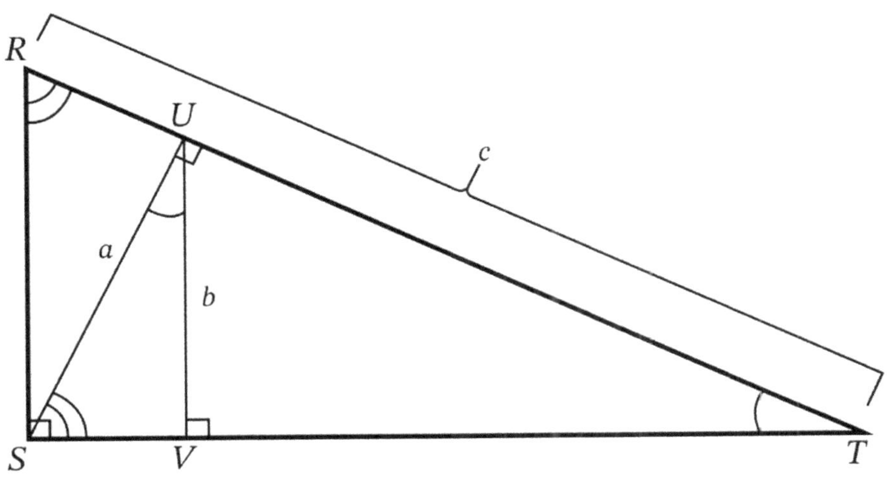

43. D: In a graph of distance traveled as a function of time, the slope of the curve indicates the speed (a scalar quantity, which cannot be negative) of the object. Thus, the first third of the graph has a positive slope corresponding to a speed of 10 ft/sec. The second third has a steeper slope corresponding to a speed of 15 ft/sec. During these first six seconds, the car travels a distance of $10 \cdot 3 + 15 \cdot 3 = 75$, or 75 feet. Thus, in the last three seconds the car must travel 75 feet, giving it a speed of 75 ft $\div$ 3 sec $=$ 25 ft/sec. So, the final third of the graph should have an even steeper slope (as it does), but that slope should still be positive, reflecting that the total distance traveled is still increasing. To put it another way, speed is scalar quantity, which cannot be negative. The student mistakenly gives the last third a negative slope (representing the vector quantity velocity, rather than speed), effectively treating the vertical axis as displacement from the starting point rather than total distance traveled.

44. B: Notice that choice A cannot be correct since $x \neq 2$ because $x = 2$ results in a zero in the denominator. Now, use algebra techniques to solve for x.

$\frac{x-1}{x+2} = \frac{6}{x+2} - \frac{x-3}{x-2}$	Write out the original equation.
$\frac{x-7}{x+2} = -\frac{x-3}{x-2}$	Subtract $\frac{6}{x+2}$ from both sides.
$(x-7)(x-2) = -(x-3)(x+2)$	Cross multiply.
$x^2 - 9x + 14 = -x^2 + x + 6$	Multiply the binomials on both sides.
$2x^2 - 10x + 8 = 0$	Move all the terms to the left side of the equation.
$2(x^2 - 5x + 4) = 0$	Factor a 2 out of each term on the left side.
$2(x-1)(x-4) = 0$	Factor the new trinomial.
$x = 1, 4$	Set the two new binomials equal to 0 and solve for x.

45. 46: Use a Venn diagram to help organize the given information. Start by filling in the space where the three circles intersect: one student enrolled in all three areas. Use that information to fill in the spaces where two circles intersect. For example, seven students are taking both Intro to Law and Latin, and one of those is also taking Biophysics, so six students are taking Intro to Law and Latin but not Biophysics. Once the diagram is completed, add the number of students from each portion of the diagram. The total number of students taking special electives is $12 + 14 + 9 + 3 + 1 + 6 + 1 = 46$.

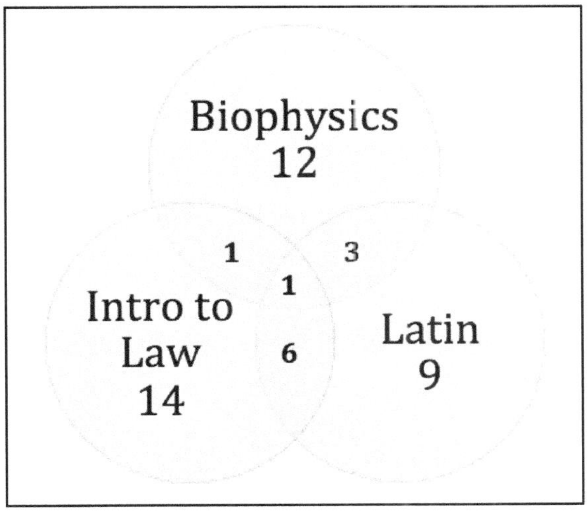

46. B, C: Let w be the number of widgets and g the number of gadgets you produce. You want to maximize your profit function $P = 8w + 32g$ subject to the availability of gears and levers. It can be seen that the feasible production levels must satisfy the following system of inequalities:

$$\begin{cases} 5w + 20g \leq 650 \text{ gears} \\ 20w + 10g \leq 1{,}200 \text{ levers} \\ w, g \geq 0 \end{cases}$$

Choice A is feasible. It satisfies the inequalities, requiring $5 \times 30 + 20 \times 25 = 150 + 500 = 650$ gears and $20 \times 30 + 10 \times 25 = 600 + 250 = 850 \leq 1{,}200$ levers, and it produces a profit of $P = 8 \times 30 + 32 \times 25 = 240 + 800 = \$1{,}040$. Thus, this solution remains a viable candidate for producing the maximum profit.

Choice B is not feasible because it violates the constraints on both gears and levers, requiring $5 \times 55 + 20 \times 20 = 275 + 400 = 675 \nleq 650$ gears and $20 \times 55 + 10 \times 20 = 1{,}100 + 200 = 1{,}300 \nleq 1{,}200$ levers. Thus, we can rule it out as a candidate for producing maximum profit.

Choice C is feasible. It satisfies the inequalities, requiring $5 \times 5 + 20 \times 30 = 25 + 600 = 625 \leq 650$ gears and $20 \times 5 + 10 \times 30 = 100 + 300 = 400 \leq 1{,}200$ levers, but it produces a profit of $P = 8 \times 5 + 32 \times 30 = 40 + 960 = \$1{,}000$, which is smaller than the profit in feasible choice A. Thus, we can rule it out as a candidate for producing maximum profit.

47. D: To solve the equation, write $x^4 + 9 = 10x^2$ in quadratic form:

$$x^4 - 10x^2 + 9 = 0$$

This trinomial can be factored as $(x^2 - 1)(x^2 - 9) = 0$. Each set of parentheses contains a difference of squares, which can be factored further as $(x + 1)(x - 1)(x + 3)(x - 3) = 0$. Use the zero-product property to find the solutions to the equation.

$$\begin{array}{cccc} x + 1 = 0 & x - 1 = 0 & x + 3 = 0 & x - 3 = 0 \\ x = -1 & x = 1 & x = -3 & x = 3 \end{array}$$

Therefore, the solutions to this equation are $x = \{-3, -1, 1, 3\}$.

48. A, C: Student A first uses the identity $\sin x° = \cos(90 - x)°$ (that is, the sine of an angle equals the cosine of the angle's complement) to replace $\sin 38°$ with $\cos 52°$. Next, the student uses the double-angle formula $\sin 2x = 2\sin x \cos x$ to rewrite $2\cos 52° \sin 52°$ as $\sin(2 \times 52°) = \sin 104°$.

Student B incorrectly believes that a product of sines equals the sine of the sum of the arguments (that is, $\sin x \sin y = \sin(x + y)$) and thus gets the false equivalence $\sin 38° \sin 52° = \sin 90°$.

Student C applies the identity $\sin x° = \cos(90 - x)°$ to replace $\sin 52°$ with $\cos 38°$. The student then applies the double-angle formula to rewrite $2\sin 38° \cos 38°$ as $\sin(2 \times 38°) = \sin 76°$. Note that choices A and C are equivalent since $\sin 76° = \sin 104°$.

49. 0.69: The area under curve can be found by evaluating the integral of the given function $f(x)$.

$$\begin{aligned} \int_1^2 \frac{1}{x} dx &= [\ln x]_1^2 \\ &= [\ln(2)] - [\ln(1)] \\ &\approx 0.69 \end{aligned}$$

50. B: One effective way to find a least common multiple is the following: first, find the prime factorization of each number. Then multiply together all the primes that appear in either number, raised to the highest powers at which they appear. So, we first find the four prime factorizations:

$$\begin{aligned} &5{,}760 = 576 \times 10 = 9 \times 64 \times 2 \times 5 = 3^2 \times 2^6 \times 2 \times 5 = 2^7 \times 3^2 \times 5 \\ &4{,}800 = 48 \times 100 = 6 \times 8 \times 10 \times 10 = 2 \times 3 \times 2^3 \times 2 \times 5 \times 2 \times 5 = 2^6 \times 3 \times 5^2 \\ &3{,}600 = 36 \times 100 = 6 \times 6 \times 10 \times 10 = 2 \times 3 \times 2 \times 3 \times 2 \times 5 \times 2 \times 5 = 2^4 \times 3^2 \times 5^2 \\ &960 = 96 \times 10 = 6 \times 16 \times 2 \times 5 = 2 \times 3 \times 2^4 \times 2 \times 5 = 2^6 \times 3 \times 5 \end{aligned}$$

Thus, $lcm(5{,}760,\ 4{,}800) = 2^7 \times 3^2 \times 5^2 = 28{,}800$ and $lcm(3{,}600,\ 960) = 2^6 \times 3^2 \times 5^2 = 14{,}400$. This gives us $lcm(5{,}760,\ 4{,}800) \div lcm(3{,}600,\ 960) = 28{,}800 \div 14{,}400 = 2$.

51. C: All three students correctly apply the properties that limits are preserved by addition, subtraction, division (in which the denominator's limit is not 0), and the taking of square roots (in which the independent variable avoids negative values near the limiting point). Student A, however, makes the common mistake of thinking that a limit of 0 in the numerator makes the limit of the whole expression fail to exist. Student B makes the common mistake of confusing $\lim_{x \to 3^-}(3 + x)$ with $\lim_{x \to -3}(3 + x)$ and then compounds it by thinking a limit of zero in the denominator makes the limit of the whole expression 0. Student C follows the same plan as Student A but draws the correct conclusion from seeing that the numerator's limit is 0.

52. B: We can find the area of the shaded part by viewing the target as five separate circles of various sizes, one inside the other. We add the areas of the three shaded circles and subtract the areas of the two white circles. The area of a circle is πr^2, so the area of the largest circle (the entire target) is $\pi(5a)^2$. We subtract the white circle inside it, add the shaded circle inside that one, and so on:

$$\begin{aligned} A &= \pi(5a)^2 - \pi(4a)^2 + \pi(3a)^2 - \pi(2a)^2 + \pi(a)^2 \\ &= 25\pi a^2 - 16\pi a^2 + 9\pi a^2 - 4\pi a^2 + \pi a^2 \\ &= (25 - 16 + 9 - 4 + 1)\pi a^2 \\ &= 15\pi a^2 \end{aligned}$$

53. D: Since it is given that $\overline{FD} \cong \overline{BC}$ and $\overline{AB} \cong \overline{DE}$, step 2 needs to establish either that $\overline{AC} \cong \overline{EF}$ or that $\Delta ABC \cong \Delta FDE$ in order for step 5 to show that $\Delta ABC \cong \Delta EDF$. The statement $\overline{AC} \cong \overline{EF}$ cannot be shown directly from the given information. On the other hand, $\Delta ABC \cong \Delta FDE$ can be determined: when two parallel lines $\overline{BC} \parallel \overline{FG}$ are cut by a transversal ($\overline{AE}$), alternate exterior angles (ΔABC, ΔFDE) are congruent. Therefore, $\Delta ABC \cong \Delta EDF$ by the side-angle-side (SAS) theorem.

54. B: The notation $\mathbb{P} \subseteq \mathbb{N} \subseteq \mathbb{Z} \subseteq \mathbb{Q} \subseteq \mathbb{R} \subseteq \mathbb{C}$ means that the set of prime numbers is a subset of the set natural numbers, which is a subset of the set of integers, which is a subset of the set of rational numbers, which is a subset of the set real numbers, which is a subset of the set of complex numbers. The set of prime numbers ($\mathbb{P}$) are natural numbers that have no factors other than 1 and itself. The set of natural numbers ($\mathbb{N}$) are all positive integers. The set of integers ($\mathbb{Z}$) are all positive and negative whole numbers, numbers without a fraction or decimal part to them. The set of rational numbers ($\mathbb{Q}$) are all numbers that can be expressed as the quotient of two integers. The set of real numbers ($\mathbb{R}$) are all numbers that can be expressed on a number line. The set of all complex numbers ($\mathbb{C}$) are all real and imaginary numbers.

55. A: Each lateral face of the pyramid is a triangle whose base has a length of h (the length of a side of the square) and whose height is the slant height of the pyramid. The slant height is the length of the hypotenuse of a right triangle whose legs have lengths of h and $\frac{h}{2}$. Thus, the slant height is $\sqrt{h^2 + \left(\frac{h}{2}\right)^2} = \sqrt{h^2 + \left(\frac{h^2}{4}\right)} = \sqrt{\frac{5h^2}{4}} = \frac{\sqrt{5}}{2}h.$

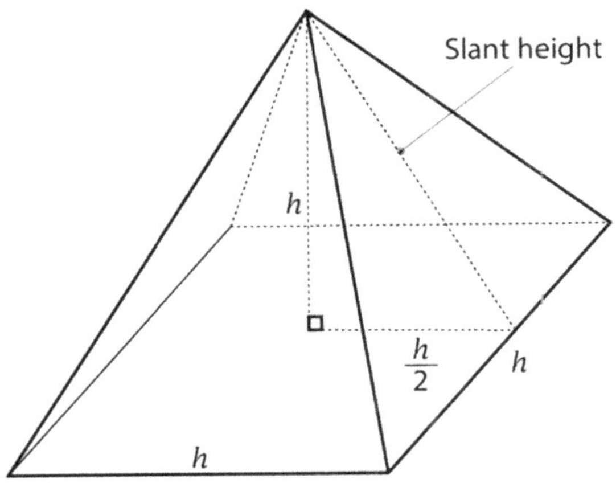

Since the area of a triangle is half the base times the height, each triangle has area $\frac{1}{2} \times h \times \frac{\sqrt{5}}{2}h = \frac{\sqrt{5}}{4}h^2$. Thus, the total area of the four lateral faces is $4 \times \frac{\sqrt{5}}{4}h^2 = \sqrt{5}h^2$ m^2. The volume of a pyramid is one-third the area of the base times the height, which, in this case, is $\frac{1}{3} \times h^2 \times h = \frac{1}{3}h^3$ m^3. Since the volume and lateral area are numerically equal, we get the equation $\frac{1}{3}h^3 = \sqrt{5}h^2$. Multiplying by 3 gives us $h^3 = 3\sqrt{5}h^2$, and dividing by h^2 (since $h > 0$) yields $h = 3\sqrt{5}$ meters.

56. D: The velocity v of the ball at any time t is the slope of the line tangent to the graph of h at time t. The slope of a line tangent to the curve $h(t) = -16t^2 + 50t + 3$ can be found using the derivative, $h'(t)$.

$$h'(t) = v(t) = -32t + 50$$

When $t = 2$, the velocity of the ball is $-32(2) + 50 = -14$. The velocity is negative because the slope of the tangent line at $t = 2$ is negative; velocity has both magnitude and direction, so a velocity of –14 means that the velocity is 14 ft/s downward.

57. B, C: Choice A makes a mistake going from the second equation to the third because $\ln(374e^{0.03t}) \neq \ln(374)\ln(e^{0.03t})$. In general, $\ln ab \neq (\ln a)(\ln b)$. The correct equivalence is $\ln ab = \ln a + \ln b$.

Choice B first divides both sides of the equation by 374 and then converts the equation from the exponential form $e^{0.03t} = 2$ to the logarithmic form $0.03t = \ln 2$ (the exponential form $e^y = x$ is equivalent to the logarithmic form $y = \ln x$). The final step divides both sides by 0.03.

Choice C first takes the common logarithm of both sides. It then simplifies the left side by rewriting the logarithm of a product, $\log(374e^{0.03t})$, as a sum of logarithms, $\log 374 + \log e^{0.03t}$. After subtracting $\log 374$ from both sides, it applies the rule $\log a^b = b \log a$ to change $\log e^{0.03t}$ to $0.03t \log e$. The final step divides both sides by $0.03 \log e$.

58. A: A pie chart shows the relationship of parts to a whole. A line graph is often used to show change over time. A box-and-whisker plot displays how numeric data are distributed throughout the range. A Venn diagram shows the relationships among sets. Therefore, a pie chart would best illustrate the percentage of a budget allocated to various departments.

59. A: We can label the vertices of the triangle A, B, and C, as in the diagram below. We see that, since the pole is 11° off vertical, the measure of $\angle ABC$ is $90° - 11° = 79°$. Further, the measure of $\angle ACB$ is $180° - 46° - 79° = 55°$. By the law of sines:

$$\frac{\text{pole length}}{\sin 46°} = \frac{96 \text{ ft}}{\sin 55°}$$

Thus,

$$\text{pole length} = \frac{96 \sin 46°}{\sin 55°} \approx 84.3 \text{ ft}$$

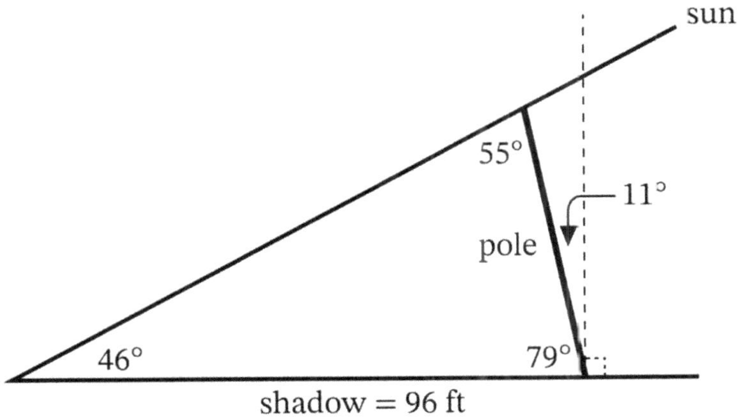

60. A: Evaluate the indefinite integral by adding 1 to each exponent and dividing each coefficient by the new exponent. Don't forget to add the constant of integration.

$$\int 3x^2 + 2x - 1\,dx$$
$$= \frac{3}{2+1}x^{(2+1)} + \frac{2}{1+1}x^{(1+1)} - x + C$$
$$= x^3 + x^2 - x + C$$

61. 4.5 GB: The expression representing the monthly charge for Plan A is $\$20 + \$15(d - 2.5)$, where d is the data used. This can be found by adding the flat rate to the price of any data used over the provided amount. Set this expression equal to the monthly charge for Plan B, which is \$50. Solve for d to find the number of GB for which the two plans charge the same amount.

$$\$20 + \$15(d - 2.5) = \$50$$
$$\$15(d - 2.5) = \$30$$
$$d - 2.5 = 2$$
$$d = 4.5$$

So, the plans have the same cost when Gillian uses 4.5 GB of data.

62. B, C: In choice A, the algebraic procedure (completing the square) is appropriate for studying a *quadratic* polynomial. In that case, it would give us a local (and absolute) extremum and the axis of symmetry. The same steps, however, do not provide any information about a *cubic* polynomial.

Choice B expands the expression using the distributive property (the FOIL method: first, outer, inner, last) to find the general form of the cubic polynomial. Since the polynomial has only odd-power terms, the function is odd, and the graphs of odd functions are always symmetric with respect to the origin. Also, in general form, the constant term gives the y-intercept.

Choice C factors x out of the second factor, expressing $f(x)$ as a product of linear polynomials (factored form). Each linear factor gives us a zero of the polynomial; and since no linear factors are repeated, each zero is simple (i.e., each has multiplicity one). Since $x = 0$ is a zero, the y-intercept is $f(0) = 0$.

63. A: Since both x and $\sin(2x)$ are nonnegative on the interval $\left[0, \frac{\pi}{2}\right]$, so is their product, $x\sin(2x)$. Thus, the area under the curve is given by the definite integral $\int_0^{\frac{\pi}{2}} x\sin(2x)\,dx$. Recall the formula for integration by parts: $\int u\,dv = uv - \int v\,du$. We let $u = x$ and $dv = \sin(2x)\,dx$. Calculate $du = dx$ and $v = -\left(\frac{1}{2}\right)\cos(2x)$. So, the area under the curve is:

$$\int_0^{\frac{\pi}{2}} \overbrace{x}^{u}\overbrace{\sin(2x)\,dx}^{dv} = \overbrace{(x)}^{u} \times \overbrace{\left(-\frac{1}{2}\cos(2x)\right)}^{v}\Bigg|_0^{\frac{\pi}{2}} - \int_0^{\frac{\pi}{2}} \overbrace{\left(-\frac{1}{2}\cos(2x)\right)}^{v} \times \overbrace{(dx)}^{du}$$

$$= -\frac{1}{2}x\cos(2x)\Big|_0^{\frac{\pi}{2}} + \frac{1}{2}\int_0^{\frac{\pi}{2}} \cos(2x)\,dx$$

$$= -\frac{\pi}{4}\cos\pi - 0 + \left(\frac{1}{4}\sin(2x)\right)\Big|_0^{\frac{\pi}{2}}$$

$$= -\frac{\pi}{4}(-1) + \frac{1}{4}\sin\pi - \frac{1}{4}\sin 0$$

$$= \frac{\pi}{4} + 0 - 0$$

$$= \frac{\pi}{4}$$

64. A: The inequality that represents this scenario is $n + 3p \leq 30$, where n is the number of notebooks and p is the number of boxes of pencils. To graph this inequality, find the intercepts of this linear inequality by making the inequality an equality and setting n and p equal to 0 and solving for the other variable.

$$(0) + 3p = 30$$
$$3p = 30$$
$$p = 10$$

$$n + 3(0) = 30$$
$$n = 30$$

The intercepts are $n = 30$ and $p = 10$. The solid line through the two intercepts represents the maximum number of each item that may be purchased. Since this inequality represents the maximum he can purchase, all first-quadrant points below the line are also possible amounts within Ben's budget.

65. D: Neighborhood A has a mean income of $70,000 and a median income of $33,000. Neighborhood B has a mean income of $52,000 and a median income of $50,000. Thus, neighborhood A has the higher mean and neighborhood B has the higher median. Should we base our interpretation on the mean or the median? To answer this question, we need some idea of the distribution of the data from a histogram, a stem-and-leaf plot, or something similar. Sketching out a double stem-and-leaf plot, we see that the incomes in both neighborhoods are positively skewed.

Neighborhood A is especially skewed because of the extreme outlier 400 (one household whose high income raises the neighborhood mean far above all the other households in the sample).

Leaf B	Stem	Leaf A
	2	6 6 7 8
9 8 6	3	2 4 7
8 5	4	2 8
3 2	5	
8 2	6	
9	7	
	⋮	⋮
	40	0

When comparing skewed data, we tend (in the absence of a clear reason to do otherwise) to take the median as a better "center" for the data than the mean. And certainly, in this case, the median is much more representative than the mean of the financial situation of most households. Thus, we conclude that neighborhood B, with the higher median income, is the more prosperous neighborhood, a conclusion supported by the stem-and-leaf plot.

66. C: The diagram reminds us that $\cos\theta$ is the x-coordinate of the point $P(\theta)$. It also shows that $P(\theta)$ and $P(-\theta)$ are symmetric with respect to the x-axis. Consequently, $P(\theta)$ and $P(-\theta)$ have a common x-coordinate, meaning $\cos(\theta) = \cos(-\theta)$ for all real numbers θ. Thus, if the point (x_0, y_0) is on the graph, meaning the equation $y_0 = \cos x_0$ is true, then $y_0 = \cos(-x_0)$ is also true, meaning the point $(-x_0, y_0)$ is on the graph. Thus, the graph of $y = \cos x$ is symmetric with respect to the y-axis.

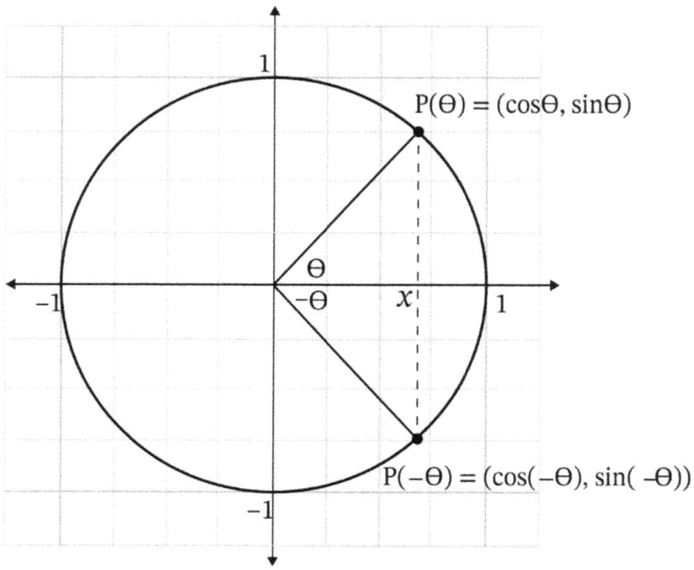

Praxis Practice Test #3

To take this additional Praxis practice test, visit our online resources page:
mometrix.com/resources719/priimath5161

How to Overcome Test Anxiety

Just the thought of taking a test is enough to make most people a little nervous. A test is an important event that can have a long-term impact on your future, so it's important to take it seriously and it's natural to feel anxious about performing well. But just because anxiety is normal, that doesn't mean that it's helpful in test taking, or that you should simply accept it as part of your life. Anxiety can have a variety of effects. These effects can be mild, like making you feel slightly nervous, or severe, like blocking your ability to focus or remember even a simple detail.

If you experience test anxiety—whether severe or mild—it's important to know how to beat it. To discover this, first you need to understand what causes test anxiety.

Causes of Test Anxiety

While we often think of anxiety as an uncontrollable emotional state, it can actually be caused by simple, practical things. One of the most common causes of test anxiety is that a person does not feel adequately prepared for their test. This feeling can be the result of many different issues such as poor study habits or lack of organization, but the most common culprit is time management. Starting to study too late, failing to organize your study time to cover all of the material, or being distracted while you study will mean that you're not well prepared for the test. This may lead to cramming the night before, which will cause you to be physically and mentally exhausted for the test. Poor time management also contributes to feelings of stress, fear, and hopelessness as you realize you are not well prepared but don't know what to do about it.

Other times, test anxiety is not related to your preparation for the test but comes from unresolved fear. This may be a past failure on a test, or poor performance on tests in general. It may come from comparing yourself to others who seem to be performing better or from the stress of living up to expectations. Anxiety may be driven by fears of the future—how failure on this test would affect your educational and career goals. These fears are often completely irrational, but they can still negatively impact your test performance.

Elements of Test Anxiety

As mentioned earlier, test anxiety is considered to be an emotional state, but it has physical and mental components as well. Sometimes you may not even realize that you are suffering from test anxiety until you notice the physical symptoms. These can include trembling hands, rapid heartbeat, sweating, nausea, and tense muscles. Extreme anxiety may lead to fainting or vomiting. Obviously, any of these symptoms can have a negative impact on testing. It is important to recognize them as soon as they begin to occur so that you can address the problem before it damages your performance.

The mental components of test anxiety include trouble focusing and inability to remember learned information. During a test, your mind is on high alert, which can help you recall information and stay focused for an extended period of time. However, anxiety interferes with your mind's natural processes, causing you to blank out, even on the questions you know well. The strain of testing during anxiety makes it difficult to stay focused, especially on a test that may take several hours. Extreme anxiety can take a huge mental toll, making it difficult not only to recall test information but even to understand the test questions or pull your thoughts together.

Effects of Test Anxiety

Test anxiety is like a disease—if left untreated, it will get progressively worse. Anxiety leads to poor performance, and this reinforces the feelings of fear and failure, which in turn lead to poor performances on subsequent tests. It can grow from a mild nervousness to a crippling condition. If allowed to progress, test anxiety can have a big impact on your schooling, and consequently on your future.

Test anxiety can spread to other parts of your life. Anxiety on tests can become anxiety in any stressful situation, and blanking on a test can turn into panicking in a job situation. But fortunately, you don't have to let anxiety rule your testing and determine your grades. There are a number of relatively simple steps you can take to move past anxiety and function normally on a test and in the rest of life.

Physical Steps for Beating Test Anxiety

While test anxiety is a serious problem, the good news is that it can be overcome. It doesn't have to control your ability to think and remember information. While it may take time, you can begin taking steps today to beat anxiety.

Just as your first hint that you may be struggling with anxiety comes from the physical symptoms, the first step to treating it is also physical. Rest is crucial for having a clear, strong mind. If you are tired, it is much easier to give in to anxiety. But if you establish good sleep habits, your body and mind will be ready to perform optimally, without the strain of exhaustion. Additionally, sleeping well helps you to retain information better, so you're more likely to recall the answers when you see the test questions.

Getting good sleep means more than going to bed on time. It's important to allow your brain time to relax. Take study breaks from time to time so it doesn't get overworked, and don't study right before bed. Take time to rest your mind before trying to rest your body, or you may find it difficult to fall asleep.

Along with sleep, other aspects of physical health are important in preparing for a test. Good nutrition is vital for good brain function. Sugary foods and drinks may give a burst of energy but this burst is followed by a crash, both physically and emotionally. Instead, fuel your body with protein and vitamin-rich foods.

Also, drink plenty of water. Dehydration can lead to headaches and exhaustion, especially if your brain is already under stress from the rigors of the test. Particularly if your test is a long one, drink water during the breaks. And if possible, take an energy-boosting snack to eat between sections.

Along with sleep and diet, a third important part of physical health is exercise. Maintaining a steady workout schedule is helpful, but even taking 5-minute study breaks to walk can help get your blood pumping faster and clear your head. Exercise also releases endorphins, which contribute to a positive feeling and can help combat test anxiety.

When you nurture your physical health, you are also contributing to your mental health. If your body is healthy, your mind is much more likely to be healthy as well. So take time to rest, nourish your body with healthy food and water, and get moving as much as possible. Taking these physical steps will make you stronger and more able to take the mental steps necessary to overcome test anxiety.

Mental Steps for Beating Test Anxiety

Working on the mental side of test anxiety can be more challenging, but as with the physical side, there are clear steps you can take to overcome it. As mentioned earlier, test anxiety often stems from lack of preparation, so the obvious solution is to prepare for the test. Effective studying may be the most important weapon you have for beating test anxiety, but you can and should employ several other mental tools to combat fear.

First, boost your confidence by reminding yourself of past success—tests or projects that you aced. If you're putting as much effort into preparing for this test as you did for those, there's no reason you should expect to fail here. Work hard to prepare; then trust your preparation.

Second, surround yourself with encouraging people. It can be helpful to find a study group, but be sure that the people you're around will encourage a positive attitude. If you spend time with others who are anxious or cynical, this will only contribute to your own anxiety. Look for others who are motivated to study hard from a desire to succeed, not from a fear of failure.

Third, reward yourself. A test is physically and mentally tiring, even without anxiety, and it can be helpful to have something to look forward to. Plan an activity following the test, regardless of the outcome, such as going to a movie or getting ice cream.

When you are taking the test, if you find yourself beginning to feel anxious, remind yourself that you know the material. Visualize successfully completing the test. Then take a few deep, relaxing breaths and return to it. Work through the questions carefully but with confidence, knowing that you are capable of succeeding.

Developing a healthy mental approach to test taking will also aid in other areas of life. Test anxiety affects more than just the actual test—it can be damaging to your mental health and even contribute to depression. It's important to beat test anxiety before it becomes a problem for more than testing.

Study Strategy

Being prepared for the test is necessary to combat anxiety, but what does being prepared look like? You may study for hours on end and still not feel prepared. What you need is a strategy for test prep. The next few pages outline our recommended steps to help you plan out and conquer the challenge of preparation.

Step 1: Scope Out the Test

Learn everything you can about the format (multiple choice, essay, etc.) and what will be on the test. Gather any study materials, course outlines, or sample exams that may be available. Not only will this help you to prepare, but knowing what to expect can help to alleviate test anxiety.

Step 2: Map Out the Material

Look through the textbook or study guide and make note of how many chapters or sections it has. Then divide these over the time you have. For example, if a book has 15 chapters and you have five days to study, you need to cover three chapters each day. Even better, if you have the time, leave an extra day at the end for overall review after you have gone through the material in depth.

If time is limited, you may need to prioritize the material. Look through it and make note of which sections you think you already have a good grasp on, and which need review. While you are studying, skim quickly through the familiar sections and take more time on the challenging parts.

Write out your plan so you don't get lost as you go. Having a written plan also helps you feel more in control of the study, so anxiety is less likely to arise from feeling overwhelmed at the amount to cover.

STEP 3: GATHER YOUR TOOLS

Decide what study method works best for you. Do you prefer to highlight in the book as you study and then go back over the highlighted portions? Or do you type out notes of the important information? Or is it helpful to make flashcards that you can carry with you? Assemble the pens, index cards, highlighters, post-it notes, and any other materials you may need so you won't be distracted by getting up to find things while you study.

If you're having a hard time retaining the information or organizing your notes, experiment with different methods. For example, try color-coding by subject with colored pens, highlighters, or post-it notes. If you learn better by hearing, try recording yourself reading your notes so you can listen while in the car, working out, or simply sitting at your desk. Ask a friend to quiz you from your flashcards, or try teaching someone the material to solidify it in your mind.

STEP 4: CREATE YOUR ENVIRONMENT

It's important to avoid distractions while you study. This includes both the obvious distractions like visitors and the subtle distractions like an uncomfortable chair (or a too-comfortable couch that makes you want to fall asleep). Set up the best study environment possible: good lighting and a comfortable work area. If background music helps you focus, you may want to turn it on, but otherwise keep the room quiet. If you are using a computer to take notes, be sure you don't have any other windows open, especially applications like social media, games, or anything else that could distract you. Silence your phone and turn off notifications. Be sure to keep water close by so you stay hydrated while you study (but avoid unhealthy drinks and snacks).

Also, take into account the best time of day to study. Are you freshest first thing in the morning? Try to set aside some time then to work through the material. Is your mind clearer in the afternoon or evening? Schedule your study session then. Another method is to study at the same time of day that you will take the test, so that your brain gets used to working on the material at that time and will be ready to focus at test time.

STEP 5: STUDY!

Once you have done all the study preparation, it's time to settle into the actual studying. Sit down, take a few moments to settle your mind so you can focus, and begin to follow your study plan. Don't give in to distractions or let yourself procrastinate. This is your time to prepare so you'll be ready to fearlessly approach the test. Make the most of the time and stay focused.

Of course, you don't want to burn out. If you study too long you may find that you're not retaining the information very well. Take regular study breaks. For example, taking five minutes out of every hour to walk briskly, breathing deeply and swinging your arms, can help your mind stay fresh.

As you get to the end of each chapter or section, it's a good idea to do a quick review. Remind yourself of what you learned and work on any difficult parts. When you feel that you've mastered the material, move on to the next part. At the end of your study session, briefly skim through your notes again.

But while review is helpful, cramming last minute is NOT. If at all possible, work ahead so that you won't need to fit all your study into the last day. Cramming overloads your brain with more information than it can process and retain, and your tired mind may struggle to recall even previously learned information when it is overwhelmed with last-minute study. Also, the urgent nature of cramming and the stress placed on your brain contribute to anxiety. You'll be more likely to go to the test feeling unprepared and having trouble thinking clearly.

So don't cram, and don't stay up late before the test, even just to review your notes at a leisurely pace. Your brain needs rest more than it needs to go over the information again. In fact, plan to finish your studies by noon or early afternoon the day before the test. Give your brain the rest of the day to relax or focus on other things, and get a good night's sleep. Then you will be fresh for the test and better able to recall what you've studied.

Step 6: Take a Practice Test

Many courses offer sample tests, either online or in the study materials. This is an excellent resource to check whether you have mastered the material, as well as to prepare for the test format and environment.

Check the test format ahead of time: the number of questions, the type (multiple choice, free response, etc.), and the time limit. Then create a plan for working through them. For example, if you have 30 minutes to take a 60-question test, your limit is 30 seconds per question. Spend less time on the questions you know well so that you can take more time on the difficult ones.

If you have time to take several practice tests, take the first one open book, with no time limit. Work through the questions at your own pace and make sure you fully understand them. Gradually work up to taking a test under test conditions: sit at a desk with all study materials put away and set a timer. Pace yourself to make sure you finish the test with time to spare and go back to check your answers if you have time.

After each test, check your answers. On the questions you missed, be sure you understand why you missed them. Did you misread the question (tests can use tricky wording)? Did you forget the information? Or was it something you hadn't learned? Go back and study any shaky areas that the practice tests reveal.

Taking these tests not only helps with your grade, but also aids in combating test anxiety. If you're already used to the test conditions, you're less likely to worry about it, and working through tests until you're scoring well gives you a confidence boost. Go through the practice tests until you feel comfortable, and then you can go into the test knowing that you're ready for it.

Test Tips

On test day, you should be confident, knowing that you've prepared well and are ready to answer the questions. But aside from preparation, there are several test day strategies you can employ to maximize your performance.

First, as stated before, get a good night's sleep the night before the test (and for several nights before that, if possible). Go into the test with a fresh, alert mind rather than staying up late to study.

Try not to change too much about your normal routine on the day of the test. It's important to eat a nutritious breakfast, but if you normally don't eat breakfast at all, consider eating just a protein bar. If you're a coffee drinker, go ahead and have your normal coffee. Just make sure you time it so that the caffeine doesn't wear off right in the middle of your test. Avoid sugary beverages, and drink enough water to stay hydrated but not so much that you need a restroom break 10 minutes into the test. If your test isn't first thing in the morning, consider going for a walk or doing a light workout before the test to get your blood flowing.

Allow yourself enough time to get ready, and leave for the test with plenty of time to spare so you won't have the anxiety of scrambling to arrive in time. Another reason to be early is to select a good seat. It's helpful to sit away from doors and windows, which can be distracting. Find a good seat, get out your supplies, and settle your mind before the test begins.

When the test begins, start by going over the instructions carefully, even if you already know what to expect. Make sure you avoid any careless mistakes by following the directions.

Then begin working through the questions, pacing yourself as you've practiced. If you're not sure on an answer, don't spend too much time on it, and don't let it shake your confidence. Either skip it and come back later, or eliminate as many wrong answers as possible and guess among the remaining ones. Don't dwell on these questions as you continue—put them out of your mind and focus on what lies ahead.

Be sure to read all of the answer choices, even if you're sure the first one is the right answer. Sometimes you'll find a better one if you keep reading. But don't second-guess yourself if you do immediately know the answer. Your gut instinct is usually right. Don't let test anxiety rob you of the information you know.

If you have time at the end of the test (and if the test format allows), go back and review your answers. Be cautious about changing any, since your first instinct tends to be correct, but make sure you didn't misread any of the questions or accidentally mark the wrong answer choice. Look over any you skipped and make an educated guess.

At the end, leave the test feeling confident. You've done your best, so don't waste time worrying about your performance or wishing you could change anything. Instead, celebrate the successful completion of this test. And finally, use this test to learn how to deal with anxiety even better next time.

Review Video: Test Anxiety
Visit mometrix.com/academy and enter code: 100340

Important Qualification

Not all anxiety is created equal. If your test anxiety is causing major issues in your life beyond the classroom or testing center, or if you are experiencing troubling physical symptoms related to your anxiety, it may be a sign of a serious physiological or psychological condition. If this sounds like your situation, we strongly encourage you to seek professional help.

Online Resources

Due to our efforts to try to keep this book to a manageable length, we've created a link that will give you access to all of your online resources:

mometrix.com/resources719/priimath5161

It's Your Moment, Let's Celebrate It!

Share your story @mometrixtestpreparation

www.ingramcontent.com/pod-product-compliance
Lightning Source LLC
LaVergne TN
LVHW062146200926
848484LV00007B/1870